ACS SYMPOSIUM SERIES **650**

Macromolecular Interactions in Food Technology

Nicholas Parris, EDITOR
Agricultural Research Service,
U.S. Department of Agriculture

Akio Kato, EDITOR
Yamaguchi University

Lawrence K. Creamer, EDITOR
New Zealand Dairy Research Institute

John Pearce, EDITOR
CSIRO Division of Food Science and Technology,
Melbourne Laboratory

Developed from a symposium sponsored
by the ACS Division of Agrochemicals at the
1995 International Chemical Congress of Pacific Basin Societies

American Chemical Society, Washington, DC

Library of Congress Cataloging-in-Publication Data

Macromolecular interactions in food technology / Nicholas Parris, editor . . . [et al.].

    p.   cm.—(ACS symposium series, ISSN 0097–6156; 650)

"Developed from a symposium sponsored by the ACS Division of Agrochemicals at the 1995 International Chemical Congress of Pacific Basin Societies, Honolulu, Hawaii, December 17–22, 1995."

Includes bibliographical references and indexes.

ISBN 0–8412–3466–3

1. Macromolecules—Congresses. 2. Food—Composition— Congresses. 3. Food industry and trade—Congresses.

I. Parris, Nicholas, 1937–   . II. International Chemical Congress of Pacific Basin Societies (1995: Honolulu, Hawaii) III. Series.

QP801.P64M33   1996
664—dc20
                                          96–36455
                                            CIP

This book is printed on acid-free, recycled paper.

PRINTED IN THE UNITED STATES OF AMERICA

Foreword

THE ACS SYMPOSIUM SERIES was first published in 1974 to provide a mechanism for publishing symposia quickly in book form. The purpose of this series is to publish comprehensive books developed from symposia, which are usually "snapshots in time" of the current research being done on a topic, plus some review material on the topic. For this reason, it is necessary that the papers be published as quickly as possible.

Before a symposium-based book is put under contract, the proposed table of contents is reviewed for appropriateness to the topic and for comprehensiveness of the collection. Some papers are excluded at this point, and others are added to round out the scope of the volume. In addition, a draft of each paper is peer-reviewed prior to final acceptance or rejection. This anonymous review process is supervised by the organizer(s) of the symposium, who become the editor(s) of the book. The authors then revise their papers according to the recommendations of both the reviewers and the editors, prepare camera-ready copy, and submit the final papers to the editors, who check that all necessary revisions have been made.

As a rule, only original research papers and original review papers are included in the volumes. Verbatim reproductions of previously published papers are not accepted.

ACS BOOKS DEPARTMENT

Contents

GEL AND FILM FORMATION

Preface

THE TWO MAJOR ATTRIBUTES OF FOOD that make it desirable are texture and flavor. Food texture is closely related to the behavior of the macromolecular food components, such as the proteins and polysaccharides.

Considerable information exists on the structure and properties of proteins and polysaccharides. This information, however, relates to the physical behavior of individual molecular species and not to behavior that results from protein–protein or protein–polysaccharide interactions. In food systems, a number of macromolecular interactions occur as a result of formulation and processing conditions. The number, type, and strength of these interactions determine the structure and properties of any particular manufactured food product. Recent developments are leading to a new understanding of the nature of these interactions and their contribution to the formulation of stable products for food and non-food uses.

This volume was developed from a symposium presented at the 1995 International Chemical Congress of Pacific Basin Societies, sponsored by the American Chemical Society Division of Agrochemicals, in Honolulu, Hawaii, December 17–22, 1995. The goal of this book is to bring together the recent findings presented by internationally prominent scientists from academia, government, and industry at the symposium and to provide a resource to stimulate scientists in the discussion of advances in this area.

Acknowledgments

The success of the symposium upon which this book is based can be attributed to its speakers. In their selection, we attempted to include experts from both industry and academia. To those experts who were missed for various reasons, we offer our apologies. We thank the authors for their excellent presentations and for the timely preparation of their chapters. We also thank the reviewers for their incisive and constructive criticism.

NICHOLAS PARRIS
Eastern Regional Research Center
Agricultural Research Service
U.S. Department of Agriculture
Wyndmoor, PA 19038

AKIO KATO
Department of Biochemistry
Yamaguchi University
Yamaguchi 753
Japan

LAWRENCE K. CREAMER
Food Science Section
New Zealand Dairy Research Institute
Private Bag 11029
Palmerston North
New Zealand

JOHN PEARCE
CSIRO Division of Food Science and Technology
Melbourne Laboratory
Graham Road
Highett, Victoria 3190
Australia

July 31, 1996

GENERAL STRUCTURAL AND FUNCTIONAL PROPERTIES

Chapter 1

Structure–Property Relationships in Foods

Vladimir Tolstoguzov

Nestlé Research Centre, P.O. Box 44, Vers-Chez-les-Blanc,
CH–1000 Lausanne 26, Switzerland

For considering food structures at the molecular level, the concept of the conformational potential of food macromolecules and the concept of structural hierarchy of foods are used. Formation of structure-property relationship of foods is mainly based on non-specific interactions of food macromolecular components that are due to partial conversion of intra- into inter-macromolecular bonds. Mixtures of proteins and polysaccharides can be regarded as the most realistic models for the study of the molecular origin of the formation of food structures at four levels: submacromolecular, macromolecular, supermacromolecular and macroscopic. Functional properties of macromolecular components of traditional and novel food systems depend on compatibility, co-solubility and complexing. Under conditions of inter-biopolymer complexing and incompatibility, biopolymer interactions affect structural, rheological and other physico-chemical properties of food systems.

The enormous diversity of food structures is based on only two kinds of macromolecules: proteins and polysaccharides. Although many food components participate in the formation of structural and textural features of foods these two biopolymers are the most important constructional materials (*1*). Moreover, there is a similarity in functional properties between individual members of these two groups of food biopolymers. Native globular proteins are usually quite similar in physico-chemical properties, such as viscosity, surface activity, conformational stability and gelation. The neutral and anionic polysaccharides used in foods are also similar in physico-chemical properties. We will discuss how is it possible to create a great number of food structures on a quite modest molecular basis using only two sorts of macromolecules as constructional materials and how is it possible to classify and describe different food sructures at the molecular level.

The diversity of food structures is based on the variety of non-specific interactions between biopolymers of both types and between biopolymers and other food components. Molecular and colloidal dispersions of food biopolymers and of their complexes with food components such as lipids, ions and surfactants participate in the formation of structural and physical properties of foods. Also water and lipids as solvents, plasticisers and fillers contribute to the structure-property relationships in foods (*1-4*).

Controlling inter-biopolymer interactions is of key significance for the development of novel processes and food products as well as for improvement of conventional food technologies. The desired composition-structure-property relationship of foods is formed during food processing. For

0097–6156/96/0650–0002$15.00/0

considering the question of the formation of food structures at the molecular level the concept of the "conformational potential" of food macromolecules seems to be a useful approach (*5,6*). The term "conformational potential" describes the capacity of biopolymers to form intermolecular junction zones that generate the desired structural, rheological and other physico-chemical properties of a given food system. From a conceptual viewpoint, the production of many foods may be regarded as the formation of intermacromolecular interactions throughout a processed food system. Intramolecular bonding potentials of the proteins and polysaccharides are involved. This means that food structures are stabilized by partial release of the intramacromolecular binding potential (e.g. by unfolding of biopolymers) and use of intramacromolecular bonds for formation of a system of intermolecular non-specific interactions.

Several physico-chemical characteristics, such as co-solubility, denaturation-renaturation, cross-linking of biopolymers and behavior of phase separated systems under heating and shearing conditions are responsible for controlling the conformational potential of food biopolymers (*5*). Objectives of this paper are to consider the incompatibility of biopolymers and the formation of inter-biopolymer electrostatic complexes as important physico-chemical factors governing structure formation in foods.

Structural hierarchy of foods

The great diversity of intermolecular interactions of two types of macromolecules is a specific feature of foods. Food systems have four structural levels of biopolymer interactions: submacromolecular, macromolecular, supermacromolecular and macroscopic. In other words, there are four levels of use of the conformational potential of food macromolecules, three microscopic levels and one macroscopic level of food sructures. They are quite similar to the hierarchy of protein structures. The language of food structures, including the three levels of macromolecular interaction, could be regarded as corresponding to the simultaneous use of the main three types of writing: alphabetic, syllabic and hieroglyphic. The fourth macroscopic level implies that all texts written are interlaced, bound and form a library. This structural hierarchy underlies an unprecedented variability of food structures.

The size of junction zones, the nature and strength of intermacromolecular forces is one of the aspects of the concept of structural hierarchy of foods. The junction zones of biopolymer aggregates and inter-biopolymer complexes are regions where segments of two or more macromolecules are joined together.

The submacromolecular level of structure formation corresponds to the size of monomers. It includes interactions between suitable functional groups, e.g. formation of disulfide, hydrophobic and hydrogen bonds of intra- and intermolecular types. This level covers both chemical and enzymatic modification of biopolymers that change structural functions, such as Maillard reactions and casein modification by rennin (chymosin).

The molecular level of structure formation corresponds to the size of macromolecules. It includes changes in the conformation, shape and size of individual macromolecules, as well as dissociation, association and aggregation of macromolecules and formation of junction zones between different macromolecules (i.e. their complexing).

The supermolecular level corresponds to colloidal dimensions and greater. It includes interactions of macromolecular aggregates with each other, with macromolecules, formation of three-dimensional networks of gels and interfacial stabilizing layers of emulsions and foams.

One more aspect of structural hierarchy of foods is colloidal dimensions of many macromolecules and supermacromolecular structures. The submolecular, macromolecular and supermacromolecular structural elements of foods are formed by non-specific interactions of biopolymers. All of them can take place successively and simultaneously and lead to formation of non-equilibrium structures of real food systems. The range of size of food structural elements, the nature and kinetics of biopolymer interactions are of great importance for the formation of food structures that are usually thermodynamically non-stable and non-equilibrium.

Model systems are usually used for the study of food structural hierarchy. A thermodynamic approach based on model systems provides valuable information on the potential behavior of food systems and on the effects of main thermodynamic variables, such as temperature, concentration, pH, etc. (4,6). Nevertheless, it is quite difficult to estimate kinetic aspects of food processing conditions and the relationship between thermodynamic and kinetic aspects of formation of food structure.

The submolecular, macromolecular and supermacromolecular levels of food structure describe the ability of a food biopolymer to perform its structural functions, that is its functionality or its functional properties in foods. Functional properties reflect the physico-chemical properties of a food biopolymer and its complexes with other system components. They affect its use in foods and contribute to the desired structure-property relationship of both the processing food systems and the final products.

The fourth level corresponds to the formation of different macroscopic (morphological) food structures by a participation of food macromolecules and other components, such as lipids and gases forming separate system phases, as well as food granules, fibers and other structural ingredients.

The fourth level of structural hierarchy reflects the macroscopic heterophase structure of foods. For instance, this includes liquid and solid suspensions, emulsions, foams, gels, filled gels and various mixtures of these dispersed systems. Heterophase foods can be regarded as complex engineering constructions. They can have one or more continuous phases and one or more dispersed phases. Foods are usually composites containing dispersed particles of different states that can be gases, liquids and/or solids.

All four structural elements almost always closely interact. Normally, for instance, two- and three-dimensional supermolecular structures formed by biopolymers are important contributory factors for physico-chemical properties of heterophase food systems.

We now turn to the nature of the main non-specific intermacromolecular interactions between the main classes of food biopolymers.

Two types of macromolecular interactions

Complexing and incompatibility of biopolymers. The two main types of interactions of proteins and polysaccharides are related either to the attraction or to the repulsion between these macromolecules. They are responsible for complexing and incompatibility of biopolymers in aqueous media, respectively (1).

Complexes can be regarded as a new type of food biopolymer whose properties differ strongly from those of the macromolecular reactants (8). Electrostatic inter-biopolymer complexes can be soluble and insoluble depending on the degree of mutual neutralization of biopolymers bearing opposite charges. Normally, maximal yield of insoluble complexes corresponds to the ratio of oppositely charged biopolymer reactants, i.e. conditions of their complete mutual neutralization. Soluble inter-biopolymer complexes are usually non-stoichiometric and charged. Insoluble complexes form a separate concentrated liquid or gel-like phase containing both interacting biopolymers. This phase was called a complex coacervate.

Limited thermodynamic compatibility (or incompatibility) means that different biopolymers are not miscible on a molecular level in any proportion. Above a certain concentration it is impossible to obtain a molecular dispersion of incompatible biopolymers in a common solvent. Incompatibility leads to phase separated systems with biopolymers concentrated in different phases (1-3).

Two types of phase separation and flocculation. The entropy difference, which is the driving force for mixing different compounds is very small for mixing biopolymers. It decreases by a factor corresponding to the degree of polymerization of the biopolymers. Therefore, molecularly homogeneous mixtures of biopolymers should only be formed when the mixing enthalpy is negative. Incompatibility, which is typical of biopolymers, can lead to phase separation of mixed biopolymer solutions and depletion flocculation of colloidal dispersed particles made up of and covered by biopolymers. The basic difference between depletion flocculation and limited thermodynamic incompatibility of biopolymers is that depletion flocculation is of a non-equilibrium nature (4-6).

Biopolymer compatibility is governed by interactions of biopolymers between each other and the solvent. Boundary conditions of phase separation are mainly determined by entropic factors given by the excluded volume of their macromolecules. The phase separation threshold for biopolymer mixtures is within the concentration range corresponding to overlapping of macromolecular volumes. When the bulk biopolymer concentration increases and approaches complete occupancy of the mixed solution space by macromolecules this solution breaks down into two liquid phases each mainly containing different biopolymers. In other words a critical concentration of the biopolymers at which their mixed solution breaks down into two phases corresponds to the transition from dilute to moderately concentrated mixed solutions [2-4]. For a rigid compact globular protein the concentration corresponding to the transition from a "dilute" to a "semi-dilute" solution is from 10 to 30%. Normally, for polysaccharides in binary solutions this critical concentration varies from 0.1% to 3%. In the case of a mixed solution of these biopolymers phase separation usually occurs at an intermediate concentration range.

Generally, proteins and polysaccharides are incompatible at a sufficiently high bulk concentration and under conditions inhibiting formation of inter-biopolymer complexes. This mainly occurs at a sufficiently high ionic strength (exceeding 0.2), pH values above the protein isoelectric point (IEP) and at total biopolymer concentrations above 3-4%.

When the attraction forces between non-identical biopolymer macromolecules are larger than between molecules of the same biopolymer two other phenomena can come into a play. These are formation of inter-biopolymer complexes and the bridging flocculation of colloidal dispersed particles made up of or covered by one of the biopolymers.

Mutual neutralization of opposite charges of the randomly mixed macro-ions and formation of the concentrated phase of an insoluble complex favor a decrease in electrostatic free energy of the system. The loss of entropy on complexing is counterbalanced by the enthalpy of interactions of macro-ions and by liberation of counter-ions and water. Electrostatic inter-biopolymer complexing of proteins with polysaccharides occurs within the pH range between the protein IEP and the pK value of the anionic (mainly carboxyl) side groups of the polysaccharides and at low ionic strengths (usually less than 0.3). Under these conditions protein molecules have a net positive charge, i.e. they behave as polycations. The carboxyl-containing anionic polysaccharides are polyanions. At pH values above the IEP, e.g. at neutral pH, positively charged subunits of oligomeric proteins are capable of forming soluble electrostatic complexes with anionic carboxyl-containing polysaccharides. Sulfated polysaccharides can also form soluble complexes at pHs above the protein's IEP.

Inter-biopolymer complexes of oppositely charged proteins form at pHs between IEPs of the proteins.

Unlike biopolymer solutions, colloidal dispersions are thermodynamically unstable. The transition from molecular to colloidal protein causes system destabilization by a polysaccharide additive to be more pronounced. The critical concentration of a polysaccharide which is required for flocculation of a protein dispersion is significantly lower than that for phase separation of protein-polysaccharide mixed solutions. For instance, it was mentioned above that phase separation of a protein-polysaccharide mixed solution usually occurs at a bulk biopolymer concentration exceeding 4%, whilst depletion flocculation can take place at concentrations of less than 0.1%.

High molecular weight and rigidity of macromolecular chains, which are typical of polysaccharides, favor both types of phase separation, i.e. formation of insoluble inter-biopolymer complexes and demixing of mixed solutions of biopolymers as well as bridging and depletion flocculation of dispersed particles coated with one of the biopolymers.

Thus, upon mixing protein and polysaccharide solutions four different types of model systems are obtainable. Two kinds of single-phase systems may be: solutions of inter-biopolymer complexes; or mixed biopolymer solutions with bulk concentrations below the co-solubility limit (i.e. the phase separation threshold) of a given biopolymer pair.

Phase separated systems result from either immiscibility of incompatible biopolymers in a solvent or from formation of an insoluble inter-biopolymer complex. Upon mixing an aqueous dispersion of protein (e.g. a protein suspension or an emulsion stabilized by protein) with a polysaccharide solution two different types of flocculation can occur.

Owing to complexing, an anionic polysaccharide can act as a flocculating agent binding dispersed particles, i.e. inducing bridging flocculation. Depletion flocculation can be regarded as a flocculation of dispersed particles in a non-wettable medium, e.g. a protein suspension in a polysaccharide solution.

All six model mixtures of food biopolymers in aqueous media differ strongly in structure and properties. In real foods these types of systems can be formed simultaneously within different phases.

Inter-biopolymer electrostatic complexes

The submacromolecular, macromolecular and supermacromolecular levels of intermolecular interaction contribute to formation of inter-biopolymer electrostatic complexes. The submacromolecular level corresponds to interactions (i) between oppositely charged (acidic and basic) side groups and (ii) between other available side groups of macro-ions. It may include formation of intermolecular ionic, coordinate, hydrophobic and hydrogen bonds and covalent crosslinks. The molecular level covers interactions between oppositely charged macro-ions.

Larger intermolecular junction zones are formed at the macromolecular level compared to the submacromolecular level of intermacromolecular interaction. Because of the rigid and compact conformation of proteins the number of oppositely charged groups capable of interacting and the length of the junction zone formed by interacting globular proteins are limited. They mainly correspond to the submacromolecular level of interaction. Linear flexible macromolecules can form more extended junction zones and favor steric adjustments between the oppositely charged interacting groups. Here the macromolecular level of interaction is more typical. Junction zone stability mainly depends on the nature and number of interacting groups (i.e. cohesion energy density). For instance, it has been shown that the enthalpy and the temperature of denaturation of the Kunitz trypsin inhibitor bound to pectin are affected by the degree of esterification of the pectin and by the protein/pectin ratio. When a small amount of the protein is bound to an anionic polysaccharide, the denaturation temperature decreases compared to that of the freely accessible protein. This obviously reflects the role of free sites on the matrix that are situated near an adsorbed protein molecule for its unfolding. As the amount of the bound protein increases, the denaturation temperature rises. Presumably, a decrease in size of the junction zones between the polypeptide and polysaccharide chains leads to a shift in equilibrium from the denatured forms to the native forms (3,5).

The composition and properties of an electrostatic inter-biopolymer complex depend on: pHs, ionic strengths; net opposite charge, size, flexibility and ratio of macro-ions; and reactivity and geometrical arrangements of their side groups. Proteins with an unordered structure (such as gelatin, casein or acid denatured globular proteins) are able to form a maximum number of contacts with an oppositely charged linear polysaccharide. At a given pH the weight fraction ratio of gelatin to anionic polysaccharide in an insoluble complex is equal to the ratio of charges of these macroreagents.

Proteins and polysaccharides usually have a number of functional side groups of different nature and with different pK values. Therefore, the formation of structurally regular junction zones between macro-ions is quite unlikely. Junction zones formed by a side-by-side packing of linear polymers may contain many individual chains, especially in the case of insoluble complexes. Since mutual adjustment of macro-ion configuration is necessary for complexing, formation of junction zones is a time consuming process and aging is typical of inter-biopolymer complexes.

The supermolecular level corresponds to the attraction and aggregation of primarily complex particles. This contributes to an increase in the extent of complex formation. Mutual compensation of reagent charges occurs within the volume of dispersed particles formed by soluble and insoluble complexes. Accordingly, solutions of "soluble" inter-biopolymer complexes can usually be regarded as metastable homogeneous systems containing associates of complex particles (8).

Reversible or irreversible complexing. Formation of electrostatic inter-biopolymer complexes is a reversible process under conditions of weak protein-polysaccharide interaction near the protein's IEP. Reversible complexing corresponds to uniform distribution of protein molecules among polysaccharide chains. This means that protein molecules may be transferred between oppositely charged polysaccharide chains and that the formation of complexes includes the exchange reaction of protein molecules between complex particles (5,9). Therefore, normally on the thermograms of inter-biopolymer complexes there is a single peak of protein thermal denaturation whose position does not coincide with that of the initial protein.

Within a narrow range of pH values near and above the protein IEP inter-biopolymer complexes can be formed by a co-operative protein adsorption on anionic polysaccharide chains (9). This results from an enhancement of protein-protein hydrophobic interactions in the vicinity of the protein IEP. Under conditions of a relatively weak protein-polysaccharide interaction free sites located near the site occupied by a protein molecule are preferred for further binding of protein molecules. The exchange of protein molecules between polysaccharide chains, reflecting the reversibility of complexing, results in a non-uniform redistribution of protein molecules among polysaccharide chains. Due to the disproportionation reaction some parts of polysaccharide chains tend to be completely covered by protein molecules, while other parts are free of protein. In the former case the composition goes to that of the electroneutral complex, while the composition of the other complex fraction approaches that of the pure polysaccharide. At pH values markedly below the protein's IEP a stronger inter-biopolymer interaction freezes the redistribution of protein molecules among the polysaccharide chains.

Strong electrostatic interaction between a positively charged protein and an anionic polysaccharide at low ionic strength and at pH's below the protein's IEP is mainly responsible for non-equilibrium complexing. When solutions of oppositely charged biopolymers are mixed under conditions of strong inter-biopolymer interaction, protein molecules act as cross-linking agents between polysaccharide chains. This forms a three-dimensional network of complex gel particles. Non-equilibrium complexes are especially typical of polyelectrolytes with a high charge density. The composition and properties of non-equilibrium complexes depend on formation conditions. Thus, inter-biopolymer non-specific complexing can be either reversible or irreversible, and cooperative or non-cooperative(*1,3,8,9*). Non-equilibrium complexing has been studied in detail for mixtures of bovine serum albumin and dextran sulfate (*9*).

Stability of complexes. Generally, electrostatic complexes dissociate when the ionic strength exceeds 0.3 and when the pH is above the protein's IEP. However, in many cases the extent of dissociation is very small. This is typical of protein denaturation in the complex and the formation of triple protein-multivalent cation (e.g. Ca ion)-anionic polysaccharide complexes. Such triple complexes are usually non-equilibrium. Their composition and properties are determined by their formation conditions.

The formation of protein-anionic polysaccharide and protein-protein complexes implies an increase in local protein concentration. Denaturation promotes aggregation of bound protein molecules and stability of the complex. In the complexed state, renaturation of the denatured protein is inhibited (*3,5*). An example is the irreversibility of thermal denaturation of the Kunitz trypsin inhibitor bound by both anionic polysaccharides and oppositely charged proteins. This is obviously of interest in oilseed protein processing since relatively small amounts of an anionic polysaccharide are able to suppress protein renaturation. Presumably, protein denaturation on the charged polysaccharide matrix leads to: protein unfolding; a higher degree of exposure of hydrophobic groups of unfolded molecules; and diminishing like net charges and mutual attraction. Aggregation of several bound unfolded polypeptide chains can be accompanied by their conformational, renaturation-like, cooperative re-arrangement. This can intensify protein-protein interactions and weaken protein-polysaccharide interactions. Aggregates may then be expelled from the matrix by other free protein molecules and leave the matrix in a non-renaturable form. This supposed "catalytic" function of anionic polysaccharides in stimulating protein-protein interactions is in agreement with the irreversible denaturation of complexes of oppositely charged proteins. This could be exploited for functional protein engineering (*5*).

Functional properties of electrostatic complexes. During electrostatic complexing the net charge of anionic polysaccharides decreases with each successively attached protein macromolecule. This successively reduces the net charge, hydrophilicity and solubility of the resultant complex. It also leads to a diminished IEP of the complex and enlarges the range of pH's where the solution is stable compared to those of the initial protein (*1,3*).

The more compact conformation of protein-polysaccharide complexes compared to those of the initial reagents is probably due to a less hydrophilic nature of junction zones formed by mutually neutralized macro-ions. A compact conformation, typical of electrostatic inter-biopolymer complexes corresponds to a relatively low viscosity of their solutions and to an increase in the critical concentration for gelation (*8*).

Complexes of anionic polysaccharides with proteins of an unordered-structure are formed by interaction of randomly distributed macromolecules centers of mass of which do not coincide. As a result the junction zones can be surrounded by non-bound hydrophilic segments of single-biopolymer chains. Presumably, the junction zones can form a compact hydrophobic core, i.e. a "hydrophobic interior" of a complex particle. This means that the structure of primarily complex particles could be similar to that of globular proteins. Segments of chains that are not incorporated into the junction zones play a key role in dictating the solubility (dispersibility), surface properties, aggregation and gelation of the complex. Their interaction can lead to aggregation of complex particles and formation of a complex gel. Complex gels (e.g. alginate-gelatin, pectin-gelatin, alginate-casein, carrageenan-casein) with properties greatly different from gels of complex components can be formed under conditions where biopolymers taken separately do not gel. This means that inter-biopolymer complexing can result in synergistic gels (1,8-10).

Reasons for using protein-polysaccharide complexes as emulsion stabilizers are their high surface activity, and their ability to form gel-like charged and thick adsorbed layers. Complexing implies intensifying intermacromolecular interactions in the adsorbed layer. This suppresses the competition between different proteins for the interfaces and cause partial unfolding of protein molecules. Mechanical strength, thickness (steric barrier) of the adsorption gel-like layer and electrostatic repulsion of emulsion droplets carrying like net charges are the most important contributory factors to higher kinetic stability of oil-in-water emulsions (1,3,9, 11).

Anionic polysaccharides forming protein-anionic polysaccharide complexes can act as protective hydrocolloids inhibiting aggregation and precipitation of like-charged dispersed protein particles. This protective action can for instance, increase the stability of suspensions of denatured proteins.

Anionic polysaccharides binding colloidal particles can act as flocculents for the precipitation of protein suspensions or emulsions stabilized by a protein at pHs below the IEP. The rate and extent of bridging flocculation depend on the composition of the particle surface, the molecular weight and the conformation of the anionic polysaccharide. Bridging flocculation is used to recover protein from dilute dispersions and for food system clarification. Inter-biopolymer complexes are also used to encapsulate flavors, lipids and other liquid and solid food ingredients.

It should be stressed that many aspects of the behavior of inter-biopolymer complexes remain unstudied. For instance, the glass transition temperature of inter-biopolymer complexes and their behavior at low moisture level have not been studied. Complexing could be of importance for the thermodynamic compatibility and functionality of biopolymers in ice-cream mixes, in systems subjected to thermoplastic extrusion, etc.

Incompatibility

Incompatibility is typical of mixtures of the following biopolymers (3,6-9):
(i) structurally dissimilar polysaccharides (12-16),
(ii) proteins and polysaccharides (17-20),
(iii) proteins belonging to different classes according to Osborne' classification (i.e. albumins, globulins, glutelines and prolamines),
(iv) proteins differing in conformation (e.g., gelatin or casein and globular proteins),
(v) native and denatured forms and aggregated and non-aggregated forms of the same protein (21-24).

A relatively high phase separation threshold is typical of mixed solutions of globular proteins. Presumably this is due to the compactness and rigidity of protein molecules. Incompatibility is strongly dependent upon the conformational state (denaturation) and the difference in hydrophilicity of globular proteins and slightly dependent upon pH, ionic strength and temperature. Protein1 - protein2 - water systems are often notable for highly asymmetric phase diagrams (i.e. the ratio of co-existing phase concentrations). This is probably due to the compactness of molecules and the self-ordered structure of highly concentrated protein phases [2,3]. The latter effect may also contribute to interfacial multilayer adsorption (i.e. of protein aggregates) in oil-in-water emulsions stabilized by proteins [3,4,6].

Compared to mixtures of different proteins, protein-polysaccharide-water systems are distinguished by fairly low phase separation thresholds. Normally, linear polysaccharides are more incompatible with proteins than branched polysaccharides. Incompatibility decreases in the order carboxyl-containing polysaccharides> neutral polysaccharides> sulfated polysaccharides. The larger the difference in molecular weight and in hydrophilicity the more pronounced the incompatibility of the biopolymers. The difference in hydrophilicity and in molecular weight between proteins and polysaccharides is of great importance for phase equilibrium in protein-polysaccharide-water systems, particularly for the asymmetry of their phase diagrams.

Biopolymer compatibility in solution may be controlled by (i) solvent composition, e.g. by addition of salt, sugars, alcohol, complexons binding Ca-ions, adjusting pH; (ii) inducing functional side groups providing specific interactions between biopolymers, e.g. a change in esterification degree of pectin in mixtures with gelatin; (iii) a change in biopolymer conformations, e.g. by protein denaturation and aggregation; (iv) an increase in molecular weight of polysaccharides; (v) dissociation and limited proteolysis of proteins, (vi) producing protein-polysaccharide hybrids or conjugates; and finely (vii) soluble complexes, e.g. protein-lipid, polysaccharide-lipid and protein-polysaccharide complexes.

Effects of incompatibility on food biopolymer functionality.

Oil-in-water emulsions stabilized by protein. Presumably, adsorbed (by oil particles) and dissolved (in the dispersion medium) molecules of the same protein do not recognize each other as being the same. This "incompatibility" between the adsorbed and dissolved protein molecules results in the specific very extended plateau found on adsorption isotherms of globular proteins (*3,4,6*). This plateau covers a wide range of bulk protein concentration: from that required for formation of an adsorbed monolayer to that needed for multilayer formation. This plateau is absent in the casein adsorption isotherm because casein is an unordered structure protein whose subunits are compatible and have a tendency to associate.

The excluded volume effect means that the molecules of one of the biopolymers do not have access to the volume occupied by the molecules of the other incompatible biopolymer. Biopolymers behave as if they were in a solution of a higher concentration. This enhances protein adsorption at the interfaces and stability of emulsions. This means that the effect of a polysaccharide added to the protein solution is not limited to the increased viscosity of the dispersion medium of the emulsion. It also leads to an increase in the protein's thermodynamic activity (real concentration) and surface activity. The critical concentration of the protein corresponding to its multilayer adsorption decreases proportionally with the excluded volume of the added polysaccharide (*3,25,26*). Presumably, formation of adsorbed multilayers results from phase separation in the aqueous dispersion medium of the emulsion and can be regarded as encapsulation of lipid droplets by a protein-rich phase (*3,4,6*). However, when the volume fraction of the protein phase increases, phase separation of aqueous continuous phase of emulsions can lead to emulsion destabilization. An example is the effect of addition of maltodextrin on the heat stability of infant formulae. Phase behavior of systems containing a polysaccharide and a protein of unordered structure, such as casein or gelatin, is notable for low compatibility, low co-solubility and a reduction of both these values with increasing temperature (*17,19,20,27*). Presumably, this causes a decrease in the heat stability of casein-stabilized oil-in-water emulsions by the addition of maltodextrin (*27*). Increasing the polysaccharide molecular weight inhances this effect (*28*).

Rheological properties of phase separated aqueous systems are mainly determined by those of the continuous phase of a system. Generally, however, the viscosity of an aqueous phase separated system is lower than that of the most viscose of a system's phases. The less viscous dispersed phase and the interfacial layers (i.e. depletion layer of low biopolymer concentration and low viscosity) can act as a lubricant in flowing water-in-water emulsions. Low interfacial tension and easy deformability of dispersed particles also affect emulsion rheology.

It should be noted that, unlike foam formation, in emulsions protein-lipid interaction may results in a reduction of conformational stability and hydrophilicity of proteins. Presumably this effect minimizes the difference between proteins in their emulsifying properties. Differences between various proteins have a more pronounced effect on foaming properties (*4,6*).

Multicomponent gels. Mixed and filled gels are the simplest models of solid foods (*10,29*). The relationship between the co-solubility limits of macromolecular components and their critical concentrations for gelation determines boundary conditions for the formation of mixed and filled gels (*1,4*). Mixed gels are obtained when two or more gelling agents are co-soluble and form separately continuous networks. Filled gels are obtained by gelation of a phase separated mixed biopolymer solution. Two or more independent interacting interpenetrating networks can be formed depending on the co-solubility of the biopolymers (*1-4,8,10, 29*).

Excluded volume effects are of special importance in food gels because the shear modulus of a gel is usually proportional to the square of its concentration. In mixed solutions each incompatible biopolymer behaves as if it was in a more concentrated solution, therefore small additions or mixing of gelling agents can result in a several-fold increase of the elastic modulus of the mixed gel. Thus, the main cause of the synergistic effects resulting from blending of gelling agents seems to be biopolymer incompatibility. Mutual concentration effects also result in an increased gelation rate of a mixed biopolymer that can also be sufficiently strong to be referred to as synergistic (*8,10*).

The mechanism of formation of three-dimensional interpenetrating networks of mixed gels under conditions of biopolymer incompatibility is an important point. The gelling agents can be mixed throughout the solvent space since there is a difference in conditions and rate of their gelation. Gelation is always related to aggregation of macromolecules, formation of a three-dimensional network of macromolecular aggregates and to a strong decrease in the excluded volume of the gelling agent macromolecules. Therefore, in a solution containing two different gelling agents, gelation of one of them results in a great decrease in excluded volume of its molecules. Accordingly, the free volume of the system accessible for macromolecules of the second gelling agent increases. This means that compared to the solution of a gelling agent, the dispersion medium of its gel is a better solvent for the other gelling agent. Gelation kinetics and structure-property relationship of gels strongly depend upon the sequence and conditions of gelation of individual gelling agents. Another consequence of the great decrease in the excluded volume effect is that insoluble components of a liquid food system can be dissolved after system gelation.

Some rheological features of filled gels are determined by the nature and state of the filler. Liquid or gel-like dispersed particles filling the gel result from phase separation of a mixed biopolymer solution. Because of co-solubility of biopolymers, both phases of a filled gel may be mixed gels. Incompatibility determines the spherical shape of filler dispersed particles and their very low adhesion to the gel continuous phase. Therefore, an increase in volume fraction of the dispersed phase can be regarded as both an increase in the amount of network defects and a decrease in the cross-section of the gel. The increase in volume fraction of the dispersed phase results in a large decrease in the elasticity modulus of the gel and can be referred as an antagonistic effect of mixing biopolymers (*10*). A change in composition of a liquid two-phase system can produce phase inversion. As a result, mechanical properties of the filled gel can be changed according to difference of the system phases. This can also be referred to either the synergetic or the antagonistic effect of gelling agent blends (*10*).

In the case of attraction (complexing) between the gel matrix and the filler, the latter can be regarded as an "active filler". Active filler acts as a multifunctional crosslinking agent that is able to decrease the critical concentration of gelation, increase the cross-linking density of a network, the elasticity modulus of the gel. For instance, weak, non-covalent associations between the gel macromolecular network and dispersed particles of filler can "induce" and "amplify" thixotropy of a gel. Under mechanical treatment, weak filled gels can reversibly turn from solid to liquid state and back. Such very weak gels are important for stabilization of food suspensions and drinks since they can become fluid when stirred, shaken or poured out of a can.

The mechanical properties of heterophase gels with two continuous phases obey the additivity law. Both the elasticity modulus and the tensile strength of composite food systems with two continuous phases are linear functions of the volume fraction of one of the phases and equal to the sum of these mechanical characteristics of constituent phases (*2,30-32*).

The non-equilibrium nature of mixed and filled gels determines time-dependence of their properties. Kinetics of system phase separation, mass exchange between the phases and their gelation rates are of great importance.

Effects of incompatibility on formation of food macrostructure.

Normally aqueous mixed solutions of different biopolymers are phase separated systems at bulk concentration of macromolecules, pH and ionic strength typical of foods. Therefore, features of water-in-water emulsions are of great significance for a better understanding of the structure-property relationships in foods (6,7).

Phase equilibrium. Membraneless osmosis. Unlike oil-in-water and water-in-oil emulsions, the phases of water-in-water emulsions are in thermodynamic equilibrium. Food system aqueous phases are usually in competition for water and ions. The asymmetry of the phase diagrams underlies water partition between phases of foods. Equilibrium between biopolymer solutions is established by diffusion of their components. When one of the phases gels, water and ion exchange between the system phases can stop before equilibrium is achieved. The reason is that the swollen gel applies a backpressure to the liquid phase (1-4,6-9, 33).

The polysaccharide phase is usually diluted compared to the co-existing protein phase. This underlies a new method for concentrating protein solutions: membraneless osmosis (33). Membraneless osmosis is of great importance for structure formation of many foods, e.g. bread like products [1,2]. Presumably, formation of a strong continuous gel-like gluten phase in classical bread technology is caused by a great increase in its concentration due to starch gelatinization. Starch is more competitive for water than gluten which is a highly hydrophobic protein. Therefore, continuous phase solidification can result from gluten dewatering at temperatures above the temperature of starch gelatinisation. This means that the solid foam structure and the form of the loaf is fixed by dewatering of the continuous gluten phase and by protein denaturation in a highly concentrated state. Accordingly, an excessive amount of water (depending on flour quality) usually makes the dough sticky and difficult to process resulting in wet and soggy bread.

Membraneless osmosis can be used for modeling food processing including the concentration of one of the system phases. For instance, on freezing of many food dispersed systems, e.g. meat products, water separation in the form of ice leads to irreversible dewatering accompanied by formation of protein aggregates and gelation of the protein phase. This irreversible cryotexturization is used for producing (by freezing and thawing the protein suspensions) foods such as kori-tofu and textured vegetable proteins.

Low interfacial tension. Low interfacial tension of water-in-water emulsions is due to the similarity in composition of their co-existing phases. This is because water is the main component of both phases. The second reason is the partial co-solubility of biopolymers in the co-existing phases (7).

Low interfacial tension between equilibrium phases and their close viscosity leads to easy deformability, dispersibility, coalescence and good spinnability of dispersed phase particles. Another characteristic feature of water-in-water emulsions is interfacial area that developed between co-existing phases with particles varying from spherical to fibrous form. Flowing water-in-water emulsions becomes anisotropic as a result of deformation of dispersed particles. Liquid filaments arising in flow can be fixed by rapid gelation of one or both phases of the system (34). The result may be either fibers or anisotropic gels filed with oriented filaments. Formation conditions of filamentous protein particles primarily depend on shear forces, kinetics of phase separation and gelation. Structure formation of a heterophase food system in flow is called spinneretless spinning [1-3,6,7]. This process is of importance for formation of food structures since, during processing, food systems are always subjected to shear treatments such as mixing, pumping, homogenization, centrifuging and extruding. Presumably, spinneretless spinning is the main mechanism of the formation of fibrous precipitates of isolated proteins and polysaccharides as well as of structure formation of thermoplastic extrusion products (1-3.30-35).

Interfacial layers in two-phase systems. An important feature of two-phase biopolymer systems is the presence of low biopolymer concentration and low-viscosity interface layers between immiscible aqueous phases. Formation of interfacial layers reflects a trend towards minimizing contact between non-compatible macromolecules. Presumably, these interfacial layers with thickness corresponding to the large size of macromolecules can concentrate non-polar particles and form a new continuous phase. For instance, interface layers seem to contribute to formation of the lipid continuous phase of low-fat spreads containing less than 30% (10,36,37). This assumption was made to explain how the continuous lipid phase encapsulates a large amount of aqueous phases containing gelatin and maltodextrin (6,7,10).

Due to biopolymer co-solubility, both phases contain gelatin and a polysaccharide in different proportions. Gelatin is more surface active at the oil-water interface than the polysaccharide. Therefore, gelatin could dominate at the interfacial surfaces of both aqueous phases. The gelatin gel surface formed by contact with a non-polar phase such as oil or air is highly hydrophobic (10,11). Therefore, the surfaces of both immiscible aqueous phases of the water-in-water emulsion used for producing spreads should be highly hydrophobic. Accordingly, lipids added to a water-in-water emulsion are concentrated at the interfacial layers and could form a continuous three-dimensional honeycomb-like structure of thin lipid layers. This honeycomb-like structure filled with immiscible aqueous biopolymer solutions can be fixed by solidification of all three system phases upon cooling. Both aqueous phases form thermoreversible gels with melting temperatures close to mouth temperature. In mouth, highly viscous melted droplets covered by thin lipid layers are formed. The sensory perception of the thermoreversible gel granules covered by a lipid layer can imitate that of both butter and mayonnaise droplets depending on size, concentration and viscosity of the dispersed particles. Thus, phase separation and formation of water-in-water emulsions with hydrophobic surfaces between the aqueous phases can control lipid functionality. Presumably, honeycomb-like lipid structures are typical of low-fat spreads, ice-cream mixes and other similar foods (6,7,27).

Conclusion

It should be stressed that the concept of the conformational potential of food macromolecules and the concept of structural hierarchy of foods seem to form a useful approach for considering molecular origins of specific structural features of a great number of foods. The study of the thermodynamic incompatibility and complexing of biopolymers, the nature of non-specific macromolecular interactions in mixed solutions and gels forms a sound basis for understanding food composition-property relationships, including synergistic and antagonistic effects. This trend is of great importance for both improving conventional food technologies and designing novel formulated foods.

Acknowledgments

The author is very grateful to Dr. Elizabeth Prior for editing the manuscript.

Literature Cited

1.   Tolstoguzov,V.B. Functional properties of protein-polysaccharides mixtures. In *Functional Properties of Food Macromolecules*; Mitchell, J.R., and Ledward, D.A. Eds.;Elsevier Appl.Sci.: London, **1986**; pp 385-415.
2.   Tolstoguzov, V.B. Some physico-chemical aspects of protein processing into foodstuffs. *Food Hydrocolloids*, **1988,** *2*, 339-370.
3.   Tolstoguzov,V.B. Functional properties of food proteins and role of protein-polysaccharide interaction. *Food Hydrocolloids*, **1991**, *4*, 429-468.
4.   Tolstoguzov,V. The functional properties of food proteins. In *Gums and Stabilisers for the Food Industry*; Phillips, G.O., Williams, P.A. and Wedlock, D.J. Eds.; IRL Press: Oxford. **1992,** Vol.6; pp 241-266.
5.   Tolstoguzov,V. Some physico-chemical aspects of protein processing into foods. *In Gums and Stabilisers for the Food Industry;* Phillips, G.O., Williams, P.A. and Wedlock, D.J. Eds.; Oxford University Press: Oxford, **1994,** Vol.7; pp 115-124.
6.   Tolstoguzov,V. Thermodynamic aspects of food protein functionality. In *Food Hydrocolloids: Structure, Properties and Functions*. Nashinari K. and Doi E. Eds.; Plenum Press: NewYork. **1994;** pp 327-340.
7.   Tolstoguzov,V. Thermodynamic incompatibility of food macromolecules, In *Food Colloids and Polymers :Structure and Dynamics.* Walstra, P. and Dickinson, E. Eds.; Royal Society of Chemistry: Cambridge. **1993,** pp 94-102.

8.   Tolstoguzov, V.B. Interactions of gelatin with polysaccharides. In *Gums and Stabilisers for the Food Industry*. Phillips, G.O., Williams, P.A. and Wedlock, D.J. Eds.; Oxford University Press: Oxford, **1990**, Vol.5; pp 157-175.

9.   Tolstoguzov, V.B., Grinberg, V.Ya., and Gurov, A.N. Some physicochemical approaches to the problem of protein texturization. *J. Agric. Food Chem*, **1985**, *33*, 151-159.

10.   Tolstoguzov,V. Some physico-chemical aspects of protein processing in foods. Multicomponent gels. *Food Hydrocolloids*, **1995**, *9*, 4.

11.   Burova,T.V., Grinberg,N.V., Grinberg,V.Ya, Leontiev,A.L. and Tolstoguzov,V.B. Effects of polysaccharides upon the functional properties of 11S globulin from broad beans. *Carbohydrate Polymers*, **1992**, *18*, 101-108.

12.   Antonov, Yu.A., Pletenko, M.G., and Tolstoguzov, V.B. Phase state of water-polysaccharide-1-polysaccharide-2 systems. Macromolecules (*Vysokomolek. Soed., USSR*), **1987**, *A29*, 2477-2481.

13.   Antonov, Yu.A., Tolstoguzov, V.B., and Pletenko, M.G. Thermodynamic compatibility of polysaccharides in aqueous-media. Macromolecules (*Vysokomolek. Soed.,USSR*), **1987**, *A29*, 2482-2486.

14.   Kalichevsky, M.T., Orford,P.D. and Ring, S.G. (The incompatibility of concentrated aqueous solutions of dextran and amylose and its effect on amylose gelation. *Carbohydrate Polymers*, **1986**, *6*, 145-154.

15.   Kalichevsky, M.T., and Ring, S.G. Incompatibility of amylose and amylopectin in aqueous solution. *Carbohydrate Research*, **1987**, *162*, 323-328.

16.   Walter,H., Johansson,G. and Brooks,D.E. Partitioning in Aqueous Two-Phase Systems:Recent Results.Review. *Analytical Biochem.*, **1991**, *197*, 1-18.

17.   Grinberg, V.Ya., and Tolstoguzov, V.B. Thermodynamic compatibility of gelatin and some D-glucans in aqueous media. *Carbohydrate Research*, **1972**, *25*, 313-320.

18.   Tolstoguzov, V.B., Belkina, V.P., Gulov, V.Ya., Titova, E.F., Belavtseva, E.M., and Grinberg, V.Ya. Phasen Zustand, Struktur und mechanische Eigenschaften des gelartigen Gystems Wasser-Gelatin-Dextran. *Staerke*, **1974**, *26*, 130-137.

19.   Antonov, Yu.A., Grinberg, V.Ya., and Tolstoguzov, V.B. Phasengleichgewichte in Wasser/Eiweiss/Polysaccharid-Systemen. 1. Systeme Wasser/Casein/saures Polysaccharid. *Staerke*, **1975**, *27*, 424-431.

20.   Antonov, Yu.A., Grinberg, V.Ya., and Tolstoguzov, V.B. Phase equilibria in water-protein-polysaccharide systems. 2 Water-casein-neutral polysaccharide systems. *Coll. Polym., Sci*, **1977** *255*, 937-947.

21.   Polyakov, V.I., Grinberg, V.Ya., Antonov, Yu.A., and Tolstoguzov, V.B. Limited thermodynamic compatibility of proteins in aqueous solutions. *Polymer Bull*, , *1*, 593-597.

22.   Polyakov, V.I., Popello, I.A., Grinberg, V.Ya., and Tolstoguzov, V.B. Thermodynamic compatibility of proteins in aqueous media. 3.Studies on the role of intermolecular interactions in the thermodynamics of compatibility of proteins according to the data of dilution enthalpies. *Nahrung*, **1986**, *30*, 81-88.

23.   Polyakov, V.I., Grinberg, V.Ya., Popello, I.A., and Tolstoguzov, V.B. Thermodynamic compatibility of proteins in aqueous medium. *Nahrung*, **1986**, *30*, 365-368.

24.   Tombs,M.P., Newsom, B.C. and Wilding, P. Protein solubility: Phase separation in arachin-salt-water systems. *Int. J. Prot. and Peptide Res.*, **1974**, *6*. 253-277.

25.   Tsapkina, E.N., Semenova, M.G., Pavlovskaya, G.E., Leontiev, A.L., and Tolstoguzov, V.B. The influence of incompatibility on the formation of adsorbing layers and dispersion of n-decane emulsion droplets in aqueous solution containing a mixture of 11S globulin from Vicia faba and dextran. *Food Hydrocolloids*, **1992**, *6*, 237-251.

26.   Pavlovskaya, G.E., Semenova, M.G, Thzapkina, E.N., and Tolstoguzov, V.B. The influence of dextran on the interfacial pressure of adsorbing layers of 11S globulin Vicia faba at the planar n-decane/aqueous solution interface. *Food Hydrocolloids*, **1993**, *7*, 1-10.

27.   Tolstoguzov,V. Applications of phase separated biopolymer systems. In *Gums and Stabilisers for the Food Industry*. Phillips, G.O., Williams, P.A. and Wedlock, D.J. Eds.; Oxford University Press: Oxford, **1996**; (in press).

28.   Cruijsen, J.M.M., van Boekel, M.A.J.S. and Walstra, P. P.Effect of malto-dextrins on the heat stability of caseinate emulsions. *Netherlands Milk & Dairy J.*, **1994,** *48*, 177-180.

29.   Tolstoguzov, V.B., Braudo,E.E. Fabricated foodstuffs as multicomponent gels. *J. Texture Studies*, **1983**, *14*, 183-212.

30.   Tolstoguzov,V.B. Creation of fibrous structures by spinneretless spinning. In *Food Structure - Its Creation and Evaluation.* Blanshard,J.M.V. and Mitchell,J.R. Eds.; Butterworths: London, **1988;** pp 181-196.

31.   Tolstoguzov,V. Thermoplastic extrusion - mechanism of the formation of extrudate structure and properties. *Food Technology International,* **1991**, pp 71-75.

32.   Tolstoguzov,V. Thermoplastic extrusion - the mechanism of the formation of extrudate structure and Properties, *J. Amer. Oil Chem.Soc.,* **1993**, *70,* 417-424.

33.   Tolstoguzov, V.B. Concentration and purification of proteins by means of two-phase systems. *Food Hydrocolloids,* **1988**, *2,* 195-207.

34.   Tolstoguzov,V.B., Mzhel'sky,A. and Gulov,V. Deformation of emulsion droplets in flow. *Coll. Polym. Sci.,* **1974**, *252*, 124-132.

35.   Tolstoguzov,V. Development of texture in meat products through thermodynamic incompatibility. In *Developments in Meat Science - 5.* Lawrie,R.A. Ed.; Elsevier Applied Science: London. **1991**; pp 159-189.

36.   Cain,F.W., Jones,M.G. and Norton,I.T. Spread. *European Patent* No 0237120 (**1987**).

37.   Kasapis,S., Morris,E.R., Norton,I.T. and Gidley, M.J. Phase equilibria and gelation in gelatin/maltodextrin systems - Part 2: polymer incompatibility in solution. *Carbohydrate Polymeres,* **1993**, *21*, 249-259.

Chapter 2

Macromolecular Interactions of Food Proteins Studied by Raman Spectroscopy

Interactions of β-Lactoglobulin, α-Lactalbumin, and Lysozyme in Solution, Gels, and Precipitates

Eunice C. Y. Li-Chan

Department of Food Science, University of British Columbia, Vancouver, British Columbia V6T 1Z4, Canada

Coagulation and gelation of food proteins are a manifestation of intermolecular interactions occurring at high solute concentrations. In order to study these interactions, analytical methodologies are required which can be applied to study the structural characteristics of proteins at these high concentrations, in solutions as well as in insoluble or gelled phases. Visible laser Raman spectroscopy is a useful tool for this purpose. Changes reflecting the microenvironment and interactions of aromatic and aliphatic residues, disulfide bonds, hydrogen bonds and relative proportions of secondary structural types can be monitored, by detailed investigation of various bands in the Raman spectrum arising from vibrational motions of the different amino acid residue side chains as well as the polypeptide backbone. This chapter illustrates a few applications of Raman spectroscopy to study interactions of food proteins in solutions, gels and coagula. Changes in protein structure during thiol- or heat-induced conversion of solutions of hen egg white lysozyme to opaque gels, and of β-lactoglobulin and α-lactalbumin into transparent gels have been monitored. Evidence of protein-protein interactions in unheated as well as heated (90 °C, 30 minutes) binary mixtures of the two whey proteins and lysozyme is also presented.

Gelation of globular proteins has intrigued many researchers for decades. Of great interest is an elucidation of the basic mechanisms to explain the formation of protein gels, as well as the great diversity of the resultant gel properties, including rheological characteristics, optical properties such as transparency or opacity, and thermo-reversibility or irreversibility. Not all proteins can form gels; some may remain in the sol phase, while others may form precipitates or coagula. Even for those proteins which have the potential to form gels, the sol-gel transition is markedly dependent on specific condition of pH, ionic strength, protein concentration and composition of other components in the system. Processing conditions are also important variables. Heat-set gels are the most common, but other processes including high-pressure treatments are increasingly being explored.

Various models have recently been proposed regarding molecular structure of food proteins which result in distinctive gel properties. However, few studies have actually investigated the gel protein structure *in situ*. The objective of this chapter is to present an overview on current hypotheses regarding the heat-induced gelation of globular proteins, and to illustrate the potential use of Raman spectroscopy as a tool to study structural characteristics of food proteins which may play a role in intermolecular interactions resulting in gel or coagulum formation. Macromolecular interactions of two major proteins in whey (α-lactalbumin and β-lactoglobulin) and a basic protein from hen egg white (lysozyme) are used as examples.

Gelation of Globular Proteins

Gelation may be defined as the formation of a continuous and well-defined, three-dimensional network of particles or polymers, accompanied by immobilization of the solvent water.

Early investigators proposed a two-step process for gelation or coagulation involving globular proteins (*1*). Intramolecular changes were postulated to lead to a conversion of protein molecules from the native to denatured state; intermolecular interactions or association of the denatured molecules would then lead to formation of a three-dimensional network. Based on this general two-step mechanism, a 4-step sequence of events was proposed in the heat-induced gelation of proteins (*2,3*):

Step I:	$N \rightarrow D \ (U)$
Step II:	$D \ (U) \rightarrow A$
Step III:	$A \rightarrow S$
Step IV:	$S \rightarrow G$

The first step may be brought about by increasing temperature to 60-80°C, resulting in an intramolecular first-order transition of compact native (N) molecules to denatured (D) or unfolded (U) states, due to opposing temperature dependencies of the entropic and enthalpic forces which stabilize the native protein structures. Non-specific interactions of exposed hydrophobic regions in the unfolded molecules are generally assumed to be the driving force for the next step, but the precise molecular mechanisms which may be responsible for formation of aggregates (A) during and after unfolding are still obscure (*3*). Hydrogen bonding, ionic interactions and disulfide crosslinks are postulated to contribute to stabilization of these soluble aggregates, which are the basic building blocks leading to strand (S) or pro-gels, and ultimately to gels (G).

The formation of globular protein gels from aggregates is postulated to fall into one of two categories - random clustering, and a linear chain or "string-of-beads" structure (*4*). The type of structure which is formed depends on both the intrinsic properties of the proteins as well as the specific heating conditions, and ultimately is reflected in the distinctive characteristics of the resulting coagulum or gel.

Molten Globule State.   Although the earliest work on gelation theory commonly assumed that the denatured state (D) consisted of almost completely unfolded molecules (U) which then formed aggregates through side-by-side associations, this assumption was challenged at least forty years ago (*5*).   In recent years, many researchers have supported the involvement of a partially unfolded state, known as the "molten globule state" or "compact denatured state" (*6-11*).

As summarized by Hirose (*10*), the conformational properties of protein molecules in the molten globule state can be characterized as follows:

(1)   polypeptide backbone with native-like secondary structure, in contrast to amino acid side chains with denaturation-like environment ;

(2)   higher degree of expansion of molecule compared to the native state, but higher degree of compactness compared to the denatured state;

(3)   increased exposure of hydrophobic groups, as reflected by binding of hydrophobic fluorescent probes;

(4)   absence of significant cooperativity in the temperature-dependent unfolding from molten globule to denatured state ;

(5)   enhancement in dynamic accessibility of the peptide N-H protons, as estimated by hydrogen exchange rates.

Thus, the molten globule state is postulated to be a unique state with partially folded conformation, having some similarities yet being distinct from both the native and the fully denatured forms. The term "globule" refers to a native-like compactness, while "molten" refers to the increased enthalpy and entropy on transition from the native structure to the new state, analogous to the melting of solid into liquid.   Interestingly, the molten globule state has also been proposed to explain the mechanism of protein folding, being the intermediate stages formed during the transition from a completely unfolded molecule to the native state. Although the molten globule state has been suggested to be involved in various functional properties of food proteins, including gelling, emulsifying and foaming properties (*10*), it is not clear whether this involvement in functionality reflects an incomplete unfolding from native to complete random coil or is a result of partial re-folding from the completely unfolded state.

Coagula versus Transparent and Opaque Gels. Differences between coagula, turbid gels and transparent gels have been suggested to be attributable primarily to the varying degree and order of association of denatured molecules or aggregates (*12*).   It is also possible that underlying differences in conformation of denatured states *per se* are responsible.   However, detailed knowledge about the conformation of globular protein molecules in heat-induced gels at the molecular level is scant (*11*).

Shimada and Matsushita (*13*) proposed that the propensity of a protein to form translucent as opposed to coagulum-type gels could be predicted from its amino acid composition, molecular weight, and net hydrophobicity.   They suggested that proteins containing above 31.5% of certain hydrophobic amino acid residues form coagulum-type gels, whereas those with less than 31.5% hydrophobic residues form translucent-type gels.   In contrast, Wang and Damodaran (*14*) presented evidence indicating that hardness or gel strength of typical globular gels is fundamentally related to size and shape of the polypeptides in the gel network, rather than to their chemical nature such as amino acid composition and distribution. They also suggested that a minimum weight-average molecular weight of 23,000 is required to form a self-supporting gel network.

In fact, it has been demonstrated that a single protein species may form either transparent or opaque gels, depending on the gelling conditions, including pH, ionic strength, protein concentration, heating rate and final temperature (9, 11, 15-16). Based on recent studies, transparent gels are believed to result from a three-dimensional network of linear polymers, the so-called "strings of beads" model, while random agglomerates of large aggregates would yield turbid or opaque gels (3-4, 9). Hence, turbid gel formation is favored by rapid heating rate, high heating temperature, pH close to the isoelectric point, and low ionic strength. Under these conditions, random or multi-site interactions of the denatured protein aggregates occur due to low repulsive forces and high attractive forces. A predominance of random protein-protein interactions resulting in extremely large molecular weight aggregates could lead to formation of precipitates or coagula, distinguished from gels by their inability to hold water. In contrast, a greater balance of repulsive to attractive forces restricts the extent of protein-protein interactions, and allows a more ordered association of partially denatured molecules into linear polymers which are then linked into a network observed as a transparent gel.

Protein Structure in Gels or Precipitates. One of the earliest proposals on the interactions leading to protein gels suggested the alignment of unfolded peptide chains and formation of extensive regions of β-sheets involved in intermolecular crosslinks (17). By measuring secondary structure profiles of the native proteins and aggregated forms, it should be possible to test this hypothesis (5). Spectral methods may be used to determine the presence of aggregation before visible precipitation occurs, to discriminate between native and denatured forms of the protein molecules. Until recently, however, most methods were unable to directly monitor protein structure in solid phases such as gels or precipitates. Fortunately, improvements in vibrational spectroscopy, mass spectrometry and laser light scattering techniques, from the viewpoint of measurement as well as interpretation, can now provide much better characterization of protein structure in solution or solids (18).

Tani et al. (11) studied soluble aggregates, the intermediate stage to transparent gels of ovalbumin, hen egg white lysoyzme and bovine serum albumin. Their investigation showed that the proteins in these linear aggregates remained partially folded, with some hydrophobic regions becoming exposed to the solvent environment; these observations were reported to be consistent with the molten globule theory. Wang and Damodaran (19) conducted circular dichroic analysis of the protein fluid expressed from protein gels upon centrifugation, to indirectly elucidate the structural state of globular proteins in gels. Gelation of bovine serum albumin involved conversion of α-helical and aperiodic structures into β-sheet conformation. Soy proteins, which contain high proportion of β-sheet in their native state, showed a reduction in the sheet content and an increase in aperiodic structure in the gel state.

Vibrational spectroscopic techniques including infrared and Raman spectroscopy can be applied for structural analysis of proteins in the solid phase, including opaque samples. The relative proportions of α-helix and β-sheet of twelve proteins in their native state and in precipitates obtained using the chaotropic salt KSCN were estimated from analysis of the Amide I band in Raman spectra (*20*). Formation of β-sheet structures with a concomitant loss of α-helical structures were observed in the precipitates, and it was hypothesized that formation of β-sheet strands may be a fundamental phenomenon in self-associating or aggregating protein systems. Clark et al. (*5, 21*) also suggested the characteristic generation of sheet structure during aggregation processes. However, based on qualitative infrared and Raman spectroscopic analysis, no obvious correlation was observed between secondary structure content and gels of varying clarity and texture. Clark and Lee-Tuffnell (*5*) suggested that while β-sheet structures appear to be characteristic components in gel formation, they may play only a limited role in intermolecular interactions; instead other bonding mechanisms involving non-specific hydrophobic interactions may be responsible for intermolecular crosslinking.

Gelation of whey proteins, in particular, α-lactalbumin and β-lactoglobulin. Whey protein isolates and concentrates are important ingredients in food products, due in many respects to their ability to form heat-induced gels (*22,23*). Numerous studies have been published on the major whey proteins, including β-lactoglobulin, α-lactalbumin and bovine serum albumin, with respect to thermal denaturation studied by differential scanning calorimetry (DSC), Fourier transform infrared (FTIR) spectroscopy, and circular dichroism (CD), as well as studies on the mechanical or rheological behavior of gels. Nevertheless, there are still many conflicting reports and confusion regarding gelling properties and structural changes of the whey protein molecules during gelation under various conditions. This may be due to the fact that very few *in situ* studies have been reported, and that the secondary and tertiary structure of the whey proteins may be dependent on protein concentration (*24,25*). Furthermore, different purity of protein preparations may lead to variable results.

Although denaturation is commonly assumed to precede gelation, the denaturation temperature of a protein is not necessarily related to its gelling behaviour. The denaturation temperature (T_d) ranges have been reported as 59-62 and 76-82 °C for α-lactalbumin and β-lactoglobulin, respectively (*3*). However, the latter is the major gelling agent in whey. The content of intramolecular disulfide bonds in the native structure of the two proteins appears to be of minor importance in heat stability or gelation, since the more thermostable yet easily gelled β-lactoglobulin molecule contains only two disulfide bridges, compared to four disulfide bridges in α-lactalbumin. Paulsson and Visser (*26*) proposed that the high proportion of β-sheet structure in β-lactoglobulin may explain its observed thermostability. However, notwithstanding this stability, thermodenaturation of β-lactoglobulin is irreversible and leads to gel formation, probably as a result of a thiol group which is capable of promoting thiol-disulfide and disulfide-disulfide interchange reactions. In contrast, although α-lactalbumin is less thermostable, it shows 80-90% reversibility as assessed by DSC after heat treatment in the temperature range 20-120°C at 10°C/min, and subsequent cooling.

β-Lactoglobulin as well as whey protein isolates may form transparent or opaque gels with different rheological characteristics, depending on the specific composition and heating conditions (16, 22-31). Generally, translucent or transparent gels were formed at neutral pH conditions in the absence of added salt ions, or in the presence of low (e.g. 20-30 mM) concentrations of NaCl. Opaque gels were formed in the presence of $CaCl_2$, higher NaCl concentrations or pH close to the isoelectric point.

Recent studies have attempted to study the protein structure of β-lactoglobulin in the soluble aggregate and gel phases. Iametti et al. (32) studied the concentration dependence of various structural parameters (measured using intrinsic and probe fluorescence, electrophoresis and gel permeation chromatography) during heating and cooling of 3.8, 8 and 16 mg/ml solutions. It was concluded that irreversible modification of the tertiary structure is not concentration dependent, while the temperature required for the occurrence of protein swelling (the inital step in formation of soluble aggregates) increases with protein concentration. Stabilization of aggregates by intermolecular disulfide bonds was only concentration-dependent at temperatures below 75°C. Loss of native structure was reported to occur more quickly at higher protein concentrations (e.g. 100 mg/ml) representative of conditions conducive to gelation (30). *In situ* measurement of CD spectra of solutions varying in concentration from 0.1-100 mg/ml indicated that secondary and tertiary structure of β-lactoglobulin are dependent on protein concentration, both before and after heating (25). A concentration-dependent increase in β-sheet was reported, which was independent of salt concentration and heat denaturation. In the absence of salt, heating of low (0.1-10 mg/ml) concentration solutions resulted in far-UV CD spectra consistent with α-helix formation. In contrast, an increase in β-sheet content was observed in the clear gels obtained by heating 70 mg/ml solutions. Near-UV CD spectra showed greater retention of tertiary structure after heating at higher concentrations. FTIR and Raman spectroscopic studies showed increasing anti-parallel β-sheet content with little change in helical content in the gels formed by heating 3.5% (33) and 15% (24) solutions.

Conflicting reports exist in the literature on the gelling behaviour of α-lactalbumin. Several studies have suggested the inability of α-lactalbumin to form a self-supporting gel under a wide variety of conditions, including 8% solutions at pH 6.8 and 80 °C for 30 minutes (34), 3-3.5 % solutions at pD 6.2 or 7.8 and 80°C for 1 hour (33), and 20% solutions at pH 6.6 and temperatures up to 95 °C (35). In contrast, other researchers have demonstrated transparent gel formation after heating 7% solutions at pH 7 and 80 °C for 3 hours (36) or after heating 15% solutions at pD 6.8 and 90°C for 30-90 minutes (24). Some of the confusion may arise from differences in the particular preparations of α-lactalbumin used in these studies, including the variability in calcium content. The holo- and apo-forms of this protein have been demonstrated to show a 20°C difference in the T_d values as well as in surface hydrophobicity as monitored by the temperature dependence of intrinsic fluorescence and binding of the fluorescence probe, 1-anilinonaphthalene-8-sulfonate or ANS (37). Small amounts of thiol-reagents such as glutathione have been shown to induce gelation of α-lactalbumin (38), and it is possible that traces of impurities in the protein preparation would contribute to this route for gelation. Increasing content of β-sheet accompanied by decrease in content of α-helix were observed in the transparent gels formed after heating at 90°C for 30-60 minutes (*24*).

Gelation of lysozyme. Hen egg white lysozyme, like α-lactalbumin, is an example of a protein which does not gel easily upon heat treatment. For example, although T_d for lysozyme solution at pH 7 was determined by DSC to be 75°C (*39*), lysozyme did not gel after heat treatment at 80 (*40*) or even 95 °C (*41*).

Formation of opaque lysozyme gels could, however, be induced by the addition of thiol reagents such as 2-mercaptoethanol or dithiothreitol, even under mild temperature conditions (*40,42*). Partial reduction of the intramolecular disulfide bonds stabilizing the native lysozyme structure was hypothesized to be a pre-requisite to gel formation, by allowing flexibility or unfolding of the lysozyme molecules, and thereby enabling protein-protein interactions through exposed groups. However, intermolecular disulfide crosslinks were not a necessary component of the resulting gel network. On the basis of the ability of gels to be solubilized in various denaturing and/or reducing solvents, it was suggested that non-covalent interactions, predominantly of the hydrophobic or hydrogen-bonding type, were involved in the opaque gel network structure. A two-step heating procedure was applied to the formation of transparent lysozyme gels (*11,15*). Linear aggregates in a clear sol were formed by initial heating at 80 °C for 20 minutes in the presence of 7.5 mM dithiothreitol, but in the absence of salt. Transparent gels were obtained by reheating in the presence of 50 mM NaCl. Far-UV CD spectra indicated only slight changes in secondary structure of the molecules in the heat-denatured, linear aggregates compared to native lysozyme. On the other hand, results of near-UV CD, intrinsic tryptophan fluorescence and ANS fluorescence probe measurements indicated exposure of some aromatic or hydrophobic regions to the solvent environment.

Influence of protein mixture interactions in gelation and aggregation. Since food proteins are commonly used as mixtures of two or more protein components, as in the case of commercial protein concentrates or isolates, there has been much interest on investigating the nature of macromolecular interactions occurring in protein mixtures before and after heating, and their influence on thermal denaturation characteristics and rheological properties of heat-induced aggregates.

DSC thermograms suggested no interactions between whey proteins (α-lactalbumin, β-lactoglobulin and bovine serum albumin) as monitored by their denaturation temperatures, although the reversibility of α-lactalbumin thermodenaturation disappeared when in mixture with β-lactoglobulin (*26*). Since heat-induced interactions were indicated by ion exchange and gel permeation chromatographic studies, it was suggested that these interactions occur not in the unfolding reactions which could be detected by DSC, but in the aggregation reaction (*26*). Bovine serum albumin and α-lactalbumin were reported to contribute to the storage modulus (G') of mixed protein gels made primarily with β-lactoglobulin (*36*). Stronger gels were formed in 1:1 (w/w) mixtures of β-lactoglobulin and bovine serum albumin heated at 80°C for 30 minutes, compared to the individual proteins (*43*). Transparent, stronger gels were also formed by addition of 3% α-lactalbumin to 6% bovine serum albumin, although the soluble aggregates formed in the heated mixture had lower molecular weights than those from bovine serum albumin alone. Protein-protein interactions to form soluble aggregates through the thiol oxidation reaction during heating were proposed to cause the formation of a finer, more uniform matrix leading to a stronger and more stable gel in these mixed systems.

The effects of glutathione and NaCl addition on the gel strength upon heating mixtures of α-lactalbumin and β-lactoglobulin were reported by Legowo et al. (38). Intensification of gelation was observed by interactions of the two proteins; in the presence of glutathione, α-lactalbumin was postulated to play a more important role in the gel formation of the mixed proteins, whereas in the absence of glutathione, a high proportion of β-lactoglobulin resulted in gel formation. The role of reactive thiol in altering intramolecular disulfide bonds and consequently in conformational changes conducive to gel formation was postulated, although the specific mechanisms of interaction between the two proteins in the gel formation were unknown.

Macromolecular interactions between negatively charged proteins and positively charged proteins have been reported to enhance functional properties including foaming (44) and aggregation phenomena or gelation (34-35, 43, 45-47). DSC thermograms under different conditions of pH and NaCl indicated that interactions between lactoferrin and β-lactoglobulin were primarily electrostatic in nature; however, interactions were demonstrated even under conditions in which electrostatic attraction were diminished (26). Insoluble precipitates were formed in mixtures of lysozyme and either α-lactalbumin or β-lactoglobulin (47). The interactions and amount of precipitation varied depending on the concentration of each protein in the mixture, the ionic strength and pH of the solution; a small amount of soluble complex was also recovered by ion exchange chromatography of the supernatant in the mixtures. Molecular modelling studies using interactive docking of the crystal structure, suggested specific electrostatic interactions between Glu-35 and Asp-53 in the catalytic site of lysozyme, with Lys-138 and Lys-141 at the dimerization site of β-lactoglobulin. For α-lactalbumin-lysozyme mixtures, however, the modelling studies suggested non-specific electrostatic interactions. In addition to the expected electrostatic interactions, preliminary investigation of solutions of the protein mixtures using high resolution nuclear magnetic resonance (NMR) suggested the involvement of hydrophobic groups in the protein-protein interactions (47).

Raman Spectroscopy to study Interactions in Food Protein Systems

Raman spectroscopy is a branch of vibrational spectroscopy which can be a useful probe of protein structure, since both the intensity and frequency of vibrational motions of the amino acid side chains or polypeptide backbone are sensitive to chemical changes and the microenvironment around the functional groups (48). Bands in the Raman spectrum arising from the amide I, amide III and skeletal stretching modes of peptides and proteins are useful to characterize backbone conformation, including changes in the secondary structure fractions. Valuable information can also be obtained on various side chains, based on stretching and bending deformations of the S-S or S-H groups in cystine or cysteine residues, CH groups of aliphatic hydrophobic residues, aromatic rings of tryptophan, tyrosine and phenylalanine residues.

An important advantage of the spectroscopic technique based on Raman scattering with visible laser excitation, is its applicability to systems containing high solute concentrations, critical to investigate *in situ* protein structural changes occurring during coagulum or gel formation. The remainder of this chapter describes some of the applications of Raman spectroscopy in the author's laboratory to study protein structure of lysozyme in opaque gels, whey proteins in clear gels, and lysozyme-whey protein binary mixtures in precipitates and gels.

Experimental - Materials and Methods. Lysozyme (L- 6876, from chicken egg white, 3X crystallized, dialyzed and lyophilized), α-lactalbumin (L-5385, from bovine milk, Type 1, containing 1-2 moles of calcium per mole protein) and β-lactoglobulin (L-0130, from bovine milk, 3x crystallized and lyophilized) were purchased from Sigma Chemical Co. (St. Louis, MO). Protein solutions were prepared in water or deuterium oxide, and pH adjusted to the specified value. Solutions were introduced into a hematocrit capillary tube (Nichiden-Rika Glass Co., Ltd) for spectral analysis. Samples receiving heat treatment were incubated in a water or steam bath at the specified temperature and time conditions; samples were cooled in an ice water bath for 5 minutes, then kept @ 4°C overnight prior to spectral analysis. Raman spectra were recorded on a JASCO NR-1100 laser Raman spectrometer with excitation from the 488 nm line of a Spectra Physics 168B argon ion laser. Details of the experimental procedures including computer-aided data analysis are described elsewhere (*24, 49*, Howell, N. K.; Li-Chan, E. *Int. J. Food Sci. Technol.* in press).

Heat or Thiol-Induced Gelation of Lysozyme (Opaque gels). Opaque white gels were formed by heating 10% lysozyme solutions (pH 7.2, 50 mM NaCl) at 80 or 100°C. Gels could also be formed at 37 or 60 °C in the presence of low concentrations of dithiothreitol (DTT). Gels formed in the presence of 100 mM DTT could be solubilized in 8 M urea, while gels formed with no or 10 mM DTT at 100 °C were insoluble (*49*).

The Raman spectrum of the unheated lysozyme solution is shown in Figure 1, and assignment of the major features in the lysozyme spectrum are given in Table I. Changes in the Raman spectrum were used to monitor secondary structural changes and involvement of amino acid side chains in the opaque gels formed in the presence of DTT or heat treatment. Marked differences were noted in spectra of lysozyme only after heat treatment at 80 or 100°C. However, in the presence of 10 or 100 mM DTT, changes were observed even after heating at 37 or 60°C. The results indicated that molecular flexibility imparted by either thiol addition (disulfide bond reduction) or high temperature treatment (disulfide interchange) is a necessary pre-requisite for the formation of lysozyme gels. Alteration in the conformation around one of the four disulfide bonds of lysozyme, from an all gauche to gauche-gauche-trans conformation, was indicated by changes in relative intensity of the S-S stretching bands at 508 and 528 cm^{-1}. Increasing intensity of the S-H band at 2580 cm^{-1} indicative of free thiol groups was observed especially after heat treatment in the

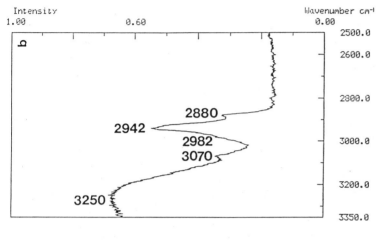

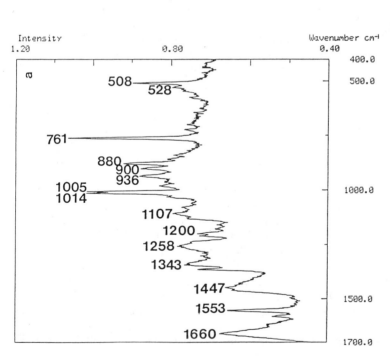

Figure 1. Raman spectrum of 10% lysozyme solution (0.05 M NaCl, pH 7.2) in the wavenumber shift regions (a) 400-1700 cm⁻¹ and (b) 2500-3350 cm⁻¹. The assignment of bands is shown in Table I. (Reproduced with permission from ref. 49. Copyright 1991 American Chemical Society).

Table I. Raman Band Assignment for Lysozyme solution (0.05 M NaCl, pH 7)

Wavenumber (cm^{-1}) [a]	Tentative Assignment[b]
508, 528	ν S-S (g-g-g and g-g-t)
538, 543, 754	Trp
626	Phe
644	Tyr
699	ν C-S, Met
761	Trp
839,859	Tyr
879	Trp
900, 936	ν C-C (α-helix)
980	ν C-C (β-sheet)
1005	Phe
1014	Trp
1030	Phe
1080, 1107, 1130 (sh)	ν C-N
1200, 1210	Tyr, Phe
1238 (sh), 1258, 1278 (sh)	Amide III (ν C-N, δ N-H, ν CH$_3$-C)
1343, 1365	Trp and δ C-H
1447, 1460 (sh)	δ C-H$_2$
1553, 1582	Trp
1620 (sh)	Tyr, Trp, Phe
1660	Amide I (ν C=O, δ N-H), δ H-O-H
2880, 2942, 2982 (sh)	aliphatic ν C-H
3070	aromatic ν C-H
3250	ν O-H and ν N-H

[a] (sh) refers to shoulder. [b] ν and δ are stretching and bending vibrations, respectively; g-g-g and g-g-t refer to the gauche (g) or trans (t) conformations of the C-S-S-C bonds of the disulfide grouping.

SOURCE: Adapted from ref (*49*)

presence of 100 mM DTT, consistent with evidence of chemical reactivity using Ellman's reagent (Table II).

For these samples having higher molecular flexibility through thiol or heat treatment, the Raman spectra also indicated changes in tertiary structure and involvement of hydrophobic interactions in gel formation. Increased exposure of aromatic residues was suggested by the decrease in intensity and sharpness of the Trp bands at 761, 879 and 1360 cm^{-1}. Increase in the intensity of the C-H stretching vibration at 2938-2942 cm^{-1} suggested increasing exposure of aliphatic side chains to the aqueous environment. Alterations in the secondary structure conformation were also noted in the Amide I and III bands. The Raman spectral analysis package of Przybycien and Bailey (20) for least squares analysis of the amide I band was applied to quantitatively estimate the secondary structure fractions (Table II). Maximum increase in β-sheet content was observed by heating at 80°C in the absence or presence of DTT, or at 37°C in the presence of 100 mM DTT, and was accompanied by a decrease in the α-helical content.

Heat-Induced Gelation of Whey Proteins (Transparent gels). Solutions of α-lactalbumin and β-lactoglobulin (15% w/v in D$_2$O, pD 6.8, 20 mM NaCl) were heated at 50, 70 or 90°C for 30, 60 or 90 minutes (24). Transparent gels were formed for the samples heated to 90 °C , while all of the other samples remained as clear solutions. Raman spectra were measured for solution and gelled samples.

Results from the spectral analysis for the unheated solutions and the transparent gels after 90°C, 30 minute heat treatment are shown in Table III. Similar changes were observed after heating at 70 and 90°C, but were more intense in the gelled (90 °C) than ungelled samples. Amide I' and Amide III' bands of the Raman spectra indicated an increase in β-sheet structure and a decrease in α-helical structure in α-lactalbumin gels, whereas increased β-sheet structure was accompanied by a decrease of turn structure in β-lactoglobulin gels. Changes in the conformation around cystinyl residues were indicated by a decrease in the intensity and broadening of the S-S stretching band near 508 cm^{-1}. These changes in disulfide bond conformation were notable especially in the case of α-lactalbumin which has four disulfide bonds, compared to two in β-lactoglobulin. Similar, though less marked, changes in the S-S stretching band were also observed in the spectra of α-lactalbumin heated at 70°C for 60 or 90 minutes. Changes in tertiary structure were observed, including the increasing exposure of buried tryptophan residues in the transparent gels indicated by the decrease in intensity and sharpness of the band at 760 cm^{-1}, for both α-lactalbumin and β-lactoglobulin gels. Involvement of tyrosine residues in gel formation was monitored by the intensity ratio of a doublet at 850 and 830 cm^{-1}, which is a good indicator of hydrogen bonding and environment of the phenolic hydroxyl group. A decrease in the intensity ratio I_{850}/I_{830} suggests an increase in "buriedness" or involvement as strong hydrogen bond donors (48); the lower intensity ratio observed in whey protein gels compared to the unheated solutions support the involvement of tyrosyl residues in hydrogen bond formation, possibly in intermolecular hydrogen bond-formation between β-sheets acting as junction zones in stabilizing the gel network, as suggested by Wang and Damodaran

Table II. Changes[a] in lysozyme sample after treatment with DTT or heat

Treatment	Observation	SH content	Secondary Structure Fractions			
			α-helix	β-sheet	turn	random
no DTT						
unheated	-	0.03	0.25	0.31	0.37	0.07
37°C, 24 h	±	0.01	0.26	0.39	034	0.03
60°C, 15 min	±	0.01	0.32	0.31	0.32	0.05
80°C, 12 min	++[b]	0.00	0.12	0.53	0.24	0.11
100°C, 12 min	++++[b]	0.06[d]	0.00	0.31	0.00	0.69
10 mM DTT						
unheated	±	0.01	0.46	0.26	0.12	0.17
37°C, 24 h	++[b]	0.86	0.44	0.24	0.28	0.03
60°C, 15 min	+[b]	1.3	0.35	0.39	0.20	0.06
80°C, 12 min	+++[c]	0.54[d]	0.00	0.63	0.20	0.17
100°C, 12 min	++++[c]	0.97[d]	0.35	0.48	0.03	0.14
100 mM DTT						
unheated	++	0.14	0.22	0.46	0.22	0.09
37°C, 24 h	++[b]	6.0	0.08	0.73	0.05	0.14
60°C, 15 min	++[b]	5.9	0.31	0.46	0.02	0.23
80°C, 12 min	+++[b]	5.4	0.05	0.75	0.10	0.10
100°C, 12 min	++++[b]	5.6	0.08	0.69	0.17	0.06

[a] Changes included visual observations, sulfhydryl content measured by Ellman's reagent and 2° structure fractions based on analysis of the Amide I band. Symbols for observations represent the following: - solution; ± slightly turbid; ++ turbid; + weak gel; ++ moderately strong gel; +++ strong gel; ++++ very strong gel.
[b] gel became dispersible after sonication.
[c] gel was not dispersible after sonication.
[d] A_{412} after reaction of supernatant with Ellman's reagent (insoluble fraction was disregarded).

SOURCE: Adapted from Ref. (*49*)

Table III. Raman band intensity[a] of whey proteins in solution and transparent gels

Raman band		Normalized intensity			
		α-lactalbumin		β-lactoglobulin	
ν (cm^{-1})	Assignment	unheated	90°C, 30'	unheated	90°C, 30'
945	α-helix	1.18 ± 0.02	0.87 ± 0.03	0.58 ± 0.02	0.53 ± 0.02
985	β-sheet	0.65 ± 0.01	0.90 ± 0.03	0.69 ± 0.02	1.01 ± 0.02
960	undefined	0.99 ± 0.02	0.92 ± 0.03	0.72 ± 0.01	0.60 ± 0.01
508	S-S (g-g-g)	0.50 ± 0.01	0.40 ± 0.03	0.24 ± 0.03	0.17 ± 0.01
760	Trp	1.02 ± 0.03	0.69 ± 0.03	0.48 ± 0.01	0.31 ± 0.01
850/830	Tyr	1.40 ± 0.10	1.11 ± 0.06	1.02 ± 0.02	0.94 ± 0.04
1408	His	0.39 ± 0.02	0.38 ± 0.04	0.39 ± 0.01	0.24 ± 0.03

[a] Values are the normalized intensity of each band, except for the ratio of normalized intensities for the Tyr 850/830 doublet, and are expressed as the mean ± standard deviation calculated from triplicate spectra, with each spectrum based on 10 scans.

SOURCE: Adapted from the data of Ref (24)

(*19*). Changes were also observed in the intensity of the band at 1409 cm^{-1}, assigned to the N-deuterated imidazolium ring of histidine; lower intensity of this band in spectra of the gels, especially of β-lactoglobulin, may indicate change in the ionization (de-deuteration) of the imidazole group. Whether this change in histidine side chains contributes to gelation or is simply a consequence of gelation is not known.

Interactions in Binary Mixtures of Whey Proteins and Lysozyme. Binary mixtures of lysozyme with either α-lactalbumin or β-lactoglobulin resulted in formation of white precipitates, whereas a clear solution was observed upon mixing α-lactalbumin and β-lactoglobulin. After heat treatment at 90°C for 30 minutes, clear gels were obtained for each of the whey proteins as well as their binary mixture while an opaque gel formed in the heated lysozyme sample. The whey protein-lysozyme binary mixtures remained as opaque solid material after heating. The Raman spectra of each of the individual proteins and their 1:1 (w/w) binary mixtures were measured before and after heating. Theoretical spectra for binary mixtures which would be expected in the absence of interactions were calculated as an average of the component spectra measured experimentally for the individual proteins. Evidence of interactions between proteins was investigated by analysis of difference spectra, expressed as the calculated average spectrum minus the experimental spectrum for each binary mixture. Figure 2a shows the spectrum in the 450-1900 cm^{-1} region for the precipitate formed in the β-lactoglobulin + lysozyme (BL) binary mixture;   interactions between the proteins are clearly demonstrated by the differences between calculated and experimental spectra. Similar changes were observed in the spectra for the precipitate formed in the α-lactalbumin + lysozyme (AL) binary mixture. In contrast, little or no interaction between α-lactalbumin and β-lactoglobulin (AB) is shown in Figure 2b.

Figure 3 shows the normalized intensity values in the experimental and calculated spectra of the three binary mixtures for three selected Raman bands, reflecting changes in (a) the S-S stretching band of cystinyl residues in all-gauche conformation, (b) the C-H bending motion of methylene groups of aliphatic residues, and (c) the environment or hydrogen-bonding status of tyrosine residues.

Both AL and BL unheated binary mixtures had lower normalized intensity of the cystinyl S-S band in the experimental spectrum than in the calculated spectrum; however, upon heating, the experimental intensity values were higher than the calculated values, particularly for the heated BL mixture (Fig. 3a). These results suggest that disulfide bonds are significantly altered by interactions between the whey proteins and lysozyme, both in the unheated precipitates and after heating. The experimental value for the unheated whey protein binary mixture (AB) was almost identical to the calculated value; after heating, the experimental value was slightly higher than the calculated value, suggesting that interactions involving disulfide bonds between the whey proteins only occur after heat treatment.

The intensity of C-H bending motion of aliphatic side chains was higher in the experimental than calculated spectra for both unheated and heated AL binary mixtures, as well as for the heated BL binary mixture (Fig. 3b). These results could suggest the involvement of aliphatic amino acid residues in hydrophobic interactions

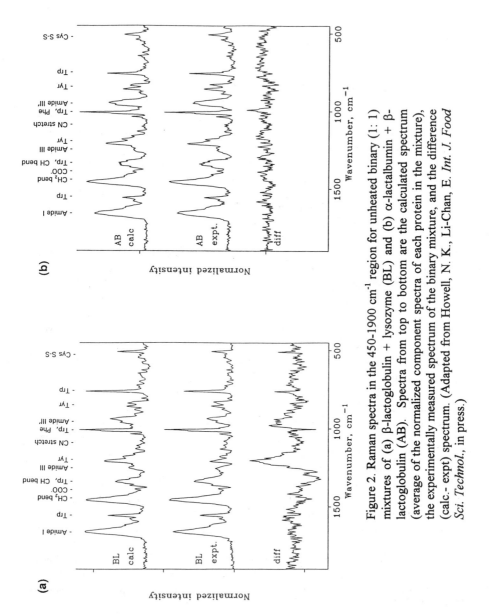

Figure 2. Raman spectra in the 450-1900 cm⁻¹ region for unheated binary (1: 1) mixtures of (a) β-lactoglobulin + lysozyme (BL) and (b) α-lactalbumin + β-lactoglobulin (AB). Spectra from top to bottom are the calculated spectrum (average of the normalized component spectra of each protein in the mixture), the experimentally measured spectrum of the binary mixture, and the difference (calc.- expt) spectrum. (Adapted from Howell, N. K., Li-Chan, E. *Int. J. Food Sci. Technol*, in press.)

in these mixtures. In contrast, experimental and calculated spectra showed similar values of the C-H bending intensity for unheated BL precipitate, as well as both the unheated and heated AB mixtures.

Unheated AL precipitate showed a much lower I_{850}/I_{830} ratio in the experimental spectrum than the calculated spectrum (Fig. 3c), suggesting either increasing buriedness or involvement of tyrosine residues as strong hydrogen bond donors in the complex. A similar, though much less pronounced, trend was also observed in the ratio for the unheated and heated AB mixtures. However, the experimentally observed values for unheated and heated BL mixtures as well as the heated AL mixture showed slightly higher intensity ratio than the calculated values, suggesting the exposure of some tyrosine residues to the polar environment.

Changes were also observed between experimental and calculated values for Raman bands assigned to α-helical (Fig. 4a) and β-sheet (Fig. 4b) structures. The unheated as well as heated AL mixtures showed lower α-helix and higher β-sheet content in the experimental than calculated spectra, while BL mixtures showed lower contents of both secondary structure types. The AB mixtures showed little change in helical content and slightly decreased β-sheet content.

Conclusions. A four-step sequence involving both intramolecular and intermolecular events has been postulated for the heat-induced gelation of globular proteins. The intramolecular transition of native molecules to denatured or unfolded intermediates is believed to be followed by the formation of soluble aggregates, which may then associate further, leading either to insoluble coagula, or to strands or pro-gels which ultimately form gels with characteristic rheological and optical properties. To elucidate the precise molecular mechanisms involved in the intermolecular interactions, it is necessary to study protein structure existing in these aggregates, coagula and gels. The results of Raman spectroscopic analysis of lysozyme and whey protein systems indicate that considerable secondary structure is retained in these proteins following gelation or coagulation; a decrease in the α-helical content and an increase in β-sheet content are common features in many, but not all, cases. Intermolecular interactions involving hydrophobic groups are reflected by changes in the Raman spectra representative of the vibrational motions of aromatic and aliphatic residues. Changes in the tertiary structure including an increased exposure of hydrophobic residues to the aqueous environment, and involvement of tyrosine residues in hydrogen bonding are also observed. Alterations in the disulfide bond stretching motions observed under some conditions suggest either reduction or interchange reactions leading to conformational changes around cystinyl residues. Further research should be focussed on correlating the results of *in situ* vibrational spectroscopic analysis with specific functional properties, to identify the distinctive interactions which lead to protein coagulation versus gelation.

Acknowledgments

Research collaborations with Mr. M. Nonaka and Dr. N. K. Howell, and financial assistance from the Natural Sciences and Engineering Research Council of Canada are gratefully acknowledged.

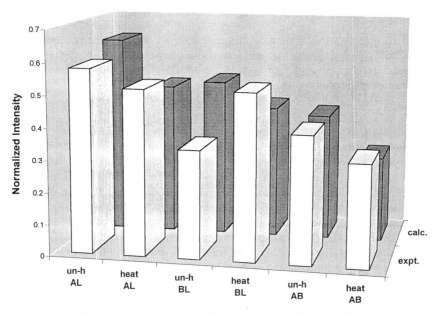

(a) 508 cm⁻¹ band (S-S stretch)

Figure 3. Normalized intensity values for the following bands in the experimental (expt.) and calculated (calc.) spectra of the binary mixtures before heating (un-h) and after heating at 90°C, 30 minutes (heat): (a) 508 cm^{-1} band assigned to S-S stretching; (b) 1455 cm^{-1} band assigned to C-H bending of methylene groups; (c) ratio of the 850 cm^{-1}/830 cm^{-1} tyrosine doublet. (Abbreviations: A, α-lactalbumin; B, β-lactoglobulin; L, lysozyme).

(b) 1455 cm^{-1} band (CH$_2$ bend)

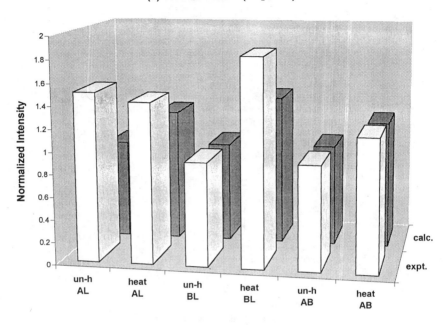

(c) 850 cm^{-1}/830 cm^{-1} (Tyrosine doublet)

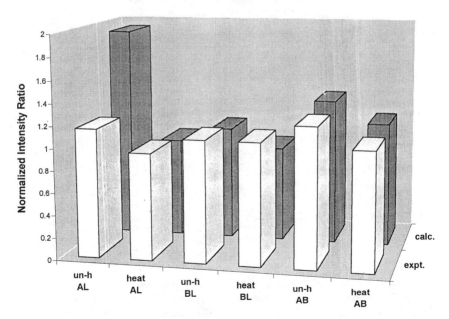

Figure 3. Continued.

(a) 934 cm^{-1} band (alpha- helix)

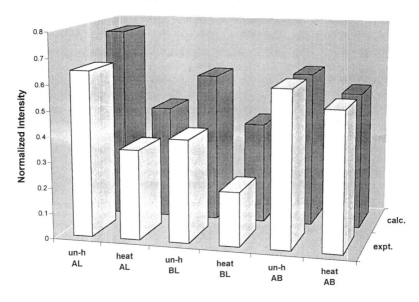

(b) 985 cm^{-1} band (beta sheet)

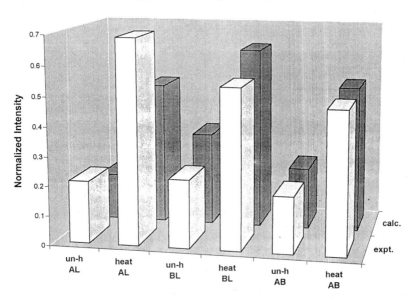

Figure 4. Normalized intensity values for the following bands in the experimental (expt.) and calculated (calc.) spectra of the binary mixtures before heating (un-h) and after heating at 90°C, 30 minutes (heat): (a) 934 cm^{-1} band assigned to α-helix; (b) 985 cm^{-1} band assigned to β-sheet. (Abbreviations: A, α-lactalbumin; B, β-lactoglobulin; L, lysozyme).

Literature Cited

1. Ferry, J. D. *Adv. Protein Chem.* **1948**, *4*, 1-78.
2. Schmidt, R. H. *Protein Functionality of Foods.* ACS Symposium Series 147; Cherry, J. P., Ed.; American Chemical Society: Washington, DC, 1981; Chapter 7, pp. 131- 147.
3. Aguilera, J. M. *Food Technol.* **1995**, *49 (10)*, 83 - 89.
4. Tombs, M. P. *Proteins as Human Foods.* Lawrie, R. A., Ed.; Butterworths: London, England, 1970; pp.126-138.
5. Clark, A. H.; Lee-Tuffnell, C. D. *Functional Properties of Food Macromolecules.* Mitchell, J. R.; Ledward, D. A., Eds; Elsevier Applied Science Publishers: New York, NY, 1986; Chapter 5, pp. 203-272.
6. Ohgushi, M.; Wada, A. *FEBS Lett.* **1983**, *164*, 21-24.
7. Ptitsyn, O. B. *Protein Folding.* Creighton, T. E., Ed.; Freeman: New York, NY, 1992; Chapter 6, pp. 243-300.
8. Ptitsyn, O. B. *Adv. Protein Chem.* **1995**, *47*, 83-229.
9. Doi, E. *Trends Food Sci. & Technol.* **1993**, *4*, 1-5.
10. Hirose, M. *Trends Food Sci. & Technol.* **1993**, *4*, 48-51.
11. Tani, F.; Murata, M.; Higasa, T.; Goto, M.; Kitabatake, N.; Doi, E. *J. Agric. Food Chem.* **1995**, *43*, 2325-2331.
12. Barbut, S.; Foegeding, E. A. *J. Food Sci.* **1993**, *58*, 867-871.
13. Shimada, K.; Matsushita, S. *J. Agric. Food Chem.* **1980**, *28*, 413-417.
14. Wang, C.-H.; Damodaran, S. *J. Agric. Food Chem.* **1990**, *38*, 1157-1164.
15. Tani, F.; Murata, M.; Higasa, T.; Goto, M.; Kitabatake, N.; Doi, E. *Biosci. Biotechnol. Biochem.* **1993**, *57*, 209-214.
16. Kitabatake, N.; Doi, E.; Kinekawa, Y.-I. *J. Food Sci.* **1994**, *59*, 769-772.
17. Astbury, W. T.; Dickinson, S.; Bailey, K. *Biochem. J.* **1935**, *29*, 2351.
18. Schein, C. H. *Protein Refolding.* ACS Symposium Series 470; Georgian, G; de Bernardez-Clark, E., Ed.; American Chemical Society: Washington, DC, 1991; Chapter 2, pp. 21-34.
19. Wang, C.-H.; Damodaran, S. *J. Agric. Food Chem.* **1991**, *39*, 433 - 438.
20. Przybycien, T. M.; Bailey, J. E. *Biochim. Biophys. Acta* **1991**, *1076*, 103-111.
21. Clark, A.H.; Saunderson, D. H. P.; Suggett, A. *Int. J. Peptide Protein Res.* **1981**, *17*, 353-364.
22. Rector, D. J.; Kella, N. K.; Kinsella, J. E. *J. Texture Studies* **1989**, *20*, 457-471.
23. Mulvihill, D. M.; Kinsella, J. E. *J. Food Sci.* **1988**, *53*, 231-236.
24. Nonaka, M.; Li-Chan, E.; Nakai, S. *J. Agric. Food Chem.* **1993**, *41*, 1176-1181.
25. Matsuura, J. E.; Manning, M. C. *J. Agric. Food Chem.* **1994**, *42*, 1650-1656.
26. Paulsson, M.; Visser, H. *Protein Interactions.* Visser, H., Ed.; VCH Press: Weinheim, Germany, 1992; Chapter 7, pp. 117-134.
27. Kuhn, P. R.; Foegeding, E. A.. *J. Agric. Food Chem.* **1991**, *39*, 1013-1016.
28. Foegeding, E. A.; Kuhn, P. R.; Hardin, C. C. *J. Agric. Food Chem.* **1992**, *40*, 2092-2097.
29. Xiong, Y. L.; Dawson, K. A.; Wan, L. *J. Dairy Sci.* **1993**, *76*, 70-77.
30. McSwiney, M.; Singh, H.; Campanella, O.; Creamer, L. K. *J. Dairy Res.* **1994**, *61*, 221-232.

31. Li, H.; Hardin, C. C.; Foegeding, E. A. *J. Agric. Food Chem.* **1994,** *42,* 2411-2420.
32. Iametti, S.; Cairoli, S.; De Gregori, B.; Bonomi, F. *J. Agric. Food Chem.* **1995,** *43,* 53-58.
33. Byler, D. M.; Purcell, J. M. *SPIE Fourier Transform Spectroscopy* **1989,** *1145,* 415-417.
34. Matsudomi, N.; Oshita, T.; Kobayashi, K.; Kinsella, J. E. *J. Agric. Food Chem.* **1993,** *41,* 1053-1057.
35. Paulsson, M.; Hegg, P.-O.; Castberg, H. B. *J. Food Sci..* **1986,** *51,* 87-90.
36. Hines, M.E.; Foegeding, E. A. *J. Agric. Food Chem.* 1993, *41,* 341-346.
37. Eynard, L.; Iametti, S.; Relkin, P.; Bonomi, F. *J. Agric. Food Chem.* **1992,** *40,* 1731-1736.
38. Legowo, A. M.; Imade, T.; Hayakawa, S. *Food Research International* **1993,** *26,* 103-108.
39. Donovan, J. W.; Mapes, C. J.; Davis, J. G.; Garibaldi, J. A. *J. Sci. Food Agric.* **1975,** *26,* 73-83.
40. Hayakawa, S.; Nakamura, R. *Agric. Biol. Chem.* **1986,** *50,* 2039-2046.
41. Hegg, P.-O. *J. Food Sci.* **1982,** *47,* 1241-1244.
42. Li-Chan, E.; Nakai, S. *Food Proteins.* Kinsella, J. E.; Soucie, W. G., Eds.; The American Oil Chemists' Society: Champaign, IL, 1989; Chapter 14, pp. 232-251.
43. Matsudomi, N.; Oshita, T.; Kobayashi, K. *J. Dairy Sci.* **1994,** *77,* 1487-1493.
44. Poole, S.; West, S. I.; Fry, J. C. *Food Hydrocolloids* **1987,** *1,* 301-316.
45. Arntfield, S. D.; Bernatsky, A. *J. Agric. Food Chem.* **1993,** *41,* 2291-2295.
46. Howell, N. K. *Biochemistry of Food Proteins.* Hudson, B.J F., Ed.; Elsevier Applied Science Publishers: New York, NY, 1992; Chapter 2, pp.35-74.
47. Howell, N. K. *Gums and stabilisers for the food industry - 7.* Philips, G. O.; Williams, P. A.; Wedlock, D. J., Eds; Oxford University Press:Oxford, England, 1994; pp. 77-89.
48. Li-Chan, E.; Nakai, S.; Hirotsuka, M. *Protein Structure-Function Relationships in Foods.* Yada, R. Y.; Jackman, R. L.; Smith, J. L., Eds; Blackie Academic & Professional, Chapman & Hall Inc.: London, England, 1994; Chapter 8, pp. 163-197.
49. Li-Chan, E.; Nakai, S. *J. Agric. Food Chem.* **1991,** *39,* 1238-1245.

Chapter 3

Factors Determining the Character of Biopolymer–Biopolymer Interactions in Multicomponent Aqueous Solutions Modeling Food Systems

M. G. Semenova

Institute of Food Substances, Russian Academy of Science, Vavilov Str. 8, 117813 Moscow, Russia

The recent results of thermodynamic studies of the basic physico chemical factors determining character of the biopolymer-biopolymer (protein, polysaccharide) interactions in aqueous medium are discussed. The emphasis is on consideration of such factors as biopolymer structure (molecular mass, size, conformation, charge) and composition of the aqueous medium (content of the low molecular weight inorganic /salts, pH/ and organic /lipophilic molecules: lipids, surface active additives or sucrose/ compounds. Correlation between the character of the biopolymer-biopolymer interactions and biopolymer functional properties in aqueous medium is demonstrated.

By now a lot of the experimental findings have indicated an interrelationship between the character of the intermolecular interactions of the same or different biopolymers in a bulk aqueous medium and biopolymer functional properties in aqueous media such as solubility, cosolubility (1,2), gelforming ability (3,4) or surface activity (5,6) and stability of emulsions or foams (7,8). For example, for the pair of 11S Globulin with Dextran it was indicated clearly that protein surface activity changes dramatically, under alteration of the protein - polysaccharide interaction character in a bulk aqueous medium (5). So, it was established that under conditions of weak attraction between protein and polysaccharide in a bulk aqueous medium a decrease of the protein surface activity was observed. In contrast, under conditions of weak exclusion between protein and polysaccharide in a bulk aqueous medium an increase of the protein surface activity was observed. Recently, it was also established that protein emulsifying capacity may be strongly dependent on intensity of the repulsion forces acting between protein and polysaccharide in a bulk aqueous medium by the example of the pair of 11S Globulin with Dextran molecules with different molecular weight under conditions of their thermodynamic incompatibility in aqueous medium (8). With low-molecular weight Dextran the better emulsifying capacity of the protein was observed under all regions of the polysaccharide protein molar ratio studied. With high-molecular weight Dextran the emulsifying capacity of the protein was improved only at low polysaccharide contents. Significant reduction in the protein emulsifying capacity was obtained due in large part to the multilayer formation at the interface as a result, probably, of liquid-liquid phase transition in a bulk aqueous medium.

In order to gain more insight into the basis of both biopolymer functionality in real food systems and quality of the food system structure, more understanding of the main factors controlling character of the biopolymer-biopolymer interactions in real food systems is required. This paper is mainly concerned with the recent results of our thermodynamic study of the some basic factors determining the interaction character between biopolymers in aqueous medium.

Thermodynamic Approach to the Study of the Interaction Character between Biopolymers

In order to elucidate the factors controlling the character of biopolymer-biopolymer interactions in an aqueous medium the thermodynamic approach may be very informative. That is determination of thermodynamic parameters of the biopolymer- biopolymer interactions in biopolymer aqueous solutions at equilibrium state of the system under distinct experimental conditions will enhance the description of the system.

In so doing we have used the chemical potentials of the components in the ternary system: solvent(1) + biopolymer (2) + biopolymer (3), which are expressed in the following forms (9):

$$\mu_2 = \mu_2{}^0 + RT[ln(m_2/m^0) + A_2m_2 + A_{23}m_3] \quad (1)$$
$$\mu_3 = \mu_3{}^0 + RT[ln(m_3/m^0) + A_3m_3 + A_{23}m_2] \quad (2)$$
$$\mu_1 = \mu_1{}^0 - (RT/m_1) \times (m_2 + m_3 + 1/2A_2m_2{}^2 + 1/2A_3m_3{}^2 + A_{23}m_2m_3), \quad (3)$$

where $\mu_i{}^0$ and m_i are the standard chemical potential and concentration (molal scale) of component-i (i =1-3), A_2 and A_3 are the second virial coefficients or interaction constant of the biopolymer (2) and biopolymer (3) respectively, A_{23} is the cross second virial coefficient for the biopolymer(2) - biopolymer (3) pair interaction, m0 is the standard-state molality, R is the gas constant and T is absolute temperature.

The sign and magnitude of the second virial coefficient provide information about the initial deviation of the macromolecular solution from ideal state at small macromolecular concentrations and reflect nature and intensity of the intermolecular pair interactions in the solutions (10). A positive value of the second virial coefficient indicates thermodynamically unfavorable interactions between biopolymers in the system, that is mutual repulsion or exclusion between biopolymers in the systems. In contrast, a negative value of the second virial coefficient indicates thermodynamically favorable interactions between biopolymers in the system, that is, mutual attraction between biopolymers in the system. For example, under thermodynamically unfavorable conditions between different biopolymers, resulting in a positive value of the interaction constant (cross second virial coefficient) (A_{23}), the chemical potentials of the biopolymers increase on mixing and this can cause phase separation of the system at high biopolymer concentrations (*1,2*).

For experimental determination of the thermodynamic parameters of the biopolymers in a bulk aqueous medium, in particular, second virial coefficients, light scattering methods may be used (*10,11*).

Effect of the biopolymer structure on the character of the biopolymer(2) - biopolymer (3) interaction in aqueous medium

The role of the structural factor in biopolymer interactions is best observed under conditions of high ionic strength in the system, so that electrostatic interactions between biopolymers are minimized. That is pairs of biopolymers with well known different structures in aqueous medium were under investigation, namely: 1) 11S Globulin - Dextran, 2) 11S Globulin - Pectinate (*1*), 3) Fibrinogen - Dextran, 4) Fibrinogen - Alginate, 5) Alginate - Dextran (Semenova,M.G.; Savilova L.B. *Food Hydrocolloids*, in press). The binary (polymer-solvent) and ternary (polymer-polymer-solvent) have been studied by light scattering method. Molecular weight, size and second virial coefficient characterizing the nature and intensity of the pair interaction of the same biopolymer molecules have been obtained in the binary solutions. Cross second virial coefficient has been determined in the ternary solutions.

The role of the biopolymer conformation. Analyzing the role of biopolymer conformation in the character of the interactions between different biopolymers we have carried out a comparison of experimentally obtained values of the cross second virial coefficients with theoretical ones calculated on the base of the size and conformational features of the interacting biopolymers. In characterizing the interactions between different biopolymers by the sign and magnitude of the cross second virial coefficient it should be noted that current theories of the second virial coefficient attest that in the case of polyelectrolytes, such as proteins and anionic polysaccharides, the value of the second virial coefficient, is mainly determined by the value of the thermodynamically excluded volume (A_{ij}^{exc}) of the macroion and by the contribution of electrostatic forces acting between macroions (A_{ij}^{electr}) (*10,12*). Thus the equation for the second virial coefficient may be written as:

$$A_{ij} = A_{ij}^{exc} + A_{ij}^{electr} \qquad (4)$$

where, i= j for the case of interactions of the same macroions.

Let us assume in a first approximation that thermodynamically excluded volume is defined only by physical volume occupied by one biopolymer molecule which is inaccessible for other biopolymer molecules. In so doing for theoretical calculations we have used following equations in accordance with conformational features of the interacting biopolymer chains (10,13):

I. Macromolecules are presented as two interacting solid spheres:

$$A_{23} = 4/3\pi N_a 10^{-3} (R_2 + R_3)^3, \tag{5}$$

where N_a is Avogadro's number; R_2, R_3 are the equivalent hard sphere radius of the biopolymer-2 and biopolymer-3, respectively.
II. Macromolecules are presented as interacting a bent cylinder and solid sphere.

$$A_{23} = 2L\pi N_a 10^{-3} (R_f + R), \tag{6}$$

where N_a is Avogadro's number; R_f is the cylinder radius and L is the total contour chain length of the cylinder.
III. Macromolecules are presented as interacting rigid rod and solid sphere.

$$A_{23} = N_a 10^{-3} [2\pi (R_2 + R_3)^2 l_2 + 4\pi/3 (R_2 + R_3)^3], \tag{7}$$

where N_a is Avogadro's number; R_2 is rod cylindrical radius an hemispherical ends of radius R_2; l_2 is half-length of the rod cylinder; R_3 is equivalent hard sphere radius.
It should be pointed out that an equivalent hard sphere corresponds to the space occupied in aqueous medium by one biopolymer which is absolutely not accessible for other biopolymers (10). For the calculation of the radius of the equivalent hard sphere in some cases it is necessary to use a penetration parameter which is specific for different conformations of macromolecules. This parameter is the ratio of the radius of the equivalent hard sphere (R_e) to the radius of gyration of the macromolecule $(<R_g^2>^{1/2})$ (10):

$$\gamma = (3/16\pi Na)^{1/3} \times (A_{ii}^{exc})^{1/3} M_w^{2/3} / <R_g^2>^{1/2}, \tag{8}$$

where A_{ii}^{exc} is the contribution of the excluded volume of the biopolymer to the second virial coefficient, M_w is weight average molecular weight of the biopolymer.
Table I presents penetration parameters experimentally obtained for biopolymers studied. The experimental values of the penetration parameter is in good agreement with the values expected theoretically for specific conformations of the biopolymers.

Table I.   Penetration Parameter of Biopolymer Molecules

Biopolymer	Experimental γ	Literature data (10)
Dextran, Mw = 40 kDa	0.75	for flexible
Dextran, Mw = 270 kDa	0.69	chain polymer
Dextran, Mw = 2000 kDa	0.54	$\gamma = 0.735$
Pectinate	0.28	for semi- rigid
Alginate	0.19	chain polymer
Fibrinogen	0.21	$\gamma = 0.3$

Tables II - V present experimental and calculated values of the cross second virial coefficients.

Table II. The Cross Second Virial Coefficients, For Pair Interactions between Globular Protein and Polysaccharide with Different Conformations

Biopolymer pair	Type of the conformation of the interacting biopolymers	A_{23} x 10^{-5}, cm^3/mol		
		Experi-mental	Theoretical sphere - sphere	Theoretical bent cylinder - sphere
11S Globulin - Dextran $M_w = 270$ kDa	globule - flexible random coil	0.31(8)	$R_e{}^{dex} = 0.7R_g{}^{dex}$ 0.29	
11S Globulin - Pectinate $M_w = 250$ kDa	globule - semi - rigid random coil	0.27(1)	$R_e{}^{pec} = R_g{}^{pec}$ 11.5 $R_e{}^{pec} = 0.6R_g{}^{pec}$ 2.8 $R_e{}^{pec} = 0.3\ R_g{}^{pec}$ 0.47	0.35

Experimentally determined cross second virial coefficients (Table II) for the interactions of 11S Globulin, with Dextran (flexible random coil conformation) and with Pectinate (semi-rigid random coil) are remarkably similar. The theoretical sphere - sphere model appears to be suitable for the description between 11S Globulin and Dextran. However, although the molecular weights of the Dextran and Pectinate are similar, the sphere-sphere model predicts a much larger value for the cross second virial coefficient than was obtained experimentally. The best coincidence of the theory with experiment is observed by using the assumption that interactions of the macromolecules of the semi-rigid chain pectinate with globular protein may be represented like interactions of the bent cylinder with hard sphere. In other words, assuming that under electroneutral conditions space occupied by a semi-rigid chain polymer to a large extent is accessible to the protein globules in contrast to the space occupied by flexible-chain biopolymer. In consequence of this a marked decrease in contribution of the excluded volume effect in interaction of the biopolymers was observed.

Increase in the order of magnitude of A_{23} were observed for interactions of the flexible-chain biopolymer. Dextran, with biopolymers in globular, rigid rod and semi-rigid rod conformations. These increases in A_{23} (Table III) indicate increases in the space occupied by biopolymer in rod-like conformations.

Table III. The cross second virial coefficient exemplifying pair interactions between flexible chain biopolymer and biopolymers with different conformation

Biopolymer pair	Type of the conformation of the interacting biopolymers	$A_{23} \times 10^{-5}$, cm³/mol		
		Experimental	Theoretical sphere-sphere	Theoretical rod - sphere
Dextran M_w=270 kDa- 11S Globulin M_w=330 kDa	Flexible random coil - Globule	0.31	$R_e^{dex} = 0.7\,R_g^{dex}$ 0.29	
Dextran M_w=270 kDa - Fibrinogen M_w=990 kDa	Flexible random coil - Rigid rod	4.4		4.8
Dextran M_w= 270 kDa - Alginate M_w= 1540 kDa	Flexible random coil - Semi - rigid random coil	12.6	$R_e^{alg} = R_g^{alg}$ $R_e^{dex} = R_g^{dex}$ 20.6 $R_e^{alg} = R_g^{alg}$ $R_e^{dex} = 0.7\,R_g^{dex}$ 17.6	

It is interesting to note, in the correlation between experimental and theoretical values of the A_{23} manifests, firstly, that 30% of the space occupied by a flexible chain biopolymer in solution may be accessible to other macromolecules and, secondly, that space occupied by semi-rigid-chain biopolymer is practically not accessible to a random coil of the flexible-chain biopolymer.

The role of the molecular weight. The next Table exhibits data about the influence of molecular weight on intensity of the thermodynamically unfavorable interactions between biopolymers using the example of the mixtures of the Dextran with different molecular weight and 11S Globulin.
A regular increase of the value of A_{23} was observed with increases of molecular weight and size of Dextran molecules.
It is interesting to note that with increases in the molecular weight of the neutral polysaccharide, Dextran, the space occupied by this macromolecule in aqueous medium appears more accessible for globular proteins.

The role of the charge. At constant ionic strength, the value of A_{23} increases significantly on going from the interaction of a charged biopolymer with a neutral one to interactions between two charged biopolymers.

Table IV. Influence of the molecular weight of Dextran on intensity of its pair interactions with 11S Globulin

Biopolymer pair	A_{23} x 10^{-5}, cm^3/mol	
	Experimental	Theoretical sphere - sphere
Dextran Mw=48 kDa - 11S Globulin	0.09	$R_e^{dex} = R_g^{dex}$ 0.03
Dextran Mw=270 kDa - 11S Globulin	0.31	$R_e^{dex} = 0.7\ R_g^{dex}$ 0.29
Dextran Mw=2500kDa - 11S Globulin	2.05	$R_e^{dex} = R_g^{dex}$ 8.7 $R_e^{dex} = 0.7\ R_g^{dex}$ 3.3 $R_e^{dex} = 0.65\ R_g^{dex}$ 1.9

Source: Reproduced with permission. Copyright 1996 Oxford University Press.

Table V. The cross second virial coefficient exhibiting pair interactions between charged biopolymer with neutral one or with charged biopolymer

Biopolymer pair		A_{23} x 10^{-5}, cm^3/mol		
		Experimental	Theoretical sphere - sphere	Theoretical cylinder - sphere
Alginate - Dextran	charged - uncharged	13	$R_e^{alg} = R_g^{alg}$ $R_e^{dex} = 0.7\ R_g^{alg}$ 18	
Alginate - Carboxymethyl-cellulose	charged - charged	2715	$R_e^{alg} = R_g^{alg}$ $R_e^{cmc} = R_g^{cmc}$ 513	
Alginate - Fibrinogen	charged - charged	1877		41

The contribution of the electrostatic interactions determines the magnitude of A_{23}, and consequently, intensity of the intermolecular interactions between biopolymers. The role of the charge in the interaction of the biopolymers is most significant when such parameters of the system as pH and ionic strength are varied.

The recent study (Wasserman, L.A.; Semenova, M.G.; Tsapkina, E.N. *Food Hydrocolloids*, in press) provides example of the effects of varying pH on the interactions between two globular proteins in solution.

Table VI shows that the decrease in the net negative charge on the proteins when the pH is lowered caused sharp weakening of the intensity of the interactions between both the same and different protein molecules. This caused a significant decrease in the magnitude of the experimentally determined second virial coefficients. This result clearly attests strengthening of the mutual attraction between protein molecules as the charge is decreased, that is a shift in a delicate balance of the intermolecular forces, apparently, from electrostatic repulsion towards hydrophobic attraction for amphiphilic protein molecules.

Table VI.   Effect of pH on interaction between protein molecules

Interaction between Ovalbumin and 11S Globulin		
pH	Experimental A_{23} x 10^4, m³/mol	Theoretical A_{23} ^{exc} x 10^4, m³/mol
7.8	70.0	13
7.0	7.3	13

Interaction between the similar protein molecules			
pH	Ovalbumin		11S Globulin
	Experimental A_{22}x10^4, m³/mol	Theoretical A_{22}^{exc} x10^4, m³/mol	Experimental A_{33}x10^4, m³/mol	Theoretical A_{33}x10^4, m³/mol
7.8	22.5	5.5	174.2	25.2
7.0	6.5	5.5	- 39.2	25.2

It is interesting to note that decreases in the ionization of similarly charged protein molecules did not caused an increase of the protein miscibility in aqueous medium as might be expected, but on the contrary appeared to increase the concentration area of protein immiscibility of the two proteins (see Figure 1, concentration area of the protein immiscibility is higher binodal curve). The main reason for this effect, probably, is strengthening of the protein self - association.

In the case when only one interacting biopolymer posses charge the character of the biopolymer interactions, seems, most controlled by enthropic contribution of the counter ions into free energy of the biopolymer interactions (*14,15*). So, as Table VII and Figure 2 manifest that under interaction of the charged globular protein 11S Globulin with neutral polysaccharide - Dextran entropy

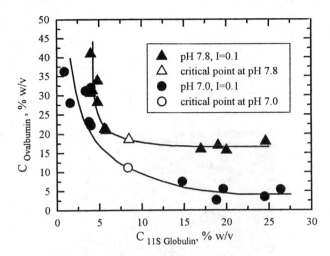

Figure 1. Effect of pH on phase diagram (binodal and critical point) of the system: Ovalbumin - 11S Globulin - Water. (Reproduced with permission. Copyright 1996 Oxford University Press.)

contribution of the counter ions, evidently, mainly determines thermodynamically favorable interactions between biopolymers in the wide concentration region at low ionic strength ($I=0.01$ mol/dm^3) in solution.

Table VII. Influence of ionic strength on interaction between charged and neutral biopolymer

Ionic strength, mol/dm^3	$(A_{22}-A_{33})\times 10^{-5}$, cm^3/mol	$A_{23}\times 10^{-5}$, cm^3/mol
0.1	1.6	3.5
0.01	4.6	- 1.8

When the ionic strength of the aqueous medium is increased, the entropy contribution of the counter ions to the free energy of the biopolymer interactions causes the take on interaction between charged and neutral biopolymer to some features of the interaction between two neutral biopolymres. The character of their interactions is controlled mainly by the effect of the excluded volume.

Thus, the charge dependence of the character of the biopolymer interaction is determined in many respects by features of biopolymer nature as well as by ionic environment in aqueous medium.

Effect of the composition of the aqueous medium on the character of the biopolymer(2) - biopolymer (3) interaction.

Among factors controlling character of the biopolymer's interaction in aqueous medium composition of the aqueous medium takes very important place.

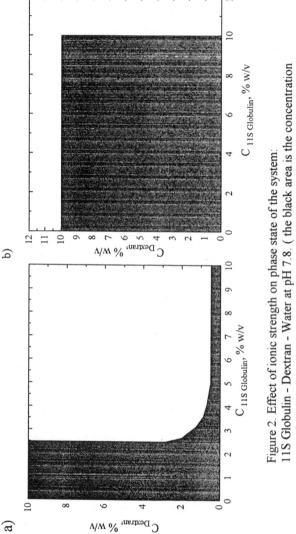

Figure 2. Effect of ionic strength on phase state of the system:
11S Globulin - Dextran - Water at pH 7.8. (the black area is the concentration area of the biopolymer miscibility),
a) I = 0.1 M;
b) I = 0.01 M.

Effect of the lipophilic molecules. By the example of the two globular proteins Table VIII indicates that effect of the nonpolar lipophilic molecules may appear as association of the protein molecules in aqueous medium both in the case of 11S Globulin and in the case of Ovalbumin. It is believed that the protein association observed is caused by increase of the protein surface hydrophobicity resulting in enhancement of the hydrophobic interactions between protein molecules. This result, possibly, is determined by increase of the effect of the excluded volume on interactions between proteins due to higher molecular weight and consequently size of the protein associates formed under the influence of lipophilic molecules (Wasserman, L.A.; Semenova M.G. *Food Hydrocolloids*, in press).

Table VIII. Effect of the nonpolar lipophilic molecule on molecular and thermodynamic parameters of the proteins

Lipophilic compound (LC) - Decane	11S Globulin		Ovalbumin		11S Globulin - Ovalbumin
	M_w, kDa	$A_{22}x10^{-3}$, cm³/mol	M_w, kDa	$A_{33}x10^{-3}$, cm³/mol	$A_{23}x10^{-3}$, cm³/mol
without LC	330	- 113	44	0.65	9.2
with LC	1900	238	294	28	388.0

Figure 3 shows that increase of the intensity of the thermodynamically unfavorable interactions between protein, as well as, possible increase of the difference in protein affinity toward solvent under nonpolar lipophilic molecule cause the increase of the concentration area of the protein immiscibility in aqueous medium.

Effect of the sucrose. Recently we have observed a significant effect of sucrose on protein solubility and cosolubility with different biopolymers in aqueous medium (Antipova A.S.; Semenova,M.G. *Carbohydrate Polymers*, in press). Let me introduce only one example of the effect of the sucrose on the character of the protein-protein interaction by the example of the Caseinate- Ovalbumin-water system. Table IX indicates that in the case of Ovalbumin the presence of 25% sucrose in solutions causes an increase of the protein second virial coefficient. This result points up increase of the thermodynamic affinity of the protein to the aqueous medium.

In the case of the Caseinate sucrose causes dissaggregation of the Caseinate complexes. Caseinate molecular weight essentially decreases at 25% the sucrose in the solution. As this takes place, however, the surface of the new protein associates with lower molecular weight appears more hydrophobic (negative values of the A2). Hydrophobic parts of the protein molecules, are, expose of from the interior of the associates to the protein surface as a result of the disaggregation in presence of sucrose.

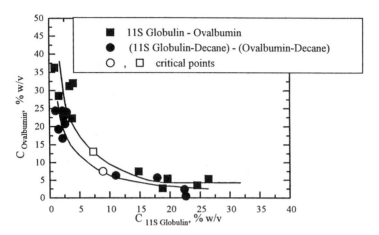

Figure 3. Effect of the nonpolar lipophilic molecule on phase diagram (binodal and critical point) of the system: 11S Globulin - Ovalbumin - Water - Decane at pH 7.0 and I= 0.1M.

Table IX. Effect of the sucrose on molecular and thermodynamic parameters of the proteins in aqueous medium

Biopolymer	Mw, kDa		$A_{ii} \times 10^4$, cm³/mol (i=2,3)	
	without sucrose	25% w/v of sucrose	without sucrose	25% w/v of sucrose
Ovalbumin	44	44	4.6	7.6
Caseinate	8600	1400	0.67	- 0.45
Biopolymer mixture	$A_{23} \times 10^4$, cm³/mol			
	without sucrose		25%w/v of sucrose	
Ovalbumin + Caseinate	- 0.56		- 7.47	

Interactions between different biopolymers becomes more favorable in the presence of the sucrose. Presumably, because of both the decrease of the excluded volume effect as a result of the Caseinate molecular weight reducing and rise of the Caseinate hydrophobicity. This change of character of the interactions between proteins under the influence of sucrose leads to the increase of the concentration of the protein's miscibility, especially under low concentration of the Caseinate molecule in the solution (See Figure 4). Upper branch of the phase diagram shifts toward higher Caseinate concentration in the system. Critical point shifts towards higher concentration of the both protein in the system.

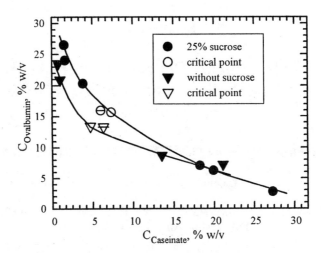

Figure 4. Effect of the sucrose on phase diagram (binodal and critical point) of the system: Caseinate-Na- Ovalbumin - Water at pH 6.6.

"Literature Cited"
1. Semenova, M.G.; Bolotina, V.S.; Grinberg, V.Ya.; Tolstoguzov, V.B. *Food Hydrocolloids*. **1990**, *3*, 447-456.
2. Semenova, M.G.; Pavlovskaya, G.E.; Tolstoguzov, V.B. *Food Hydrocolloids*. **1991**, *4*, 467-479.
3. Plashchina, I.G.; Semenova, M.G.; Braudo,E.E.; Tolstoguzov, V.B. *Carbohydrate Polymers*. **1985**, *5*, 159-179.
4. Braudo, E.E. *Food Hydrocolloids*. **1992**, *6*, 25-43.
5. Pavlovskaya, G.E.; Semenova, M.G., Tsapkina, E.N.; Tolstoguzov, V.B. *Food Hydrocolloids*. **1993**, *7*, 1-10.
6. Dickinson, E. *J.Chem. Soc. Faraday Trans.* **1992**, *88*, 2973 - 2983.
7. Tsapkina, E.N.; Semenova, M.G., Pavlovskaya, G.E., Leontiev, A.L.; Tolstoguzov, V.B. *Food Hydrocolloids*. **1992**, *6*, 237-251.
8. Dickinson,E.; Semenova, M.G. *J. Chem. Soc. Faraday Trans.* **1992**,*88*,849-854.
9. Edmond, E.; Ogston, A.G. *Biochem. J.* **1968**, *109*,569-576.
10. Tanford, C. *Phisical Chemistry of Macromolecules*; Wiley and Sons: NewYork and London, **1961**.
11. Kratochvill, P.; Sudelof, L.O. *Acta Pharm. Suec.* **1986**, *23*, 31-46.
12. Nagasawa, M.; Takahashi, A. in Huglin,M.B. *Light scattering from polymer solutions*; Academic press: London, **1972**, 671-719.
13. Ogston, A.G. *J.Phys. Chem.* **1970**, *74*, 668 -669.
14. Khoklov, A.R.; Nyrkova, I.A. *Macromolecules*. **1992**, *25*, 1493 - 1502.
15. Piculell, L.; Lindman, B. *Advancea in Colloid and Interface Science*. **1992**, *41*, 149 - 178.

Chapter 4

Use of Nonlinear Regression for Analyzing β-Lactoglobulin Denaturation Kinetics in Skim Milk

D. J. Oldfield[1], Harjinder Singh[1], M. W. Taylor[1], and K. N. Pearce[2]

[1]Department of Food Technology, Massey University, Palmerston North, New Zealand
[2]New Zealand Dairy Research Institute, Private Bag 11029, Palmerston North, New Zealand

Five different methods were used to calculate the kinetic parameters for the denaturation of β-lactoglobulin (genetic variant A). The activation energy (E_a) and pre-exponential term ($\ln(k_0)$) determined by nonlinear regression (NLR) had smaller confidence intervals and more degrees of freedom than those obtained by the commonly used two-step method, which uses linear regression. Of the methods investigated, NLR was the preferred method for analysing denaturation kinetics.

Heat-induced denaturation of whey proteins in skim milk is important in determining the functional properties of the final milk product. A number of researchers have used kinetics to quantify whey protein denaturation in milk (*1-5*). The rate of protein loss as a function of temperature, time and concentration is typically described by the general rate equation 1 and the Arrhenius equation 2.

$$-\frac{dC}{dt} = k_n \, C^{\,n} \qquad (1)$$

where
C = protein concentration (g kg^{-1}),
t = time (s),
k_n = rate constant ((g kg^{-1})$^{(1-n)}$ s^{-1}), and
n = reaction order.

0097–6156/96/0650–0050$15.00/0

$$k_n = k_0 \, e^{-\frac{E_a}{RT}} \tag{2}$$

where
k_0 = pre-exponential term $((g\ kg^{-1})^{(1-n)}\ s^{-1})$,
E_a = activation energy $(kJ\ mol^{-1})$,
R = universal gas constant $(8.314 \times 10^{-3}\ kJ\ mol^{-1}\ K^{-1})$, and
T = temperature (K).

Whilst equations 1 and 2 are used to quantify denaturation reactions they are not necessarily correct mechanistic descriptions of what actually occurs. Though they often fit the observations well. The procedure most commonly used to determine kinetic parameters has been described as a two-step method (6) and uses linear regression. The first step is to determine the concentration dependence of the rate at a fixed temperature. The temperature dependence of the rate constant is then found using the Arrhenius equation 2. The advantages of the two-step method are the ease of use and the small number of calculations required. Any non-uniform distribution of the variances of the raw data and the rates are not taken into account, and logarithmic transformation may produce trends in the residuals (7, 8). It has been suggested that better estimates of the kinetic parameters can be obtained by nonlinear regression (NLR) of concentration, temperature and time data (9). With advances in computing technology, methods such as NLR, which require more computational power, can be within easy reach of a researcher, namely through statistical software packages or one's own written software.

In this study, the two-step (linear regression) method was compared with NLR, using β-lactoglobulin genetic variant A (β-lg A) denaturation in milk as an example. The precisions of the kinetic parameters, E_a and k_0, were calculated for each method so that comparisons could be made.

Materials and Methods

Heat Treatment of Milk. Raw whole milk was obtained from the No. 1 dairy farm, Massey University, Palmerston North, New Zealand. The milk was separated at 40°C using a hermetic milk separator (Alfa-Laval, Sweden), and the resultant skim milk was stored at 5°C. The skim milk was then processed on a pilot-scale UHT plant (Type D, Alfa-Laval, Sweden). For each run, the skim milk was heated by direct steam injection (DSI) to the required temperature; DSI allowed for a rapid step change in the temperature. A range of temperatures (100-130°C) and a range of holding times (3-160 s) were investigated. After the holding tube, a flash vessel was used to reduce the milk temperature to approximately 65°C, effectively ending any further denaturation. The milk was collected and further cooled to 20°C in an ice bucket.

Protein Analysis of Milk. Milk samples were sealed in disposable plastic centrifuge tubes (13.5 ml, part No. 344322, Beckman, Palo Alto, CA) and placed in a Type 80 Ti rotor (Beckman, Palo Alto). Ultracentrifugation was carried out in a Beckman L8-80M centrifuge at 50,000 rpm (average 175,000 g) for 1 h at 20°C. After centrifugation, the top of the tube was cut open and all the supernatant was carefully removed. The supernatants were then analyzed for native β-lg A by polyacrylamide gel electrophoresis (native-PAGE) under non-dissociating conditions (10). The gels were scanned using a laser densitometer (Molecular Dynamics, Sunnyvale, CA), and the integrated intensities of the β-lg A protein bands were calculated by a software program, ImageQuant (Molecular Dynamics). A concentration standard of β-lg A (Sigma No. L-7880, lot 13H7020, Sigma Corp., St. Louis, MO) was run on the gel to convert the band intensities to concentration units of g kg^{-1}.

Statistical Analysis of Kinetic Data. The concentration/time data were analyzed by a variety of methods using the statistical software package SPSS (version 4.0.1, SPSS Inc., Chicago, IL). Five different methods of data analysis based on the general rate equation 1 and the Arrhenius equation 2 were used (11).

The concentration/time data of native β-lg A, determined by native-PAGE, were used to test the different methods. As a break in the Arrhenius plot is observed at 90°C (1, 3), a temperature region of 100-130°C was chosen to avoid this break. The data used were unweighted so that all the data points were treated equally by the program.

(i) Two-step Method (Linear Regression). The first step was to determine the rate constant from the integrated general rate equation. The resulting integrated equations are given in equations 3 and 4 (1).

When $n \neq 1$,

$$\left(\frac{C_t}{C_0}\right)^{1-n} = 1 + (n-1)\, k_n\, C_0^{n-1}\, t \qquad (3)$$

where
$C_t = $ concentration of undenatured protein at t = t (g kg^{-1}), and
$C_0 = $ concentration of undenatured protein at t = 0 (g kg^{-1}).

When $n = 1$ (first order reaction),

$$\ln\left(\frac{C_0}{C_t}\right) = k_1\, t \qquad (4)$$

where

k_1 =   rate constant for first order reaction (s^{-1}).

The left-hand side of the equation was plotted against time and k_n was calculated from the slope. The second step was to plot $\ln(k_n)$ against $1/T$ so that the Arrhenius equation could be used to calculate E_a and k_0 from the slope and intercept, respectively. Calculation of k_0 by equation 2 involved finding the intercept at $1/T = 0$ or when $T \to \infty$. As this point is a long way from where the data points lie on the Arrhenius plot, a small change in E_a causes a large change in k_0 (7). Linear regression of equation 2 assumed that the two parameters could be determined independently.

(ii) Two-Step Method with Adjusted Intercept (Linear Regression). The Arrhenius equation was modified so that a reference temperature was used in the exponential term (equation 5). This creates a new pre-exponential term, k_{ref}, which is related to k_0 by equation 6 (7).

$$k_n = k_{ref} \, e^{-\frac{E_a}{R}\left(\frac{1}{T} - \frac{1}{T_{ref}}\right)} \tag{5}$$

where

k_{ref} =   reference pre-exponential term ($(g\ kg^{-1})^{(1-n)}\ s^{-1}$), and
T_{ref} =   reference temperature (K).

$$k_0 = k_{ref} \, e^{-\frac{E_a}{R\,T_{ref}}} \tag{6}$$

By placing the intercept in the middle of the data set, k_{ref} becomes independent of the slope (E_a), and the estimated error in the pre-exponential term will be reduced (equation 5). The reference temperature chosen was 115°C, being in the middle of the temperature range 100-130°C. Thus the kinetic parameters were renamed k_{115} and T_{115}.

(iii) Scale Transformation (Linear Regression). Scale transformation of the concentration/time data was carried out by substituting equation 2 into equation 3 and rearranging (equation 7). Plotting the left-hand side against $1/T$ yields E_a and k_0 from the slope and intercept, respectively (7, 8).

$$\ln\left[\frac{\left(\frac{C_t}{C_0}\right)^{1-n}-1}{(n-1)\ C_0^{n-1}\ t}\right] = -\frac{E_a}{R}\frac{1}{T} + \ln k_0 \tag{7}$$

(iv) Nonlinear Regression (NLR). Equation 2 was substituted into equation 3 and then rearranged to make C_t the subject (equation 8).

When $n \neq 1$,

$$C_t = C_0\left[1 + (n-1)\ k_0\ C_0^{n-1}\ t\ e^{-\frac{E_a}{RT}}\right]^{\left(\frac{1}{1-n}\right)} \tag{8}$$

NLR was used to fit the concentration/time data directly into equation 8. This treats all concentrations values as of equal absolute accuracy. The statistical program SPSS uses a Levenberg-Marquardt algorithm (12-13). The equation allows for the program to fit the following parameters: E_a, k_0, reaction order (n) and initial concentration (C_0). Suitable starting estimates of the parameters, determined by the two-step method and native-PAGE analysis, were put into the program.

(v) NLR with Adjusted Intercept. As in the two-step method with an adjusted intercept, the Arrhenius term in the NLR equation can be modified using a reference temperature (equations 9 and 10). This reduced the possibility for uncertainty in E_a to influence the estimate of the intercept.

When $n \neq 1$,

$$C_t = C_0\left[1 + (n-1)\ k_{ref}\ C_0^{n-1}\ t\ e^{-\frac{E_a}{R}\left(\frac{1}{T}-\frac{1}{T_{ref}}\right)}\right]^{\left(\frac{1}{1-n}\right)} \tag{9}$$

When $n = 1$ (first order reaction),

$$C_t = C_0\ e^{\left[k_{ref}\ e^{-\frac{E_a}{R}\left(\frac{1}{T}-\frac{1}{T_{ref}}\right)}\right]} \tag{10}$$

Results

The loss of native β-lg A, determined using native-PAGE, is shown in Figure 1. The concentration of native β-lg A decreased with increasing heating time and increasing temperature.

The kinetics of β-lg A denaturation, using the five different methods outlined above, were investigated as to their precision in calculating the kinetic constants.

Two-step Method (Linear Regression). The apparent reaction order (n) for the denaturation of β-lg A was found to be 1.5, using linear regression analysis on equation 3, and this value was used in all the linear regression methods. The value of n is in agreement with Dannenberg and Kessler (*1*), but other researchers have reported $n = 2$ (*2-3*). Using the two-step method, equation 3 was plotted for $n = 1.5$ (Figure 2). The rate constants and their 95% confidence intervals were determined from the slopes (Table I). The 95% confidence intervals (C.I.), degrees of freedom (d.f.) and standard errors were different at each temperature. A plot of $\ln(k_n)$ against $1/T$ yielded a straight line (Figure 3), and least squares linear regression was used to calculate E_a and $\ln(k_0)$. The calculated values of E_a and $\ln(k_0)$ were 48.36 (kJ mol^{-1}) and 12.29, respectively. These values are in general agreement with earlier studies on β-lg A denaturation in skim milk; E_a values in the range 54-36 kJ mol^{-1} have been reported (*1-3*).

Table I. Rate Constants Obtained from the Two-step Method for β-lg A Denaturation in Skim Milk

Temperature (°C)	Rate constant (k_n) ((g kg^{-1})$^{-0.5}$ s^{-1}) × 10^3	95% C.I.	d.f.	Standard error
100	34.3	± 7.5	6	3.1
110	62.9	± 3.1	7	1.3
120	77.6	± 5.8	5	2.3
130	115.8	± 15.0	5	5.9

Comparison of Methods. Table II shows the kinetic parameters derived from different methods of data analysis and their corresponding 95% C.I. The values of E_a and k_0 were slightly different between the methods, but all were within the 95% confidence bounds calculated. Statistically, the two-step method produced the least precise results, with only 2 degrees of freedom, and NLR produced the most precise results.

Adjusting the method to find the intercept reduced the error in the pre-exponential term of the Arrhenius equation (equations 2 and 5) (Table II). The 95% C.I. for $\ln(k_0)$ obtained by the two-step method was ± 8.14. By placing the intercept in the middle of the data set, the 95% C.I. of $\ln(k_{115})$ was reduced to ± 0.24, and there was a further slight improvement in the NLR model, ± 0.12. The

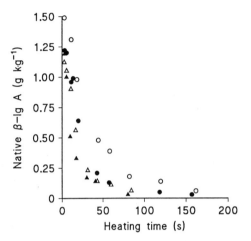

Figure 1. Loss of native β-lg A determined using native-PAGE, in heated skim milk. Heating temperatures: (○) 100°C, (●) 110°C, (△) 120°C and (▲) 130°C.

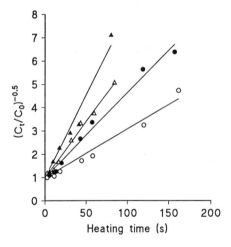

Figure 2. Denaturation of β-lg A in skim milk as a 1.5 order reaction. Heating temperatures: (○) 100°C, (●) 110°C, (△) 120°C and (▲) 130°C.

correlation between E_a and k_0 was high (r = 0.998) in the NLR model without an adjusted intercept, and non-existent (r = -0.061) in the adjusted intercept model. Because the correlation between E_a and k_0 was reduced, errors in E_a had less effect on k_0. This was observed as a reduction in the 95% C.I. of the pre-exponential term, but those of E_a remained constant at ± 7.07 (kJ mol^{-1}).

Figure 4 shows the predicted levels of native β-lg A calculated from the kinetic

Table II. Kinetic Parameters E_a and k_0 and their Respective 95% C.I. Derived from Different Methods of Data Analysis for β-lg A Denaturation in Skim Milk Heated at 100-130°C

Method of analysis	Kinetic parameters ± 95% C.I.			d.f.
	E_a (kJ mol^{-1})	$\ln(k_0)$	$\ln(k_{115})$	
Two-step	48.36 ± 26.36	12.29 ± 8.14		2
Two-step with adjusted intercept	48.40 ± 13.89	13.89 ± 8.17	-3.18 ± 0.24	2
Scale transformation	49.65 ± 12.55	12.55 ± 3.44		26
NLR	54.23 ± 7.07	13.89 ± 2.19		24
NLR with adjusted intercept	54.23 ± 7.07	13.89 ± 2.19	-2.91 ± 0.12	24

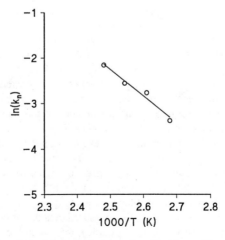

Figure 3. Arrhenius plot of β-lg A denaturation in skim milk, where k_n is in units of $((g\ kg^{-1})^{-0.5}\ s^{-1})$.

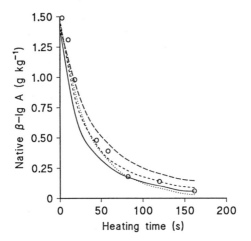

Figure 4. Comparison of predicted and experimental levels of β-lg A denaturation in skim milk heated at 100°C. (○) experimental data. Predicted values from different methods: (-) two-step, (— —) two-step with adjusted intercept, (----) scale transformation and (·····) NLR with adjusted intercept.

parameters. As the NLR and NLR with adjusted intercept methods estimated the same numerical values for the parameters (Table II), the NLR curve in Figure 4 represents both methods. When compared with the experimental data, the two-step method and the two-step method with adjusted intercept did not predict the native protein levels as closely as scale transformation and NLR. The initial concentration of β-lg A calculated by NLR, $C_0 = 1.51 \pm 0.11$ (g kg^{-1}), was in agreement with the initial concentration determined from an unheated sample by native-PAGE, 1.53 ± 0.25 (g kg^{-1}).

Discussion

Comparison of methods. The commonly employed two-step method has some limitations with respect to statistical analysis of the data. The variability in the rate constants are generally ignored when calculating E_a and k_0, and there is less utilization of the raw data, resulting in the kinetic parameters being associated with only a small number of degrees of freedom, which results in large confidence intervals (7, 8).

The method of linear regression, used to calculate the rate constants from the concentration/time data, assumes that the variance is evenly distributed over the raw data. Therefore each data point is assumed to have the same variance or absolute error; quite often it is the relative error that is constant. The result is that data points furthest away from the intercept have the largest influence, when they are often the least accurate. These points may have larger errors than the rest of the data for two possible reasons: the samples are usually taken at longer holding times where other reactions may interfere with the reaction under study; the measured concentrations are relatively low and may be at the limits of detection and accuracy of the analytical method employed.

As the rate constants from the two-step method all had different standard errors (Table I), the assumption of least squares regression does not hold true. Therefore, whenever linear regression is used to obtain the kinetic parameters from the Arrhenius plot, the variability of the rate constants should be calculated. If the standard deviations are not approximately equal, then the rate constants might need to be weighted (*8*).

A further disadvantage of the two-step method is that a lot of information (degrees of freedom) about the raw data is lost when it is transformed into a rate constant (Table II). Only a few data points are used in the Arrhenius plot, from all the concentration/time data collected.

NLR was found to be an improvement over the two-step method, and NLR with an adjusted intercept provided kinetic parameters with the highest level of precision. Compared with using the rate constants to determine E_a and k_0, greater accuracy and precision were provided when all the concentration/time data were used in either scale transformation or NLR. Both methods can be considered as one-step procedures, which process the raw data into the kinetic parameters without going through the intermediary rate constant calculations. These one-step methods had smaller and more realistic estimates of error, and more degrees of freedom associated with E_a and k_0, compared with the two-step method (Table II).

Improvement in the precision of the pre-exponential term was achieved by adjusting the intercept in the Arrhenius equation, so that it was centred in the temperature range used. This had little or no effect on E_a, but reduced the correlation between E_a and k_0, so that the error in the pre-exponential term k_{115} decreased (Table II).

Reaction Order. The reaction order determined was 1.5 ± 0.4 by linear regression, and 1.1 ± 0.3 (95% C.I) by NLR. The difference in the reaction orders was caused by the two methods treating the data differently. NLR gives all the data the same weighting, but linear regression is influenced by the larger numerical values (Figure 2), which lie at the longer holding times. The reaction order of β-lg has been reported as 1, 1.5 and 2 by numerous authors (*1-4, 14-16*). The different reaction orders may result from the variety of methods used and compositional variations in the media. Because of the complex nature of the reactions occurring in milk during heating, the apparent order of denaturation may not be very significant in terms of defining the denaturation mechanism, and therefore should be regarded as an empirical constant.

Conclusions

Of the methods tested, NLR with the adjusted intercept provided kinetic parameters with the highest level of precision. The method utilized all the raw data, thus providing a large number of degrees of freedom, and considered all the kinetic parameters at the same time when trying to obtain the best fit. As the kinetic equations themselves are nonlinear, use of NLR would appear to be the most suitable method of analysis.

Acknowledgments

The authors would like to thank the New Zealand Dairy Board for financial assistance, Prof. A. C. Cleland for helpful discussions, and Mr. G. Groube for operating the UHT plant.

Literature Cited

1. Dannenberg, F.; Kessler, H. G. *J. Food. Sci.*, **1988**, *53*, 258-263.
2. Hillier, R. M.; Lyster, R. L. J. *J. Dairy Res.*, **1979**, *46*, 95-102.
3. Manji, B.; Kakuda, Y. *Can. Inst. Food Sci. Technol. J.*, **1986**, *19*, 163-166.
4. Dalgleish, D. G. *J. Agric. Food Chem.*, **1990**, *38*, 1995-1999.
5. Resmini, P.; Pellegrino, L.; Hogenboom, J. A; Andreini, R. *Ital. J. Food Sci.*, **1989**, *1(3)*, 51-62.
6. Hill Jr, C. G.; Grieger-Block, R. A. *Food Technol.*, **1980**, *34(2)*, 56-66.
7. Wilkinson, S. A. *PhD Thesis*; Massey University, Palmerston North, New Zealand, 1981.
8. Arabshahi, A.; Lund, D. B. *J. Food Process Eng.*, **1985**, *7*, 239-251.
9. van Boekel, M. A. J. S.; Walstra, P. In *Heat-induced Changes in Milk*; Fox, P. F., Ed.; 2nd edn; International Dairy Federation: Brussels, 1995; pp 22-50.
10. Singh, H.; Creamer, L. K. *J. Food Sci.*, **1991**, *56*, 238-246.
11. Levenspiel, O. *Chemical Reaction Engineering*; 2nd edn; John Wiley & Sons: New York, NY, 1972; pp 8-86.
12. Bates, D. M.; Wats, D. G. *Nonlinear Regression Analysis and its Applications*; John Wiley & Sons: New York, NY, 1988; pp 80-82.
13. Pázman, A. *Nonlinear Statistical Models*; Kluwer Academic Publishers: Dordrecht, Netherlands, 1993; pp 124-125.
14. de Wit, J. N.; Swinkels, G. A. M. *Biochim. Biophys. Acta*, **1980**, *624*, 40-50.
15. El-Shazly, A.; Mahran, G. A.; Hofi, A. A. *Milchwissenschaft*, **1978**, *33*, 166-170.
16. Relkin, P.; Launay, B. *Food Hydrocolloids*, **1990**, *4*, 19-32.

Chapter 5

Particle Sizes of Casein Submicelles and Purified κ-Casein

Comparisons of Dynamic Light Scattering and Electron Microscopy with Predictive Three-Dimensional Molecular Models

H. M. Farrell, Jr.[1], P. H. Cooke[1], G. King[1], P. D. Hoagland[1],
M. L. Groves[1], T. F. Kumosinski[1], and B. Chu[2]

[1]Eastern Regional Research Center, Agricultural Research Service,
U.S. Department of Agriculture, 600 East Mermaid Lane,
Wyndmoor, PA 19038
[2]Department of Chemistry, State University of New York
at Stony Brook, Long Island, NY 11794

The colloidal complexes of skim milk, the casein micelles, are thought to be composed of spherical submicellar aggregates which are the result of protein-protein interactions. Studies on submicellar particles will help to elucidate the overall structure and function of these colloidal particles. Dynamic light scattering (DLS) studies on caseinate solutions in a Pipes-KCl buffer at pH 6.75 and in the absence of calcium ions showed a bimodal distribution with comparable scattering intensity contributions using the CONTIN method of analysis. Based on the weight fraction of submicelles and that of the aggregates, the casein solution is overwhelmingly in the submicellar form with radii in the 10 nm range. Similar results were obtained for purified κ-casein, the protein responsible for colloid stability; its weight average radius was found to be 12 nm. Electron microscopy (EM) using uranyl acetate-negative staining showed that both whole caseinate and κ-casein occur as irregular spherical particles and the number average sizes for their radii in this buffer were 7.7 and 8.8 nm, respectively, which are in agreement with DLS data. To develop a molecular basis for the reasons why these two different groups of particles appear so similar, the EM data were compared with previously developed three dimensional molecular models for submicellar caseinate and κ-casein. Good correlations for the shapes and sizes of these casein aggregates can be drawn from the models and structural alterations may be predicted for generation of new functional properties.

Casein constitutes the main protein component in milk and forms stable colloidal particles of approximately spherical shape, known as casein micelles which consist of about 93% protein and 7% inorganic material (principally calcium and phosphate). Four major proteins (α_{s1}-, α_{s2}-, β- and κ-casein) are present in the

0097–6156/96/0650–0061$15.00/0

micelles of bovine milk; among these phosphoproteins, the α_{s1}-, α_{s2}-, and β-caseins are sensitive to calcium ions and become insoluble at the calcium levels encountered in milk (1, 2, 3). κ-Casein is of particular interest as it remains stable at calcium concentrations up to 400 mM and can (through protein-protein interactions), stabilize the other calcium-sensitive casein components against coagulation. The κ-casein fraction is also unique in that because of its disulfide bonding pattern, it forms stable polymers within whole casein which range from dimers to octamers and above (4).

The native casein micelles are polydisperse in size, usually ranging in diameter from 100 nm to ~600 nm in bovine milk. It is believed that casein micelles are composed of a large number of subunits (submicelles) with diameters of 10-20 nm, held together by the colloidal calcium phosphate (1, 5). The exact supramolecular structure of the casein micelle is not yet clear, although various models have been proposed ranging from a structure consisting of discrete subunits to a more or less homogeneous sphere surrounded by a hairy surface layer, to the porous gellike structure (for a recent review, see reference 2)

The self-association behavior of the individual calcium sensitive casein components in aqueous solution has been well studied (5). Reduced (SH) κ-casein has been thought to exhibit soaplike micellization, i.e. follows a monomer⇌micelle model (6). The self association of κ-casein is somewhat insensitive to temperature and yields a complex with an average aggregation number of about 30 and a radius of 11 nm (6). κ-Casein polymers formed in the presence of reducing agents, have a high specific volume (6.5 or 5.7 ml/g calculated from the intrinsic viscosity or the sedimentation coefficient, respectively). To accommodate the high hydration and aggregation number, a model consisting of a κ-casein shell and a hollow core has been proposed (6, 7). Two contrasting models have been introduced, one for unreduced κ-casein micelles in a simulated milk ultrafiltrate by Thurn et al. (8) and one for reduced κ-casein by deKruif and May (9). All of these models differ somewhat from those derived from earlier electron microscopy studies of κ-casein (10, 11). Because of differences in methods of preparation, the degree of disulfide bonding could influence κ-casein structure, and also the properties of whole caseins as well-a result of the central importance of the κ-casein to the whole system (1-3).

It is known that the caseins not only exhibit self-association, but also interact with each other to form associated structures. The association between the different caseins which governs the physico-chemical behavior and the size of the submicelles formed appears to be not in a fixed stoichiometric manner with respect to the four individual caseins involved, meaning that submicelles of variable composition can be formed. However, small-angle X-ray studies (12) and gel permeation chromatography experiments (13) show that the overall sizes and shapes of submicellar whole caseinates may be somewhat similar to those reported for purified κ-caseins (7, 8, 9). The present study compares the overall physical properties of sodium caseinates and κ-casein as determined by light scattering and electron microscopy with three dimensional models (14, 15) to develop a molecular basis for the divergent aggregations which apparently converge to similar particle sizes and properties.

Methods And Materials

Materials. Whole casein typed α_{s1}-BB, β-AA, κ-AA was prepared as previously described (16). This procedure involves centrifugation at 100,000 g to remove residual lipid. κ-Casein was isolated from whole casein following the method of McKenzie and Wake (7). The κ-casein was also made lipid-free by centrifugation at 100,000 g. Polyacrylamide gel electrophoresis was according to Weber and Osborn (18) with minor modifications and at 7.5% polyacrylamide.

The whole casein was reduced and alkylated (RCM) in 8M urea essentially by the method of Schechter et al. (19). Samples, however, were dialyzed exhaustively in the dark at 4°C for 3 days against distilled, deionized water. The samples were adjusted to pH 7.0, centrifuged at 20° 100,000 × g for 20 min to remove aggregated material and lyophilized. Amino acid analysis showed >96% conversion to S-carboxymethyl cysteine. Polyacrylamide gel electrophoresis confirmed the absence of polymeric κ-casein bands (4).

Solution Preparation and Methodologies for Light Scattering. The caseinate samples were dissolved in a Pipes-KCl buffer (25 mM sodium piperazine-N,N'-bis(2-ethanesulfonate), pH 6.75, containing 110 mM KCl). The solutions were stirred for 3-5 minutes and then filtered using a low protein type Millipore filter (Millex-GV) with a nominal pore size of 0.22 μm. The casein concentrations used were in the range of 0.8 to 4.0 mg/ml.

For submicellar caseinates, a standard laboratory-built light scattering spectrometer (20) capable of both time-averaged scattered intensity and photon correlation spectroscopy measurements in an angular range of 15-140 deg was used. Intensity correlation function measurements were carried out in the self-beating mode by using a Brookhaven BI 2030AT 136-channel digital correlator. A Spectra-Physics model 165 argon ion laser operated at 488 nm was employed as the light source. All light scattering measurements were carried out at 25.0°C. The CONTIN method (21) was used for the data analysis of dynamic light scattering results.

For dynamic light scattering of κ-casein, glassware and quartz cells were cleaned with ultrapure water obtained from a Modulab Polisher HPLC Laboratory Reagent Grade Water System. Samples were prepared as described above. Dynamic light scattering was measured with a Malvern System Model 4700c equipped with a 256 channel correlator. Light at 488 nm was provided by a Spectra Physics Model 2020 5W laser. The ATTPC6300 computer supplied by Malvern was enhanced with a Sota 386si High Performance Accelerator card. Solutions were maintained at 25°C with filtered water in the goniometer chamber bath. Dynamic light scattering measurements were made at several angles using a small aperture in the photomultiplier detector. Photon count rates were kept in the low range by controlling the size of the laser beam. The data were processed by Malvern Automeasure v 4.12 software. Multi-angle analysis by Marquardt minimization was carried out with the Malvern software (22). The performance of the system and analysis software was tested on Latex Beads, 91 and 455 nm diameter (Sigma Chemical Co.).

Electron Microscopy of Caseins. Samples of whole casein, reduced carboxymethylated (RCM) whole casein and κ-casein were prepared for electron microscopy by dissolving the casein in PIPES-KCl buffer (25 mM piperazine-N-N'-bis (2-ethanesulfonic acid) pH 6.75, containing 80 mM KCl). The samples were made up to be 30 to 35 mg/ml and were passed through 0.45 μm filters. The filtrates were adjusted to 25 mg/ml with filtered buffer. Here the starting concentrations are much larger than those used in DLS, because of the dilutions attendant to the microscopy samples. Thin support films of amorphous carbon were evaporated on strips of cleaved mica and mounted on 400 mesh copper grids. All procedures were carried out in a water bath with samples and reagents at 37°C. Aliquots (10 μL) of caseins in buffered solution were placed on freshly prepared support films and suspended for 30-60 seconds over a water bath at 37°C; then the sample-side of the grid was washed with a controlled stream of 10 to 15 drops of buffered solution from a disposable Pasteur pipette containing 1% glutaraldehyde at 37°C. This was done to physically stabilize the composition of monomers in the form of complexes and to trap the equilibrium structures, while subsequently reducing the protein concentration to produce a discontinuous monolayer of casein particles. The adsorbed κ-casein particles or submicelles were then washed with a similar controlled stream of 5 to 10 drops of 2% uranyl acetate solution at 37°C for negative staining. Excess uranyl acetate solution was absorbed from the grid surface into Whatman #1 filter paper, and grids were allowed to air dry at room temperature.

Images of caseinate structures in randomly selected fields on grids were recorded photographically, at instrumental magnifications of 88,000 × using a Zeiss Model 10B electron microscope (Thornwood, NY) operating at 80KV, and 97,000 × using a Philips Model CM12 scanning-transmission electron microscope (Rahway, NJ) operating at 60 KV. Photographic prints were prepared at a magnification of 442,500 × and the circumferences of individual submicelles were traced onto transparent overlays. The overlays were digitized and circular diameters of the tracings were calculated and plotted using Imageplus software and a Dapple Microsystems digital image analyzer (Sunnyvale, CA).

Construction of Casein Aggregates by Molecular Modeling. The model κ-casein aggregate structure employed the energy minimized κ-casein model which was previously reported by this laboratory (15). Aggregates were constructed using a docking procedure on an Evans and Sutherland (St. Louis, MO) PS390 interactive computer graphics display driven by Sybyl molecular modeling software (Tripos, St. Louis, MO) on a Silicon Graphics (Mountainview, CA) W-4D35 processor. The criterion for acceptance of reasonable structures was determined by a combination of experimentally determined information and the calculation of the lowest energy for that structure. At least ten possible docking orientations were constructed, energy minimized, and assessed for the lowest energy in order to provide a reasonable sampling of conformational space.

Energy Minimization – Molecular Force Field. A full description of the concepts behind the use of the molecular force fields (including relevant equations) was given in previous communications (14, 23). In these calculations for protein-protein interactions, a Kollman Force field was employed (24). This force field uses electrostatic interactions calculated from partial charges derived by Kollman and coworkers. The following calculations were also used: a distance dependent dielectric constant, a united atom approach with only essential hydrogens, a cutoff value of 0.8 nm for all non-bonded interactions, and a conjugate gradient technique.

RESULTS AND DISCUSSION

Size of Casein Submicelles and κ-Casein as Determined by Dynamic Light Scattering. When the non-protein phase of the native casein micelles is removed by dialysis, the disintegration of the whole micelle occurs, leading to the formation of small spherical protein complexes called submicelles which consist of α_{s1}-, α_{s2}-, β-, and κ-caseins in an approximate ratio of 4:1:4:1 (1-3). Here we report our results obtained by dynamic light scattering on the average hydrodynamic size and the distribution of characteristic relaxation rates of casein submicelles which are formed by protein-protein interactions from sodium caseinate in a Pipes buffer (pH = 6.75, 110 mM KCl) in the absence of calcium ions. The casein concentration used was 8.15×10^{-4} g/ml. Figure 1 shows the results of the CONTIN analysis of the dynamic intensity data obtained at different scattering angles and at 25°C. The most striking feature of the plots given in Figure 1 is that instead of a unimodal distribution as expected for casein submicelles from small-angle X-ray scattering (*12*) and gel permeation chromatography (*13*), a well separated bimodal distribution has been consistently observed regardless of the scattering angle used, i.e., the scale of the probe window employed. The observation that, by varying the scattering angle from 15° to 90°, the intensity ratio of the two peaks in Figure 1 was dependent only weakly on the magnitude of the scattering vector, seems to rule out the possibility that the slow component monitored was caused by any dust contaminants. As seen from Figure 1, both the fast and slow species, in particular the former, have relatively narrow polydispersity, since the corresponding normalized variances are about 0.02 and 0.10, respectively. Based on this argument we may take the average diffusion coefficient (z-average, as deduced by the CONTIN analysis) to characterize each fraction. The average fast and slow diffusion coefficients plotted as a function of squared magnitude of scattering wave vector showed that the D_f values obtained at different scattering vectors, although somewhat disperse, were randomly distributed around an average value and showed essentially no q-dependence, as justified by the fact that for the fast component the relation $qR_h \ll 1$ holds. In contrast, the slow component exhibited a mild, but noticeable q^2-dependence, because $qR_h \geq 1$, particularly at higher scattering angles. Thus, D_f and D_s (by extrapolation to zero scattering angle) were found to be $(2.7 \pm 0.3) \times 10^{-7}$ and $(3.3 \pm 0.2) \times 10^{-8}$ cm^2/s, respectively. By using the Einstein-Stokes relation (*25*), the corresponding R_h values were calculated to be 8.8 ± 1 nm and 74 ± 4 nm, respectively. Apparently, the fast component with a large average diffusion coefficient corresponds to the casein submicelles, thus supporting the previous observations on the submicellar size (*12,13*).

At the present moment we have no conclusive answer to the nature of the slow component. The question arises as to whether these larger particles are aggregates of submicelles, or whether they are associated structures related to only some of the casein components. It is important to note that the complex formation between the four casein components yields structures (casein submicelles) having a smaller size than that of individual casein components, e.g., α_{s1}- and β-caseins. This means that as a result of the coexistence of all caseins in solution the complexes behave differently as compared to complexes of any single casein component. Pepper and Farrell (13) reported that for soluble whole casein in solution, changes in protein concentration may result in a variation of association behavior. For example, with increasing casein concentration the components of whole casein form submicelles. But, with decreasing casein concentration the κ-casein dissociates from the other casein components, forming associated structures independently. However, a single peak appeared on the gel chromatogram after the SS bond of κ-casein was reduced by dithiothreitol. We have performed dynamic light scattering measurements on the above casein solution in the presence of dithiothreitol (DTT). By comparing Figs. 1c (no dithiothreitol added) and 1d (the dithiothreitol concentration was 0.1 mM), no noticeable differences were detected. Finally RCM (reduced carboxymethylated) casein yielded virtually identical results (25).

As noted above the casein solution shows an angular dependence of scattered intensity which can be ascribed to the presence of the slow component. Because of its relatively large size (~75 nm radius), the intraparticle interference associated with the slow component accounts for the decrease in scattered intensity with increasing q. By correcting for interparticle interference (25) the weight fraction of each component can be estimated. Figure 2 clearly shows that the slow component does exist, but in a very small amount. Thus, the submicelles are dominating in quantity in agreement with previous findings (5, 13).

κ-Casein samples were also subjected to dynamic light scattering analysis in the same buffers as whole caseins. Samples were analyzed for particle size at 30, 60, 90 and 120°. Data analysis by the Malvern multi-angle program also showed two peaks analogous to those seen for whole casein. The results are given in Figure 3 for the weight average distribution. The smaller sized particles centering about 11.8 nm radius account for greater than 98% of the weight fraction of the material, while the larger sized particles centering about 110 nm radius account for less than 1%. Thus both whole casein and κ-casein exhibit similar bimodal particle size distributions, but the κ-casein particles are larger on the average.

Electron Microscopy of Whole Caseins. The dynamic light scattering results obtained above show that both submicellar casein and κ-casein exhibit similar overall particle size distributions. Parallel conclusions can be drown from comparisons of gel permeation chromatography results (13). Although the overall shape of a particle can be calculated from these methods, the specific distribution of voids and surface deformations can not be detected because of the relatively low resolution of the scattering experiments which are the average structure of many particles. With this in mind, transmission electron microscopy experiments (TEM) were carried out on whole casein under submicellar conditions (absence of Ca^{2+}) and on purified κ-casein for comparison of the previously reported 3D models with

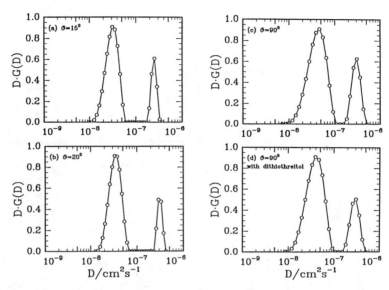

Figure 1. Relaxation rate distributions obtained at the indicated scattering angles by using the CONTIN analysis of the intensity correlation function. The peak area represents the scattered intensity contribution. (a)-(c): the whole casein solution in a Pipes-KCl buffer (pH=6.75 and total KCl=110 mM) containing no dithiothreitol. (d) the same casein solution with 0.1 mM dithiothreitol. For all solutions the casein concentration was 8.15×10^{-4} g/ml. (Reproduced with permission from reference 25. Copyright 1995 Academic.)

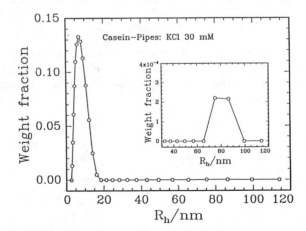

Figure 2. A plot of weight fraction distribution versus hydrodynamic radius obtained by the CONTIN analysis. The corresponding relaxation rate distribution is shown in Figure 1a. The inset shows an enlarged distribution profile for the large-size fraction. (Reproduced with permission from reference 25. Copyright 1995 Academic.)

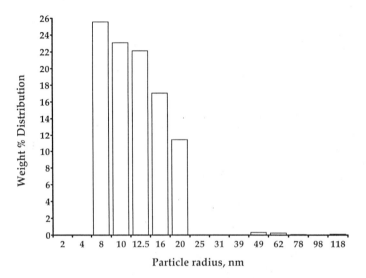

Figure 3. A plot of weight fraction distribution versus hydrodynamic radius obtained by the MALVERN analysis of κ-casein solutions. The corresponding experimental conditions are given except κ-casein in Figure 1. Concentration was 4.4 × 10^{-3} g/ml.

respect to predicted outlines and surface topology of the individual particles. RCM was chosen for study by electron microscopy because light scattering data showed little or no differences between native and RCM whole caseins. In addition, because the procedure for S-carboxymethyl derivatization (*19*) uses 8 M urea and was modified to include slow (3 day) dialysis at 4°C in the dark, a more uniform submicellar caseinate preparation with a stoichiometry of κ-:α_{s1}-:α_{s2}-:β-casein of 1:4:1:4 for computer enhanced electron microscopy was expected.

The lyophilized RCM whole casein sample was suspended in 25 mM PIPES pH 6.75 with 80 mM KCl and processed for TEM as described in Methods and Materials. A typical example of the total field of view is shown in Figure 4. A large number of particles with diameters from 10 to 20 nm were found in agreement with the dynamic light scattering results which yielded a Stokes radius of approximately 9.0 nm for native or for RCM casein submicelles. A few larger particles, possibly due to non-fixative associated cross links, occur in the total field. When 500 particles were measured and fitted to a Gaussion function the number average radius was found to be 7.7 ± 1.4 nm. Most previous studies have indicated a cauliflower-like appearance for both micelles and submicelles, with the greatest detail for submicelles being shown in the topographical images of Kimura et al. (*26*). The question here is whether or not the 3D models built from the monomeric caseins (*12, 14*) could give rise to the observed topographical details. Procedures were devised to compare the 3D models with TEM representations at comparable scales without losing the detail of the models, but also without overextending the resolution provided by the TEM samples. The rationale behind the methods developed to make the procedure more objective has been given in detail elsewhere (*27*). Briefly 200 submicellar structures were selected from 6 total fields such as Figure 4, enlarged 2× over Figure 4, and 15 particles were selected which were thought to be similar to the predicted 3D submicellar structures by one of the authors (HMF). A montage of these 15 pictures is shown in Figure 5. The submicellar particles display at least three topographical shapes: ellipsoidal, circular and rhomboid. Selected particles of the montage of Figure 5, were then photographically enlarged to bring the scale up to low resolution images generated by the modeling graphics. To reduce the resolution of the models, van der Waals surfaces of the asymmetric 3D energy minimized submicelle structure were calculated with a density of one dot/0.01 nm^2, and projected on a black background. These low resolution models were used for comparison with the photographically enhanced representations of the TEM. This approach seems justified since in work with crystalline proteins whose overall dimensions are known, negative staining with uranyl salts can achieve a high TEM resolution of up to 3.0 nm (*28*), thus the 3D models can be considered to be of low resolution if in agreement with TEM. Particle images were analyzed by Fourier transforms and compared with particle free backgrounds. These comparisons demonstrated the maximum resolution to be at least 3 nm for all of the particles used for enhancement.

The actual 3D orientation of the submicellar particle on the thin amorphous carbon grid could not be determined so the 3D models were rotated in the computer

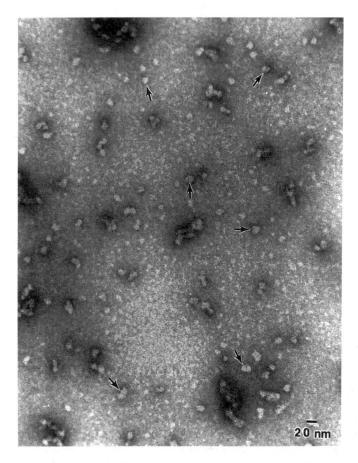

Figure 4. Negatively stained submicelles of modified whole casein. Most single particles have diameters ranging from 15-20 nm and some asymmetry in shape; bar = 20 nm.

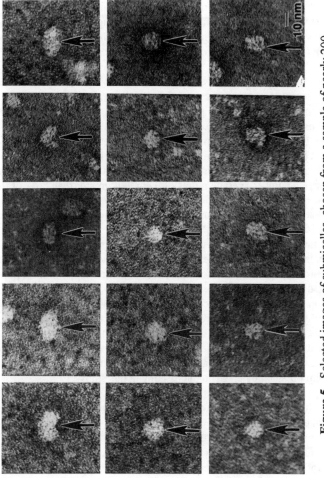

Figure 5. Selected images of submicelles, chosen from a sample of nearly 200 single particles representing three types of profiles: ellipsoidal (top row), circular (middle row) and rhomboidal (bottom row); bar = 10 nm.

for comparison with the micrographs rather than the opposite. All images compared in each TEM and 3D model are at the same scale in the figure. In Figure 6 (a) the backbone structure is shown to provide an orientation of structure in 3D space. As can be seen, this orientation of the model results in positioning of the κ-casein portion (i.e. horse and rider) at the center of the upper surface of the structure. Here the calculated van der Waals image (b) and the best representation (c) which corresponds to this bean-shaped van der Waals surface shows close agreement with respect to size, overall shape, and rugosity of surface. By rotation of this van der Waals image by 90° about the y axis (image is assumed to be in the x-y plane) the van der Waals 3D asymmetric model (d) and another TEM representation (e) are surprisingly good.

Electron Microscopy of Purified κ-Casein. When subjected to transmission electron microscopy using uranyl acetate as a negative stain, the purified κ-casein appears to occur primarily as single particles with only a few multiple particles in each field. Close examination of the micrographs shows that like whole casein, some particles have a bean-like shape and some are spherical. A typical field is shown in Figure 7. Size distributions of the κ-casein particles were made. The sample included 1500 particles from 20 fields and gave a Gaussian but narrow distribution with a number average radius of 9 nm, but greater than 86% of the particles counted gave an particle radius of 8.8 ± 0.9 nm.

Previous studies for whole κ-casein by rotary shadowing yielded diameters in the ranges of 10 to 15 nm (*11*) and 18 to 20 nm (*10*) for κ-casein particles but without statistical analysis. The particles observed in this study appear somewhat more uniform in distribution, but are in the range of the values previously reported. The rather uniform particle size observed by electron microscopy is somewhat at odds with the distribution of polymers observed for the κ-casein preparation on SDS-PAGE, in the absence of reducing agents (*4*). However by gel permeation chromatography these complexes are quite highly associated and in fact do not form reversible associating systems in the absence of reducing agents (*13*). The Stokes radius determined by gel chromatography for these maximally-associated κ-casein particles was 9.4 nm. This value is in good agreement with the average value of 8.8 nm obtained by electron microscopy in this study.

Investigations from this laboratory have generated a monomer model for κ-casein (*15*). This monomer model was used to assemble disulfide linked tetramers with an asymmetric arrangement of disulfides (11-88, 11-11 and 88-11). This tetrameric species is shown in Figure 8. The calculated radius of gyration for this particle is 4.7 nm which converts to a hydrodynamic radius of 6.1 nm somewhat smaller than the experimentally observed particles, and at 76,000 κDa, smaller than the experimental molecular weight which has been reported to range from 150,000 to 600,000 (*29, 30*). To simulate more fully the experimental data, two structures similar to Figure 8 were docked. The octamer, whose tetramers were at a 90° angle to each other, yielded the best structure after energy minimization (-200 kcal/mole of a stabilization energy). The calculated Stokes radius for this molecule was 8.0

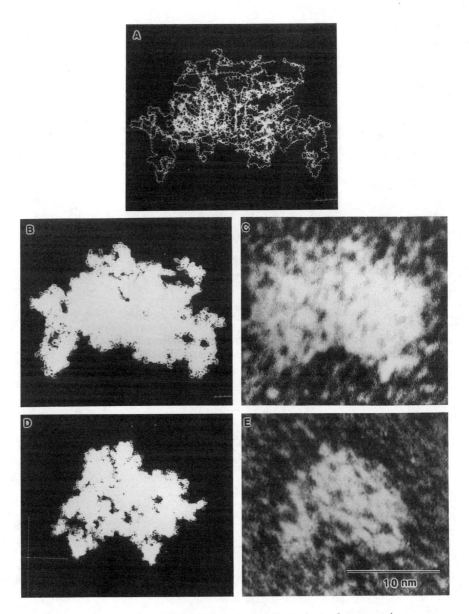

Figure 6. Comparison of matched shape and dimensions of asymmetric submicelle 3D model with photographically enlarged image enhanced representations of submicelles. (TEM bar = 10 nm; Molecular Model ½ axis = 5nm). (A) top: backbone structure for asymmetric model; (B) left: van der Waals dot surface of model; (C) right: enlargement of image enhanced micrograph; (D) left: van der Waals of (A) rotated 90° about *y* axis with enlargement of image enhanced representation of the submicelle particle (E).

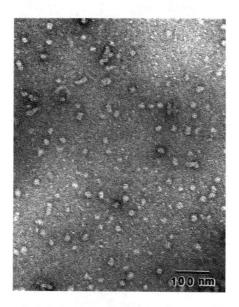

Figure 7. Transmission electron micrograph of a general field of negatively stained (uranyl acetetate 2%) κ-casein.

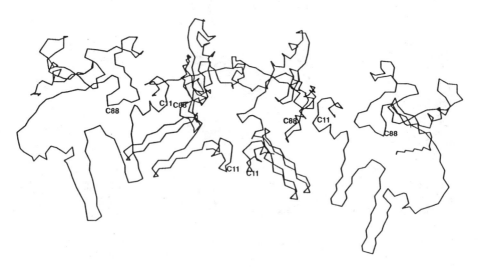

Figure 8. Three dimensional model for κ-casein tetramer linked by an asymmetric arrangement of disulfides, 11-88, 11-11, 88-11.

nm which is more in agreement with the experimental data but in the low range for aggregation number (*29, 30*).

The same procedures described above (Figure 6) were used for comparison of the TEM results for κ-casein to the 3D model of the octamer. Figure 9 shows the best example of the corresponding calculated van der Waals dot surface for the κ-model and best photographically enlarged TEM representations. The orientation of the upperhalf-surface of the octamer in the 3D model in space (Figure 9a) is similar to that shown in Figure 8 for the tetramer prior to docking its lower partner at 90°. The κ-casein structure exhibits surface features with long ridges 3 to 5 nm in length as found for the submicellar caseins. In addition the κ-casein 3D images on rotation (Figure 9b) may also be considered spherical as well as bean shaped (as were the submicellar images). The κ-casein 3D images still contain extended tails for the κ-casein macropeptide not visible in the TEM. It has been suggested (*2*) that these structures are observable hydrodynamically but are collapsed by the negative stain procedures in electron microscopy.

General Conclusions

The results obtained permit the following conclusions to be drawn for whole casein. For soluble whole casein in a Pipes-KCl buffer at pH = 6.75, DLS studies revealed a bimodal distribution either in the relaxation rate space or in the size space. The casein solution, on the weight fraction basis, remains overwhelmingly in the submicellar form with radii in the 10 nm range (8.8 ± 1 nm). RCM whole casein is in the same size range as its parent casein. RCM casein as studied by TEM has a size distribution (7.7 ± 1.4 nm) comparable to that obtained by DLS and results are compared to other studies in Table 1.

Table 1. Summary of Physical Data on Whole Caseins

	M.W.	*Radius (nm)*[a]	*Method*	*Reduction*
Pepper (*13*)		9.4	GPC[b]	None
Kumosinski (*12*)	285,000	8.0	SAXS[c]	DTT[d], hours
Chu (*25*)		8.8	DLS[e]	None
Chu (*25*)		8.8	DLS	RCM/DTT, hours
This Study		7.7	EM[f]	RCM

[a]Radius varies by type of measurement.
[b]Gel permeation chromatography.
[c]Small-angle X-ray scattering.
[d]Dithiothreitol.
[e]Dynamic light scattering (weight average).
[f]Electron microscopy.

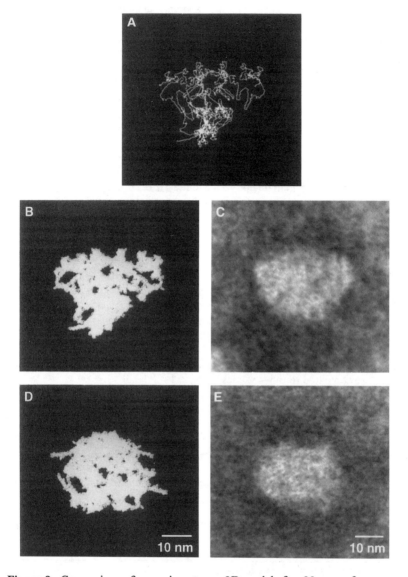

Figure 9. Comparison of κ-casein octamer 3D model after 20 psec of molecular dynamics with photographic enlargement of image enhanced representation of κ-casein. (TEM bar = 10 nm; Molecular Model bar = 10 nm). (A) Top: backbone structure for octamer of κ-casein; (B) left: calculated van der Waals surface of (A); (C) right: photographic enlargement of image enhanced EM representation of a κ-casein particle; (D) rotation of 90° about x axis of (B) van der Waals model and photographic enlargement of image enhanced EM representation of a κ-casein particle (E) showing a comparable structure.

The TEM patterns photographically enlarged are reminiscent of the surface shapes predicted by Kimura et al. (*26*). The results from TEM yield overall more structural information although the radii obtained for the particles are smaller than those from DLS, in line with possible stain penetration of their open porous structures.

κ-Casein, as purified from bovine milk, exhibits rather uniform distribution of particles 9.6 nm ± 2.5 nm radius as disclosed by DLS (number average). TEM yielded a value of 8.8 ± 0.9 nm. These data are at odds with some physical chemical data for κ-casein (Table 2).

Table 2. Summary of Physical Data on κ-Caseins

	MW	Radius (nm)[a]	Method	Reduction
Vreeman (*7*)	600,000	11.1[b]	Sedimentation	1 week, 2-ME[c]
Slattery (*30*)	600,000	11.2[d]	Sedimentation	1 hour, 40 mM DTT
deKruif (*9*)		14.7[b]	SANS[e]	5 mM DTT
Pepper (*13*)		9.4[d]	GPC[f]	None
Thurn (*8*)	2,000,000	7.0[b,g]	SANS[e]	None
This study		8.8[d]	EM[h]	None
		9.6[d]	DLS[i]	None

[a]Radius type varies with method.
[b]DEAE purified K-I casein.
[c]2-Mercaptoethanol.
[d]Whole κ-casein.
[e]Small-angle neutron scattering.
[f]Gel permeation chromatography.
[g]Internal "submicellar" particle of larger aggregate.
[h]Electron microscopy.
[i]Dynamic Light Scattering, number average; weight average = 11.8 nm.

It must be noted that the more recent data on the size distributions of κ-casein were not taken on the type of κ-casein used here. All of this data (*7, 8, 9*) was collected on samples which were reduced, purified on DEAE media in urea, dialyzed, lyophilized and then redissolved and reduced to varying extents (Table 2). The process of further isolation may yield particles with different degrees of aggregation. In this study the average radius for bovine κ-casein particles was 8.8 ± 0.9 nm which is similar to the value of 7.7 nm found for whole casein submicelles by identical electron microscopy techniques. Few higher order particles were observed in these studies. An overall view of κ-casein, as purified from bovine milk, is that of a series of disulfide-bonded polymers, ranging from monomers to octamers and above. These polymers are most likely distributed among three possible arrangements (11-88, 88-88, and 11-11) according to Rasmussen et al. (*31*). The apparent heterogeneity of the individual κ-casein chains however is overcome through protein-protein interactions which yield rather uniform particles of 8.8 nm as revealed by electron microscopy or 9.6 nm by light scattering. The shape and size of the κ-casein particles are mediated in part by the rigidity of its disulfide bonds which may cause the particles to be larger than those found for both RCM and native caseins. Thus, as noted above, purification procedures may produce protein-protein interactions in κ-casein which are not present in submicellar structures; alternatively some submicelles could be pure κ-casein.

The overall size, shape and perhaps rugosity of the 3D models are in line with the photographically enlarged representations of the TEM images of both whole casein and purified κ-casein. Whether or not the porosity predicted by the 3D model occurs cannot be support by the TEM since the depressions or pores are below resolution. Good correlation, however, between the images generated for the dot surfaces of the model and the TEM appears to exist in different spatial orientations of the models. However, when taken together with SAXS, SANS and hydration values (*7, 9, 12*) all of the data point toward open, highly hydrated and rugose structures for both κ- and whole caseins. The κ-casein 3D models built with a hydrophobic core and hydrophilic exterior confirm the conceptual model of deKruif and May (*9*) and portray internal cavities and pores necessary to accommodate the experimentally observed hydrations (*6, 7*).

As emphasized in previous papers on the 3D models of submicellar casein (*12, 14*), it must be kept in mind that these structures represent working models. They are not the final native structures but are presented to stimulate discussion and to be modified as future research unravels the nature of these non-crystallizable proteins. Inspection of a recent drawing of the casein micelle by Holt (*2*) demonstrates how structures such as those presented here could be further aggregated into the casein micelle. Continued dialogue and research in this area may stimulate the new concepts necessary to bring together divergent views and to finally produce an accurate micelle model. It is hoped that this work is a further step in that direction.

Literature Cited

1. Farrell, H. M., Jr., and Thompson, M. P. In *Caseins as Calcium Binding Proteins*; Thompson, M. P., Ed.; Calcium Binding Proteins; CRC Press: Boca Ratan, FL, 1988, Vol. 2, pp 150.
2. Holt, C. *Advances in Protein Chem.* **1992**, *43*, 63-151.
3. Swaisgood, H. *Developments in Dairy Chem.* **1992**, *1*, 51-95.
4. Groves, M. L., Dower, H. J., and Farrell, H. M., Jr. *J. Protein Chem.* **1991**, *11*, 21-28.
5. Schmidt, D. G. In *Developments in Dairy Chemistry*; 1 Fox, P. F., Ed.; Applied Science Pub. Ltd.: Essex, England, 1982, pp 61.
6. Vreeman, H. J., Brinkhaus, J. A. and van der Spek, C. A. *Biophys. Chem.* **1981**, *14*, 185-193.

7. Vreeman, H. J., Visser, S., Slangen, C. J., and Van Riel, J. A. M. *Biochem. J.* **1986**, *240*, 87-97.
8. Thurn, A., Blanchard, W. and Niki, R. *Colloid and Polymer Sci.* **1987**, *265*, 653-666.
9. deKruif, C. G., and May, R. P. *Eur J. Biochem.* **1991**, *200*, 431-436.
10. Parry, R. M. Jr., and Carroll, R. J. *Biochim. Biophys. Acta.* **1969**, *194*, 138-150.
11. Schmidt, D. G., and Buchheim, W. *Neth. Milk Dairy J.* **1976**, *30*, 17-28.
12. Kumosinski, T. F. King, G. and Farrell, H. M., Jr. *Journal of Protein Chemistry*, **1994**, *13*, 701-714.
13. Pepper, L., and Farrell, H. M., Jr. *J. Dairy Sci.* **1982**, *65*, 2259-2266.
14. Kumosinski, T. F., King, G. and Farrell, H. M., Jr. *Journal of Protein Chemistry*, **1994**, *13*, 681-700.
15. Kumosinski, T. F., Brown, E. M., and Farrell, H. M., Jr. *J. Dairy Sci.* **1993**, *76*, 2507-2520.
16. Kakalis, L. T., Kumosinski, T. F. and Farrell, H. M., Jr. *Biophys. Chem.* **1990**, *38*, 87.
17. McKenzie, H. A., and Wake, R. G. *Biochim. Biophys. Acta.* **1961**, *47*, 240-242.
18. Weber, K., and Osborn, M. *J. Biol. Chem.* **1969**, *244*, 4406-4409.
19. Schechter, Y., Patchornik, A., and Burstein, Y. *Biochemistry* **1973**, *12*, 3407-3413.
20. Zhou, Z. and Chu, B. *J. Colloid Interface Sci.* **1988**, *126*, 171-180.
21. Provencher, S. W. *Comput. Phys. Commun.* **1982**, *27*, 119-123.
22. Cummins, P. G., and Staples, E. J. *Langmuir* **1987**, *3*, 1109-1112.
23. Kumosinski, T. F. and Farrell, H. M., Jr. In *Protein Functionality in Food System*; Chapter 2, Hettiarachy, N.S. and Ziegler, G.R., Eds.; Marcel Dekker Inc.: New York, NY, 1994, pp 39-77.
24. Kollman, P. A. *Ann Rev. Phys. Chem.* **1987**, *38*, 303-331.
25. Chu, B., Zhou, Z., Wu, G. and Farrell, H. M., Jr. *Journal of Colloid Interface Science*, **1995**, *170*, 102-112.
26. Kimura, T., Taneya, S. and Kanaya, S. *Milchwissenschaft*, **1979**, *34*, 521-524.
27. Kumosinski, T. F., Uknalis, J. J., Cooke, P. H. and Farrell, H. M. Jr. *Lebens. Wiss. Technol.* **1996**, (In press).
28. Unwin, P. N. T. *Journal of Molecular Biology*, **1975**, *98*, 235-242.
29. Swaisgood, H. E., Brunner, J. R., and Lillevik, H. A. *Biochemistry* **1964**, *3*, 1616-1623.
30. Slattery, C. W., and Evard, R. *Biochim. Biophys. Acta.* **1973**, *317*, 529-538.
31. Rasmussen, L. K., Højrup, P., and Petersen, T. E. *Eur. J. Biochem.* **1992**, *207*, 215-222.

GEL AND FILM FORMATION

Chapter 6

Effects of Divalent Cations, Phytic Acid, and Phenolic Compounds on the Gelation of Ovalbumin and Canola Proteins

Susan D. Arntfield

Department of Food Science, University of Manitoba, Winnipeg, Manitoba R3T 2N2, Canada

Gelation of globular proteins can be modified by non protein components. The formation of thermally induced networks from globular proteins requires a balance between repulsive electrostatic forces and attractive forces supplied by hydrogen bonds, covalent bonds and hydrophobic interactions. Non protein components can influence gel characteristics by modifying these forces. Magnesium and calcium ions improved the firmness of ovalbumin gels due to increased crosslinking between proteins at pH values above the isoelectric point. Phytic acid or a combination of calcium and phytic acid increased the strength of canola gels only at pH values below the isoelectric point where there was direct interaction between the phytic acid and the protein. Phenolic compounds inhibited network formation, particularly at pH values above the isoelectric point, through interference with hydrophobic interactions.

The use of proteins in food systems is designed to do more that simply supply the amino acids necessary for nutrition. The ability to contribute to functional properties often determines if and where particular proteins are used. One such property is gel formation. This is an important property because the structure or texture of many food products depend on the ability of proteins to form networks.

Principles of Protein Network Formation

The mechanism responsible for network formation associated with the heat denaturation of globular proteins is an area that has been under investigation for a number of years (1-2). Current thinking indicates that the protein is first denatured by heat, but the unfolded protein retains a globular shape (3-5). These unfolded proteins will interact to form strands or soluble aggregates. It is interactions between these strands of globular proteins that leads to network formation (3-5). The need for protein unfolding can be clearly demonstrated using differential scanning calorimetry (DSC) to measure the heat flow during the endothermic transition associated with protein denaturation and comparing this transition to the development of structure as measured by an increase in the storage modulus (G') as a function of temperature using a Bohlin rheometer (Figure 1). It is clear that denaturation must precede structure development. Furthermore, there seemed to be a relationship between the difference in temperatures, and therefore times, associated with protein denaturation and structures development and the properties of the network formed (6). The greater the delay prior to structure development, the more likely the resulting structure would be a three dimensional network rather than an aggregated mass.

In addition, the conditions that promoted good network formation or a gel-like product required that there was a delicate balance between protein-protein and protein-solvent interactions. Specific interactions involved included: electrostatic interactions which supplied a repulsive force between proteins and promoted protein-solvent interactions (7); disulfide bonds (for some proteins) which were involved in protein-protein interactions during network formation (8); hydrophobic interactions which were also involved in network formation (9-10) and hydrogen bonding which seemed to play a role by forming protein-protein interactions which helped to strengthen networks (10).

Impact of Non Protein Components on Network Formation

A knowledge of the requirement for protein denaturation and a balance of protein-protein and protein solvent interactions does not supply all the answers in terms of protein behaviour and the potential for network formation. This is particularly true for systems containing non protein material. Three different examples will be given to indicate the role of small molecular weight non protein components on the thermal gelation of globular proteins. Included are an investigation of the effects of the divalent cations Ca^{+2} and Mg^{+2} on the thermal gelation of ovalbumin, the effects of phytic acid on the thermal gelation of the storage protein from canola and the effects of phenolic compounds on the thermal gelation of canola protein.

Effects of Ca^{+2} and Mg^{+2} on Ovalbumin Gelation. The rationale for investigating the effects of these cations on the thermal gelation of ovalbumin is related to the practice of removing lysozyme, a small molecular weight protein with antibacterial properties, from egg albumen prior to its use as a food ingredient (11-12). Lysozyme is normally removed using a cation exchanger and as a result divalent cations, including Ca^{+2} and Mg^{+2}, are also removed. To return the albumen to its original status as much as possible the cations can be added back. The level of addition, however, is not well defined as the values for these minerals in egg albumen is highly variable ranging from 78-89 mg/100 g (dry basis) and 72-

84 mg/100 g (dry basis) for Ca^{+2} and Mg^{+2}, respectively (13-14). In this study, salt combinations were examined at a ratio of 1 to 5 molar ratio of Mg^{+2} to Ca^{+2}. This ratio was chosen to approximate that normally encountered in egg white. The salt combinations used included $MgCl_2$ + $CaCl_2$, $MgSO_4$ + $CaCl_2$, $MgCl_2$ + calcium lactate $[Ca(C_3H_5O_3)_2]$, and $MgSO_4$ + $Ca(C_3H_5O_3)_2$. The total salt concentration ranged from 0 to 0.01M and the effects were examined at pH 5, 7 and 9. In all cases, gels were prepared by heating a solution of 10% ovalbumin in the appropriate medium to 90°C and then cooling immediately to room temperature using an ice bath. Firmness was taken as the initial slope of the curve monitoring deformation as a function of time based on a penetration test using an Ottawa Texture Measuring System.

At pH 5, which is closse to the 4.6-4.8 isoelectic point of ovalbumin (15), there were no effects due to salt type or concentration. At pH 7, the addition of salts results in stonger gels but the gel strength decreased as the salt concentration increased (Figure 2). It appeared that there was a certain level of salt which provided a bridging function resulting in an improvement in network strength. Levels that were too high resulted in increased protein-protein interactions such that the balance between protein-protein and protein-solvent interactions was skewed towards protein-protein interactions and aggregates rather that networks resulted. In comparison to this concentration effect, the effect of specific salt combinations was relatively small. Consistently higher firmness values were seen for the $MgSO_4/CaCl_2$ combination with the highest firmness value being obtained with a salt concentration of 0.005M. The $MgCl_2/CaCl_2$ combination also produced firm gels at salt concentrations of 0.001 and 0.005M. Interestingly, when the level of cation binding was determined using equilibrium dialysis (16) followed by analysis for the Ca^{+2} and Mg^{+2} ions using Ca^{+2} specific and divalent specific electrodes, it was the observed that the highest level of binding was also associated with the $MgSO_4/CaCl_2$ combination. This added support to the role of the cations as serving as crosslinks in promoting protein-protein interactions. Increased binding at the 0.01M salt levels, however, was also responsible for the weakening of the network.

At pH 9, increased salt concentration resulted in increased gel firmness (Figure 3). At this pH, the net charge on the protein is higher than at pH 7 so that there is more opportunity for the salt to bind and from crosslinks between proteins before the protein-protein interactions reach a point where aggregation is promoted. Evidence of a situation promoting aggregation was not obtained at the highest salt concentrations examined in this study. As was the case at pH 7, salt type had little effect on the firmness values obtained, although gels obtained with combinations of $MgCl_2/CaCl_2$ and $MgCl_2/Ca(C_3H_5O_3)_2$ were a bit firmer. The effect of cation binding on these results was not as well correlated as was the case at pH 7. While high levels of cation binding were seen with the $MgCl_2/Ca(C_3H_5O_3)_2$, the binding of cations for the $MgCl_2/CaCl_2$ systems was similar to those for the other salt combinations.

It is clear from this study that the addition of divalent cations to ovalbumin, prior to the formation of heat induced gels, results in improved gel firmness. The actual firmness values depend on the pH, in that weak aggregates were obtained

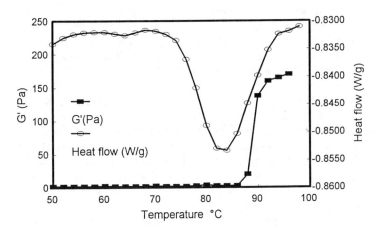

Figure 1. Relationship between protein denaturation, as monitored using differential scanning calorimetry, and the development of rheologically significant structure, as reflected by an increase in G', for 6% canola protein isolate at pH 8.

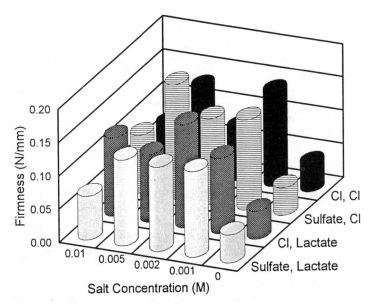

Figure 2. Effects of salt concentration and type on firmness of 10% ovalbumin gels at pH 7. Salt combinations at a ratio of 1:5 included: $MgCl_2/CaCl_2$ (Cl, Cl); $MgSO_4/CaCl_2$ (Sulfate, Cl); $MgCl_2/Ca(C_3H_5O_3)_2$ (Cl, Lactate); and $MgSO_4/Ca(C_3H_5O_3)_2$ (Sulfate, Lactate).

at pH 5, while stronger gels were obtained at pH 7 and 9. At pH 7, the maximum gel firmness was obtained with salt concentration of 0.002 to 0.005M while at pH 9, the gels were firmer with 0.01M salt, the highest level used in the study. In both cases the improvement in gel strength was a result of the role of the cations in providing a crosslink between proteins.

Effect of Phytic Acid on the Gelation of Canola Protein. Phytic acid, a compound used for storage of phosphorous, is found in most plant material including the oilseeds soybean and canola. In view of an ongoing interest in promoting utilization of canola protein for human consumption, it was desirable to get a better understanding of how natural contaminants could affect the behaviour of these proteins. Although a large portion of the phytic acid can be removed during protein isolation procedures, it is important to know how critical that level of residual phytic acid may be in terms of protein gelation properties.

In studies looking at the interactions between phytic acid and protein, it has been shown that the nature of the binding mechanism is highly dependent on pH and the presence of cations (17-18). Below the isoelectric point of the protein, a direct electrostatic interaction between the positively charged protein and the negatively charge phytic acid has been reported. Above the isoelectric point, a divalent cation is necessary for phytic acid to bind to the protein. A ternary complex of protein, phytic acid and cation has been reported (17-18).

Levels of phytic acid ranging from 0-5%, corresponding to levels that could reasonably be expected in canola protein products, were examined with and without the inclusion of 0.01M $CaCl_2$ at pH 5, 7 and 9. Canola protein was isolated using a method which included a salt extraction and precipitation by dilution (19). The level of phytate naturally associated with this protein isolate is less than 0.4%. Gel were formed in the Bohlin rheometer by heating at 2°C/min to 95°C, holding at this temperature for 2 min and cooling to 25°C at a rate of 2°C/min as described previously (6). Gel properties, measured using the Bohlin rheometer, included the storage modulus G' and the tan delta value (loss modulus G"/G'). An increase in the G' value reflects an increase in the magnitude of the elastic component in the gel, whereas a decrease in the tan delta value reflects an increase in the relative elasticity of the network.

For gels formed in the presence of phytic acid but no $CaCl_2$, the G' values obtained were pH dependent. At pH 5, which is below the isoelectric point of approximately 7 for the 12S canola protein (20-21), the addition of low levels of phytic acid increased the G' values but as more phytic acid was included, the G' value decreased (Figure 4). This situation was similar to that seen for the cations binding to ovalbumin. A certain degree of protein-protein interaction, resulting from the crosslinking action of the phytic acid improved the network, but as the degree of crosslinking increased, aggregation was promoted. At pH 7 and 9, the level of phytic acid did not significantly impact the G' value (Figure 4). This was not entirely surprising because there was no phytic acid binding at these pH values, as would be expected due to the lack of the divalent cation necessary for the formation of the ternary complex. The fact that similar results were obtained in the presence of 0.01M $CaCl_2$, however, was surprising (Figure 5). At pH 5, direct

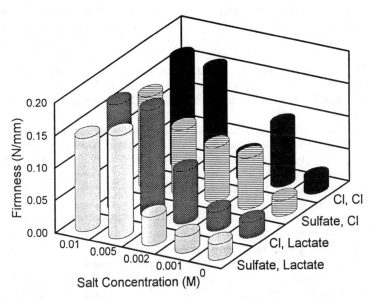

Figure 3. Effects of salt concentration and type on firmness of 10% ovalbumin gels at pH 9. Salt combinations at a ratio of 1:5 included: $MgCl_2/CaCl_2$ (Cl, Cl); $MgSO_4/CaCl_2$ (Sulfate, Cl); $MgCl_2/Ca(C_3H_5O_3)_2$ (Cl, Lactate); and $MgSO_4/Ca(C_3H_5O_3)_2$ (Sulfate, Lactate).

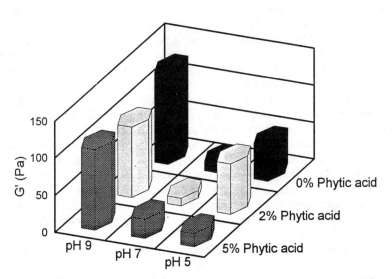

Figure 4. Effects of pH and phytic acid on the G' modulus (at a frequency of 1Hz) for gels from 10% canola isolate.

electrostatic interactions between phytic acid and protein resulted in increased G' values with the addition of 2% phytic acid but G' decreased when the phytic acid level was increased to 5%. Variations in the G' values at pH 7 and 9 were not significantly different. In addition, there was no evidence of phytic acid binding at pH 9. This implied that there was no formation of a ternary complex with the low phytic acid and $CaCl_2$ concentrations investigated in this study. Without any binding it was not surprising that the phytic acid/$CaCl_2$ combination did not have an effect on gel firmness.

The tan delta values were unaffected by the presence of phytic acid at all pH values, although the presence of Ca_2 and/or higher pH values resulted in lower tan delta values (Figure 6), indicating the formation of networks with a higher degree of crosslinking (7). The increased G' values in the presence of phytic acid at pH 5 did not promote the formation of better networks, only more interactions between proteins.

Several conclusions can be reached based on these data. First, the effect of phytic acid on the ability of the canola protein to form heat induced networks is pH dependent, giving increased G' values only at pH 5, below the isoelectric point for the canola protein. Although this indicated increased strength in the network, it did not promote the formation of a well crosslinked three dimensional network as the tan delta value was high and did not change in the presence of phytic acid. At higher pH values, there was no evidence of phytic acid binding, even in the presence of Ca^{+2}, and therefore phytic acid had no impact on the networks formed. It would appear that there is no benefit to be gained in terms of the gelation properties of canola protein by reducing the phytic acid levels to zero.

Effect of Phenolics on the Gelation of Canola Protein. Phenolic compounds are another example of naturally occurring minor components found in plants, including canola. Phenolics have traditionally been a concern because of the colors and flavors they can produce in food products (22-23). The major phenolic compound in canola is sinapic acid which may be present as is or in an esterified form as sinapine. In earlier work, it was found that at elevated pH values, the sinapic acid is converted to the lignan thomasidioic acid (24). As a result, when examining the impact of sinapic acid on the gelation properties of canola protein, it must be remembered that thomasidioic may be present.

Isolated canola protein (19) was again used as the protein source. Sinapic acid levels ranged from 0-2% (w/w based on the protein content) and represented levels that could be reasonably expected in canola products. Gels were prepared using 10% protein at pH 7 and 8.5 with and without the inclusion of 0.01 M NaCl. G' and tan delta values were measured on gels formed in the Bohlin rheometer by heating at 2°C/min to 95°C, holding at this temperature for 2 min and cooling to 25°C at a rate of 2°C/min.

Inclusion of a mixture of sinapic acid and thomasidioic acid, which were the forms present at pH 7 when sinapic acid was included, or thomasidioic acid, the only species present at pH 8.5, resulted in decreased G' values (Figure 7). Using equilibrium dialysis, followed by analysis of the unbound phenolic, it was determined that with or without salt thomasidioic acid, and not sinapic acid, bound

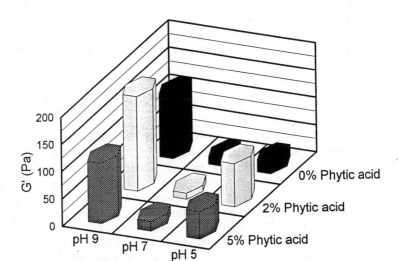

Figure 5. Effects of pH and phytic acid on the G' modulus (at a frequency of 1Hz) for gels from 10% canola isolate in the presence of 0.01M CaCl₂.

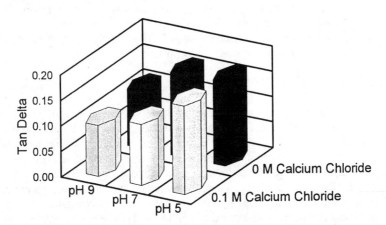

Figure 6. Effects of pH and the presence of 0.01M CaCl₂ on the tan delta values (at a frequency of 1Hz) for gels from 10% canola isolate. Data from samples with and without phytic acid included.

to the protein at pH 7. Based on increased binding in the presence of salt, the type of interaction involved in this association appeared to be hydrophobic rather than electrostatic. This interaction, however, interfered with the crosslinking necessary for good network formation. At pH 8.5, and no salt, a similar trend was observed even though the level of binding at this pH was minimal. In the presence of NaCl, however, different results were obtained in that an increase in the G' value was observed in the presence of 2% thomasidioic acid. Since the addition of either phenolics alone or NaCl alone at this pH resulted in decreased G' values, it appeared that the combination of the two resulted in an interaction between NaCl and the thomasidioic acid such that the formation of a protein network was comparable to that when neither compound was present.

A similar trend was seem when examining the impact of phenolics and NaCl on the tan delta values for the canola gels (Figure 8). At pH 7, with and without salt, and at pH 8.5, without salt, significantly higher tan delta values were obtained if phenolics were present. This indicated a decrease in the relative elasticity of the network and resulted from a decrease in protein-protein interactions contributing to the G' value. At pH 8.5, and in the presence of NaCl, low tan delta values were associated with the gel obtained in the presence of 2% thomasidioic acid. The increase in the G' value observed above resulted not just in increased interactions between proteins, but in interactions which promoted the development of a well crosslinked network.

From this study, it is clear that the effect of phenolics on the gelation properties of canola protein depends on the pH and whether or not NaCl in present. In most situations, the presence of phenolics, including both sinapic acid and thomasidioic acid, resulted in reduced network strength and a reduction in the relative elasticity of the networks. At pH 8.5, the presence of NaCl counteracted the influence of the phenolics, leaving the protein to form a network approaching that obtained when neither compound was present.

Conclusions

A number of general conclusions can be made on the impact of non protein components on the gelation of globular proteins based on the data obtained in these three studies. Strength of gels and degree of crosslinking can be improved or diminished by the inclusion of non protein components, depending on the specific compound added and the interaction that takes place between the protein and that compound. The pH of the system determines the balance between protein-protein and protein-solvent interactions by altering the role of electrostatic interactions. If non protein components increase the level of protein-protein interaction through the formation of crosslinks and at the same time reduce the protein-solvent interactions by reducing the net charge on the protein, network strength can be improved. If the extent of interaction is high or if protein-protein interactions are already favored due to the proximity of the pH in relation to the isoelectric point of the protein, however, this can lead to aggregation rather than the development of a well crosslinked network. Because hydrophobic interactions play a role in the initial formation of a network, interference with these interactions, as was the case when phenolic compounds were included, results in diminished network strength and reduced crosslinking within the network.

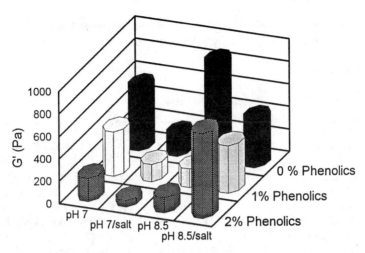

Figure 7. Effects of phenolics, pH and the presence of 0.1M NaCl on the G' modulus (at a frequency of 1Hz) of gels from 10% canola protein isolate. Phenolics added as sinapic acid, but convert to a mixture of sinapic acid and thomasidioic acid at pH 7 and thomasidioic acid alone at pH 8.5.

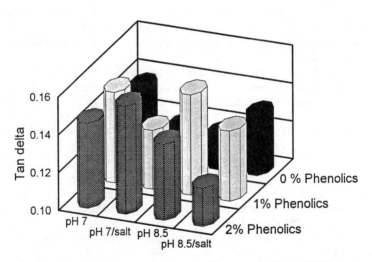

Figure 8. Effects of phenolics, pH and the presence of 0.1M NaCl on the tan delta value (at a frequency of 1Hz) of gels from 10% canola protein isolate. Phenolics added as sinapic acid, but convert to a mixture of sinapic acid and thomasidioic acid at pH 7 and thomasidioic acid alone at pH 8.5.

Acknowledgements

The contributions of Virginia Savoie, Amy Wong, Maria Rubino, Céline Nadon and Aniko Bernatsky are gratefully acknowledged.

Literature Cited

1. Ferry, J.D. *Adv. Protein Chem.* **1948**, *4*, 1-78.
2. Flory, P.J. *Faraday Disc. Chem. Soc.* **1974**, *57*, 7-18.
3. Tombs, M.P. *Faraday Disc. Chem. Soc.* **1974**, *57*, 158-164.
4. Nakamura, T.; Utsumi, S.; Mori, T. *J. Agric. Food Chem.* **1984**, *32*, 349-352.
5. Clark, A.H.; Lee-Tuffnell, C.D. In *Functional Properties of Food Macromolecules*; Mitchell, J.R.; Ledward, D.A., Eds.; Elsevier Applied Science: New York, NY, 1986. pp 203-272.
6. Arntfield, S.D.; Murray, E.D.; Ismond, M.A.H.; Bernatsky, A.M. *J. Food Sci.* **1989**, *54*, 1624-1631.
7. Arntfield, S.D.; Murray, E.D.; Ismond, M.A.H. *J. Texture Stud.* **1990**, *21*, 295-322.
8. Arntfield, S.D.; Murray, E.D.; Ismond, M.A.H. *J. Agric. Food Chem.* **1991**, *39*, 1378-1385.
9. Arntfield, S.D.; Murray, E.D.; Ismond, M.A.H. *J. Agric. Food Chem.* **1990**, *38*, 1335-1343.
10. Léger, L.W.; Arntfield, S.D. *J. Amer. Oil Chem. Soc.* **1993**, *70*, 853-861.
11. Durance, T.D.; Nakai, S. *J. Food Sci.* **1988**, *53*, 1096-1102.
12. Li-Chan, E.; Nakai, S.; Sim, J.; Bragg, D.B.; Lo, K.V. *J. Food Sci.* **1986**, *51*, 1032-1036.
13. Cotterill, O.J.; Glauert, J.L. *Poultry Sci.* **1979**, *58*, 131-134.
14. USDA. *Composition of Foods*; Agricultural Handbook No. 8-1; USDA Agricultural Research Service: Washington, DC, 1976.
15. Parkinson, T.L.; *J. Sci. Food Agric.* **1966**, *17*, 101-111.
16. Damodaran, S; Kinsella, J.E. *J. Biol. Chem.* **1981**, *256*, 3394-3398.
17. Cheryan, M. *CRC Crit. Rev. Food Sci. Nutr.* **1980**, *13*, 297-334.
18. Prattley, C.A.; Stanley, D.W.; van de Voort, F.R. *J. Food Biochem.* **1982**, *6*, 255-271.
19. Ismond, M.A.H.; Welsh, W.D. *Food Chem.* **1992**, *45*, 125-127.
20. Schwenke, K.D.; Raab, B.; Linow, K.-J.; Pähtz, W.; Uhlig, J. *Die Nahrung* **1981**, *25*, 271-280.
21. Léger, L. *Thermally Induced Gelation of the 12S Canola Globulin*; M.Sc. Thesis; University of Manitoba: Winnipeg, MB, 1992.
22. Larsen, L.M.; Olsen, O.; Ploger, A. and Sørensen, H. In *Proceedings of 6th International Rapeseed Congress*; Paris, France, 1983, pp 1477-1582.
23. Andersen, H.R.; Sørensen, H. In *Advances in the Production and Utilization of Cruciferous Crops*; Sørensen, H., Ed.; Marinus Nijhoff, Norwell, MA, 1985, pp 208-217.
24. Rubino, M.I.; Arntfield, S.D.; Charlton, J.L. *J. Amer. Oil Chem. Soc.* **1995**, *72*, 1465-1470.

Chapter 7

The Role of α-Lactalbumin in Heat-Induced Gelation of Whey Proteins

N. Matsudomi and T. Oshita

Department of Food Chemistry, Faculty of Agriculture,
Yamaguchi University, Yamaguchi 753, Japan

The effects of addition of α-lactalbumin (α-La) to β-lactoglobulin (β-Lg) or bovine serum albumin (BSA) were investigated to elucidate the gelling properties of whey protein. An β-Lg concentration of 4% (w/v) was required for the formation of a self-supporting gel following heating at 80°C for 30 min in 100 mM sodium phosphate buffer (pH 6.8). For BSA, it was below 4% (w/v). Solutions of α-La, even up to a protein concentration of 8% (w/v), did not gel under the same conditions. The addition of α-La to β-Lg or BSA significantly enhanced gel hardness; addition of α-La above 3% caused a substantial increase of gel hardness of their mixtures, although the α-La itself did not form gels. A 0.2% solution of β-Lg heated for 30 min at 80°C formed soluble aggregates, however, a similar α-La solution did not. When a mixture of both diluted solutions was heated, they interacted to form soluble aggregates through a thiol-disulfide interchange reaction, an interaction that was critical in formation of the gel network and markedly enhanced the hardness of gels. Similarly, addition of α-La to BSA induced formation of soluble aggregates through the same mechanism, to form a finer and more uniform network leading to transparent gels. It was concluded that the enhancing effect of α-La on the gel hardness of β-Lg or BSA was due to the formation of a specific soluble aggregate, and that such interactions contribute to the properties of whey protein gels.

Whey is the soluble fraction of milk that is separated from the casein curd during cheese manufacturing. The main proteins in whey are β-lactoglobulin (β-Lg), α-lactalbumin (α-La), immunoglobulins, and bovine serum albumin (BSA). Protein products derived from whey proteins, such as concentrates and isolates (approximately 70 and 90% protein, respectively), have important commercial applications in foods, a major one being their tendency to form irreversible heat-induced gels. The gel-forming ability of whey proteins is one of the most important functional properties in food system. An understanding of the mechanism for the gelation of whey proteins is essential to a variety of applications of these proteins in foods. Gelation of whey protein products and the constituent proteins has been

0097–6156/96/0650–0093$15.00/0

reviewed many times in the past (*1-4*). However, the mechanisms responsible for the formation of three-dimensional network of whey protein are not fully understood because of the complicated interactions among the different proteins of whey. β-Lg and BSA are the major gelling proteins. These proteins have different structures and molecular properties (*5*), and possess different gelling properties (*6, 7*). It is possible that interactions among the constituent proteins of whey may significantly affect the gelling properties of whey protein. The interactions of β-Lg or BSA with α-La in heat-induced gelation were investigated to elucidate the strong gelling properties of whey protein.

Gelation of Whey Proteins

Heat-induced gelation of globular proteins depends on many factors, such as pH, temperature, protein concentration, heating rate, ionic strength, and presence of specific ions. The minimum concentration of protein required for gel formation is an important criterion of the gel-forming ability of specific proteins (*8, 9*). The gel hardness of the constituent proteins of whey was measured as a function of protein concentration (Figure 1). The hardness of the β-Lg, BSA and whey protein isolate (WPI) gels increased exponentially with increasing protein concentration. The minimum concentration required for the formation of a self-supporting gel was different for each of these proteins; for β-Lg, it was 4% (w/v), and for WPI, 5%(w/v). The BSA gels were tansparent, had a smooth texture, and exhibited good water-holding capacity with little syneresis, and a BSA concentration below 4% (w/v) was sufficient for the formation of a self-supporting gel. The α-La did not gel under this condition (heating at 80°C for 30 min in 0.1 M phosphate buffer, pH 6.8) even up to a protein concentration of 8% (w/v). WPI used in this study contained about 60% β-Lg. If the hardness of WPI gels depended simply on the content of β-Lg, the gel hardness of 10% WPI would be close to that of 6% β-Lg. The hardness of the gel made from 10% WPI was approximately 80 g, and that of 6% β-Lg was about 22 g. Such gelling properties for WPI is considered to have reflected the interactions between β-Lg and the other proteins in WPI during gel formation. Therefore, the interactions of β-Lg or BSA with α-La, the main protein components of WPI, were investigated for their gel-forming abilities.

Gel Hardness of Protein Mixture

The hardness of the gels made from the mixture of 6% (w/v) β-Lg or BSA with various protein concentrations of α-La was measured (Figure 2). The addition of α-La to β-Lg or BSA enhanced gel hardness. Addition of α-La above 3% caused a substantial increase of gel hardness of their mixtures, although the α-La itself did not form gels, reflecting the interaction of these proteins during gelation. The heat-induced interactions were demonstrated from the changes in gel hardness of mixtures of β-Lg and α-La with various protein concentrations, the total protein concentration being kept constant at 8% (w/v) (Figure 3). The addition of 6% α-La to 2% β-Lg caused a significant increase in the gel hardness, while each protein did not form individually a self-supporting gel at these protein concentrations. This result implies enhancement of the gelation of β-Lg by α-La, and that β-Lg interacted with α-La at a molecular level during gelation process. Such an interaction of β-Lg (or BSA) with α-La may be the reason for the strong gelling properties observed in whey protein.

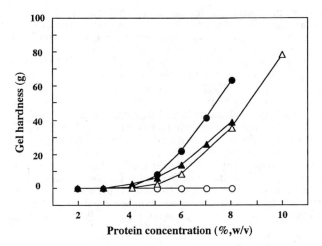

Figure 1. Effect of protein concentration on gel hardness. (●), β-Lg; (○), α-La; (▲), BSA; (△), WPI.
(Reprinted with permission from ref. 18. Copyright 1996 Japan Society for Bioscience, Biotechnology, and Agrochemistry)

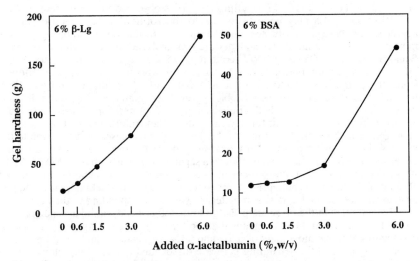

Figure 2. Hardness of gels caused by heating mixtures of 6% β-Lg (or BSA) and α-La with various concentrations.
(Reprinted with permission from ref. 18. Copyright 1996 Japan Society for Bioscience, Biotechnology, and Agrochemistry)

Stabilization of Mixed Protein Gels

The major forces involved in the formation and stabilization of the gel matrix were followed by the solubilization of protein from the gel against the five different solvent buffers [buffer A, 0.1 M sodium phosphate buffer, pH 6.8; buffer B, 0.5% SDS in buffer A; buffer C, 8 M urea in buffer A; buffer D, 10 mM dithiothreitol (DTT) in buffer A; and buffer E, 10 mM DTT in buffer C] (Figure 4A, B). Buffer C caused the gels from β-Lg or from BSA to be solubilized to about 60%. Buffer D solubilized the β-Lg gel considerably but had little effect on the BSA gel. All protein gel samples were almost completely solubilized by buffer E. These results suggest that noncovalent bondings, such as hydrophobic interaction and hydrogen bondings, were essential for forming and stabilizing the gels from BSA and β-Lg alone, and that disulfide bonding may also have contributed considerably to the interactions involved during their heat-induced gelation process, although disulfide bonding may not be important in the BSA gel compared to the β-Lg gel. In the gels from the mixture of β-Lg and α-La or from the mixture of BSA and α-La, the extent of solubilization of protein was much lower than in their single protein systems against all buffer systems except buffer E. These results indicate strengthening and stabilization of the gel matrix from the heat-induced interactions of β-Lg (or BSA) with α-La during gel formation.

Protein-Protein Interactions

Aggregation of denatured or partially denatured globular proteins in solution has been recognized as an integral part of the heat-induced gelation process. Soluble aggregates of polymerized molecules are formed during the early stages of heat-induced gelation of proteins, and subsequent polymerization results in the formation of a rigid gel network (10, 11). Gelling properties observed in the mixture of β-Lg (or BSA) and α-La may have been due to the formation of specific soluble aggregates by these protein-protein interactions. The formation of soluble aggregates in heated solutions of individual and mixture was analyzed by gel filtration (Figures 5 and 6) on a TSK Gel G-3000SW column (0.75 x 30 cm), at a flow rate of 0.5 mL/min of the elution buffer of 0.1 M sodium phosphate buffer, pH 6.8. Elution from the column was monitored at 280 nm. While a 0.2% solution of β-Lg or BSA heated for 30 min at 80°C formed soluble aggregates, a similar α-La solution did not. However, when a mixture of both diluted solutions (Figure 5 for β-Lg/α-La mixture, Figure 6 for BSA/α-La mixture) was heated, they interacted to form soluble aggregates and the peak area of the soluble aggregates increased in proportion to the decrease in the peak area of α-La monomer. These fractions, as indicated by the underlines in Figures 5 and 6 were collected separately and analyzed by SDS-PAGE. As shown in Figure 5D, the protein in Peak 1 could not enter the gel and in the absence of 2-mercaptoethanol (ME), remained on top of the stacking gel, while this protein was dissociated almost completely in bands corresponding to the monomers of α-La and β-Lg in the presence of ME. On the other hand, in the absence of ME, most of the protein in Peak 2 remained at the top of the separating gel, a part of it dissociating into the monomers of α-La and β-Lg, and bands which could not enter the stacking gel and the separating gel were dissociated almost completely in each protein (monomers of α-La and β-Lg) by reduction with ME (Figure 5E). The SDS-PAGE analysis indicated that α-La had been incorporated into those soluble aggregates formed from the mixture of β-Lg and α-La, which were not formed from either β-Lg or α-La alone. It is suggested that β-Lg formed soluble aggregates with α-La mainly through a thiol-disulfide interchange reaction during heating. Similarly, addition of α-La to BSA induced formation of soluble aggregates through the same mechanism (Figure 6).

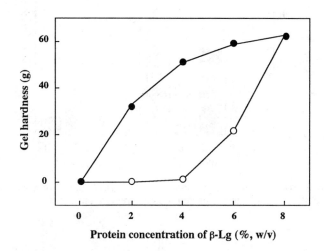

Figure 3. Effects of protein concentration of β-Lg on hardness of gels made from β-Lg alone (○) and from a mixture of β-Lg and α-La (●). The mixture of β-La and α-La was prepared by mixing β-Lg and α-La at various concentrations, the total protein concentration being kept constant at 8%.
(Reprinted with permission from ref. 18. Copyright 1996 Japan Society for Bioscience, Biotechnology, and Agrochemistry)

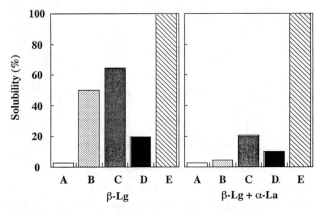

Figure 4-A. Solubilization of gels made from 6% β-Lg and from mixture of 6% β-Lg and 3% α-La by various protein-denaturing reagents.

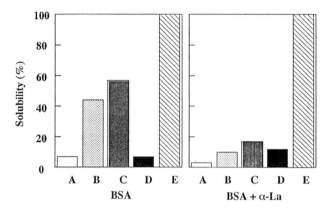

Figure 4-B. Solubilization of gels made from 6% BSA and from mixture of 6% BSA and 3% α-La by various protein-denaturing reagents.

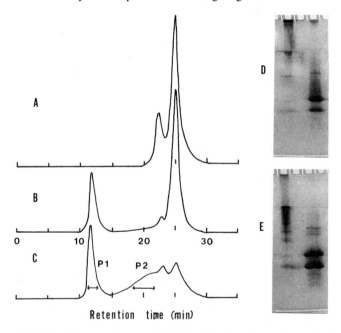

Figure 5. HPLC elution patterns and SDS-PAGE of products formed by heating a mixture of α-La and β-Lg. (A), mixture of α-La and β-Lg before heating; (B), mixture after α-La and β-Lg had been separately heated; (C), heated after α-La and β-Lg had been mixed; (D), SDS-PAGE of Peak 1; (E), SDS-PAGE of Peak 2. The right lane of each gel was treated with ME.
(Reprinted with permission from ref. 18. Copyright 1996 Japan Society for Bioscience, Biotechnology, and Agrochemistry)

Native β-Lg has one thiol group and 2 disulfide bonds per monomer (*12*), α-La having no thiol group but 4 disulfide bonds per monomer (*13*). By heating (30 min at 80°C) above the thermal transition point for β-Lg (78°C) and α-La (77°C) at pH 6.65 (*14*), the thiol group of the β-Lg molecule would easy participate in reduction of the disulfide bond in the α-La molecule. The thiol groups content in heated solution of the mixture of β-Lg and α-La was much lower than the mixture of the β-Lg and α-La solutions (0.2%) when heated individually (Figure 7). The changes in the thiol groups in the mixture containing β-Lg and α-La heated for different times at 80°C confirmed that the reactivity of thiol group in β-Lg may have been facilitated by the coexistence of α-La, and that interaction product of β-Lg and α-La would be formed through the thiol oxidation or thiol-disulfide interchange reaction during heating.

Nature of Soluble Aggregate as a Progel

The apparent molecular weight average of soluble aggregate obtained from the mixture of β-Lg and α-La or from the mixture of BSA and α-La was determined from the ratio of the peak area of laser light scattering output (LS) to that of the refractive index (RI), as described by Takagi and Hizukuri (*15*). Chromatograms on the TSK column of each heat-treated protein solution were detected by a low-angle laser light scattering photometer (LS-8, Toyo Soda Co.) and a precision differential refractometer (RI-8000, Toyo Soda Co.). The apparent molecular weights of soluble aggregates formed from the BSA alone and the mixture of BSA and α-La in the presence of N-ethylmaleimide (5 mM) on heating are given in Table I. The apparent molecular weights of soluble aggregate formed from the β-Lg or BSA alone were about 1.5 times those obtained from their mixtures with α-La. The difference in the molecular size of the soluble aggregate obtained from the mixture of BSA and α-La and from BSA alone was consistent with the difference in the mobilities of these aggregates on SDS-PAGE without ME (Figure 6D, E), indicating that the molecular size of soluble aggregate from the mixture is smaller than that formed by heating BSA alone. In the presence of NEM, which blocks thiol groups, a soluble aggregate was formed from the BSA solution after heating (30 min at 80°C) but not for β-Lg. The apparent molecular weights of soluble aggregates obtained from the BSA alone and the mixture of BSA and α-La, in the presence of NEM, were much greater than those of the aggregates formed without NEM, indicating that the apparent molecular weights of disulfide-induced aggregates were smaller than those of aggregates formed by noncovalent bondings. These results suggest that polymerization of β-Lg or BSA alone in the presence of α-La might be limited by disulfide bond formation between β-Lg (or BSA) and α-La. The addition of NEM greatly decreased the hardness of β-Lg and BSA gels (*7*), reflecting the importance of intermolecular disulfide bonds. Kato et al. (*16*) reported a correlation between decreases in the molecular size of soluble aggregate and increases in the gel hardness of dry-heated dried egg white. On the basis of these studies, a decrease in the size of aggregates following the addition of α-La to β-Lg (or BSA) solutions may play an important role in the formation of stronger, more stable gels by providing a finer gel network.

The molecular weight distribution curves of heat-induced aggregates were derived from the elution patterns of the aggregates monitored by LS and RI, according to the method of Kato and Takagi (*17*). Chromatograms of the soluble aggregates make it possible to estimate the molecular weight of the fraction eluted at a particular retention time and its relative population from the output of the refractometer (RI). The molecular weight distribution curves thus obtained are shown in Figure 8. The curve of aggregate formed from β-Lg alone showed a wide distribution of molecular weights (Figure 8A). In the mixture of β-Lg and α-La, the curve of the aggregate

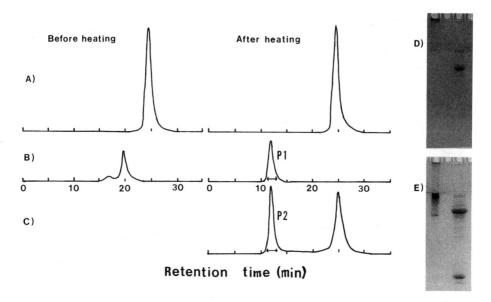

Figure 6. HPLC elution patterns and SDS-PAGE of products formed by heating a mixture of α-La and BSA. (A), α-La; (B), BSA; (C) mixture of α-La and BSA; (D), SDS-PAGE of Peak 1; (E), SDS-PAGE of Peak 2. The right lane of each gel was treated with ME.

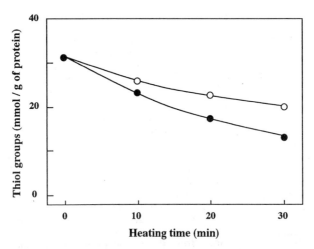

Figure 7. Changes in the thiol group content in a mixture containing β-Lg and α-La (●), following heating for the indicated times at 80°C, and those in a mixture with 1:1 ratio of the β-Lg and α-La solutions heated individually (○).

Table I. Apparent Molecular Weight of Soluble Aggregates Formed by Heating

protein system	M.W.(x10^{-6})	protein system	M.W.(x10^{-6})
β-Lg	5.4		
β-Lg + α-La	4.0		
		NEM-modified	
BSA	5.2	BSA	9.6
BSA + α-La	3.3	BSA + α-La	9.8

Protein solution (0.2%) was heated for 30 min at 80°C in the absence or presence of N-ethylmaleimide (5 mM).

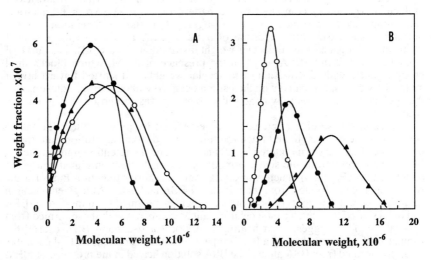

Figure 8. Molecular weight distribution curves of soluble aggregates formed by heating at 80°C for 30 min. In Panel A: (○), 0.2% β-Lg; (▲), mixture of 0.2% β-Lg and 0.1% α-La; (●), mixture of 0.2% β-Lg and 0.2% α-La; In Panel B: (●), 0.2%BSA; (○), mixture of 0.2% BSA and 0.2% α-La; (▲) 0.2% BSA with NEM.

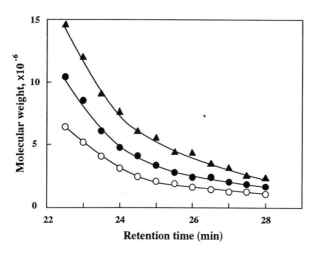

Figure 9. Relationship between retention time and molecular weight of soluble aggregates formed by heating at 80°C for 30 min. Symbols are as in Panel B of Figure 8.

indicated a narrower molecular weight distribution which shifted toward lower molecular weights as compared to the pattern of aggregate from β-Lg alone, suggesting a more restricted progress of aggregation and the formation of more uniform strands. Similarly, addition of α-La to BSA resulted in the curve of aggregate having a narrow molecular weight distribution (Figure 8B). The curve of aggregate formed from BSA alone in the presence of NEM, which blocks thiol groups, gave a wide distribution of molecular weights and shifted toward higher molecular weights, compared with the pattern obtained without NEM, indicating that the aggregates formed only by noncovalent bonding was composed of non-uniform strands.

The nature of soluble aggregates was estimated by plotting the relationship between molecular weight and retention time derived from the elution patterns of heat-induced aggregates monitored by LS and RI, since the retention time on the gel filtration column reflects the size and mode of aggregate fraction eluted at a given retention time (Figure 9). There was a large shift to shorter retention times for the soluble aggregates formed from the mixture of BSA and α-La. At a given retention time, the molecular weights of heat-induced aggregate formed from mixture of BSA and α-La shifted to a smaller molecular weight as compared with those formed from BSA alone. This suggests that heating a mixture of α-La and BSA caused the formation of aggregate having a less compact structure, as compared to the soluble aggregate formed from BSA alone. The BSA solution heated in the presence of NEM formed more compact soluble aggregates compared to those formed without NEM. A correlation between decreases in molecular size of the soluble aggregate and increases in the gel hardness of dry-heated dried egg white has been reported by Kato et al. (16). The expanded structure of heat-induced soluble aggregate formed from the mixture of β-Lg and α-La or from the mixture of BSA and α-La might be responsible for the formation of a strong and stable gel network which holds an adequate amount of water.

Conclusions

The interactions among main protein constituents in whey protein were investigated to elucidate the mechanisms of heat-induced gelation of whey protein. α-La and β-Lg (or BSA) molecules interact directly with each other to form a gel network through disulfide bonding during the course of gelation. The expanded structure of uniform soluble aggregate formed from the mixture with α-La might be correlated to the formation of a firm and stable gel network. An interaction of such a soluble aggregate might significantly affect the formation and stabilization of the resultant thermal gel and play a major role in defining the gelling ability of whey protein.

Literature Cited

1. Paulsson, M.; Hegg, P.-O.; Castberg, H. B. *J. Food Sci.* **1986**, *51*, 87-90.
2. Mulvihill, D. M.; Kinsella, J. E. *Food Technol.* **1987**, *41*, 102-111.
3. Katsuta, K. *Nippon Shokuhin Kogyo Gakkaishi.* **1990**, *37*, 73-81.
4. Mangino, M. E. *Food Technol.* **1992**, *46*, 114-117.
5. Brunner, J. R. In *Food Protein*; Whitaker, J. R.; Tannenbaum, S. R., Eds.; AVI Publishing Company: Westport Connecticut, **1977**, pp 175-208.
6. Yasuda, K.; Nakamura, R.; Hayakawa, S. *J. Food Sci.* **1986**, *51*, 1289-1292.
7. Matsudomi, N.; Rector, D.; Kinsella, J. E. *Food Chem.* **1991**, *40*, 55-69.
8. Hegg, P.-O. *J. Food Sci.* **1982**, *47*, 1241-1244.
9. Katsuta, K.; Rector, D.; Kinsella, J. E. *J. Food Sci.* **1990**. *52*, 516-521.
10. Ma, C. Y.; Holme, J. *J. Food Sci.* **1982**, *47*, 1454-1459.
11. Nakamura, T.; Utsumi, S.; Mori, T. *J. Agric. Food Chem.* **1984**, *32*, 349-352.
12. Whitney, R. M.; Brunner, J. R.; Ebner, K. E.; Farrell, H. M.; Josephson, R.V.; Moor, C. V.; Swaisgood, H. E. *J. Dairy Sci.* **1976**, *59*, 785-815.
14. Larson, B. L.; Rolleri, G. D. *J. Dairy Sci.* **1955**, *38*, 351-360.
15. Takagi, T.; Hizukuri, S. *J. Biochem.* **1984**, *95*, 1459-1467.
16. Kato, A.; Ibrahim, H.R.;Takagi,T. *J. Agric. Food Chem.* **1990**,*38*, 1868-1872.
17. Kato, A.; Takagi, T. *J. Agric. Food Chem.* **1987**, *35*, 633-637.
18. Matsudomi, N.; Oshita, T.; Sasaki, E.; Kobayashi, K.*Biosci. Biotech. Biochem.* **1992**, *56*, 1697-1700.

Chapter 8

Laser-Light-Scattering Properties of Heat-Induced Ovalbumin Gels

Yoshinori Mine

Department of Food Science, University of Guelph, Guelph, Ontario N1G 2W1, Canada

Multi-angle laser photometer and dynamic light scattering spectrophotometer were used for characterizing the aggregation behaviour and gel network structure of elastic gels and less elastic ones of ovalbumin. Ovalbumin formed opaque and less elastic gels in 50 mM phosphate buffer (pH 7.0). However, succinylated ovalbumin formed transparent and elastic gels. Ovalbumin protein formed high molecular and polydisperse aggregates upon heating. Succinylation of the protein contributed to formation of lower molecular weight polymer with a narrow molecular size distribution during subsequent heat gelation. The root mean square and hydrodynamic radii of the protein polymers indicated that the elastic gel consisted of a more extended protein polymer network compared to that of the less elastic gel. The aggregation behaviour of denatured ovalbumin appeared to play a crucial role in the subsequent gelling process of the protein.

Elucidation of the protein-structure function relationship in food proteins can result in useful practical information in food and protein chemistry. Egg white proteins display multiple functional properties such as foaming and gelling properties. Although numerous methods exist to study protein structure, only limited information is available concerning the relationship between protein structure and functionalities of food proteins. In general, a protein gel network is formed via noncovalent cross-linkages such as hydrophobic interactions, or electrostatic and hydrogen bonds interactions and less frequently, by covalent interactions such as disulfide bonds to form a three-dimensional network (1). It is believed that the gelling properties of proteins could be affected by factors determining the state of aggregates found during the subsequent heat for gelation. The mechanism, however, underlying the formation of the three-dimensional network of protein gels is not fully understood. Ovalbumin forms heat-induced opaque gels, whereas transparent gels are formed by succinylation of the amino groups of the protein (2). Transparent gels obtained by two-step heating method were firm and elastic with high-water holding capacity, while the opaque gel obtained by one-step heating was soft and less elastic (3). Absolute characterization of these protein aggregates is neccessary to understand these phenomena. There is little information regarding the size of aggregated protein polymers until the recentstudies of Kato *et al.* (*4*). Thus, it is of great interest to elucidate the relationship between the heat-induced gelling properties of ovalbumin and the aggregation behaviour of these proteins.

The development of laser light scattering techniques have enabled an advance in polymer science for characterizing the aggregation behaviour and networkstructure of polymer gels. Light scattering is an ideal means to determine the actual molecular weight and aggregate formation (*5*). In this study, the author demonstrate the technique for characterizing protein aggregation behaviour and gel structure upon heating *in situ*. Multi-angle laser scattering coupled with size exclusion HPLC and dynamic light spectrophotometry allows us to obtain not only an absolute molecular weight and the root mean square (RMS) radius of protein polymers but also the hydrodynamic radius of the protein gel network *in situ*.

Materials and Methods

Materials   Ovalbumin was purchased from Sigma Chemicals, St. Louis, MO. Other chemicals used were ACS grade from Sigma. Ovalbumin was succinylated according to the procedure described by Groninger (*6*). Succinic anhydride was added at a 50:1 protein : anhydride weight ratio. The extent of succinylation was determined from the free amino contents by the method of Concon (*7*) using dinitrobenzene sulfonate (DNBS).

Preparation of Gels   A protein solution (8.0%) in 50mM phosphate buffer (pH 7.0) was injected into an aluminum cup (5mm inner diameter, 2.5mm height). The cup was filled, completely degassed, and sealed with a silicone sheet on which a steel plate was tightly fixed with clips. The cups containing the protein solutions were heated at 80° C for 20 min., followed by rapid cooling to room temperature by immersion in tap water.

Creep Analysis of Gels   Creep behaviour under compression was analyzed with a Rheoner RE-3305 (Yamaden, Co Ltd., Tokyo) interfaced with a computer (NEC PC 9801 VM, NEC, Tokyo). The creep curves were analyzed according to the procedure described by Kamata *et al.* (*8*).

Light Scattering Measurements   Light scattering was measured on a DAWN DSP-F multi-angle laser light scattering photometer (MALLS) (Wyatt technology, Santa Barbara, CA) using a 632.8 nm laser and dynamic light scattering spectrophotometer (DLS-7000, Otsuka Electronics, Tokyo). The MALLS was coupled with HPLC system consisting of a Shodex KW-804 and KW-803 columns at a room temperature. A specific refraction index increment (dn/dc) of 0.186 was obtained for the dialyzed protein solution using a Wyatt OPTILAB DSP interferometric differential refractometer.

Equations   The MALLS photometer measures the intensity of the scattered light with the use of photodiodes placed at specific angles (θ) relative to the laser beam. This light intensity is converted to the Rayleigh ratio as follows:
Rayleigh Ratio:

$$R_\theta = I_\theta r^2 / (I_0 V) \qquad (eq.1)$$

Where:

R_θ = Rayleigh ratio

I_θ = Scattered intensity

I_0 = Incident beam intensity
V = Scattering volume
r = distance between scattering volume and detector

From the Rayleigh ratio, we can use the Zimm equation to directly determine molecular weight provided that a number of physical properties are known.
Zimm equation:

$$R_\theta/K^*c = MP(\theta) + 2A_2cM^2P^2(\theta) \qquad \text{(eq.2)}$$

Where:

R_θ = the Rayleigh ratio

K^* = an optical constant, $4\pi^2n_0 2(dn/dc)^2\lambda_0^{-4}N_A^{-1}$, where n_0 is the refractive index of the solvent at the incident radiation wavelength (nm), N_A is Avogadro's number and dn/dc is the differential refractive index increment solvent-solute solution with respect to a change in solute concentration.

c = the concentration of solute molecules in the solvent

$P(\theta)$ = the theoretically derived form factor. $P(\theta)$ is a function of the molecule's z-average size, shape, and structure.

A_2 = the second virial coefficient

M = the weight average molecular weight

Since we are using multi-angle detection here, we can extrapolate the value of q to 0. In the limit of θ->0, $P(\theta)$->1 and equation (2) becomes:

$$R_{\theta\to0}/K^*c = M + 2A_2cM^2 \qquad \text{(eq.3)}$$

If A_2=0, then

$$M = R_\theta/K^*c \qquad \text{(eq.4)}$$

Using the Debye Plot:

$$R_\theta/K^*c \text{ vs. } Sin^2(\theta/2) \qquad \text{(eq.5)}$$

We can see that at an angle of θ=0, we can read the molecular weight (M) directly off of the y-axis of the plot. It depends on the RMS radius $<r^2>$ independent of molecular conformation (5).

Using a *continuum model* (9) that the wavelength of light is usually much larger than the average distance between neighboring cross-links, self-correlation function is derived as the following linear equation:

$$C(\tau) \sim \alpha(kT/K)exp(-Dq^2\tau)$$

where:

$C(\tau)$= self correlation function
k= Boltzmann constant
T= absolute temperature
K= modulus of elasticity
D= diffusion coefficient

We produce the Stokes-Einstein relation for the diffusion coefficient of polymers in solution.

$$D = kT/6\pi\eta_0 a$$

where:

η= liquid viscosity
a= hydrodynamic radius of the polymer

Results and Discussion

Creep-Analysis of Ovalbumin Gels   By using succinic anhydride, 28% of the lysine of ovalbumin was modified.   The gel formation of the succinylated ovalbumin (Su-ovalbumin) protein was compared to that of ovalbumin.   Ovalbumin formed an opaque gel at pH 7.0 upon heating at 80° C for 20 min.   On the other hand, the Su-ovalbumin gel gave a transparent gel under the same condition (The inset of Fig. 1). The typical creep and creep-recovery curves of these two gels were obtained to clarify the difference in the gel characteristics of a transparent gel and an opaque one.   The creep compliance of the transparent gel was more than twice that of the opaque gel, indicating that the transparent gel was more soft and more deformable.   Differences in the rheological properties of these gels were also observed clearly by the residual strain after the removal of the stress.   The Su-ovalbumin gels showed lower residual strain than ovalbumin gels.   These results suggested that the Su-ovalbumin gels were more elastic, while those of ovalbumin were likely to be more plastic or less elastic.   Creep curves obtained were further analyzed with a four-element mechanical model as shown in Table 1.   Each parameter:P, stress, E_0, instantaneous modulus, E_1, retarded elastic modulus: η_1, retarded viscous modulus, and η_n, Newtonian viscous modulus was calculated.   The P and E_0 value of ovalbumin gels was over twice than that of Su-ovalbumin gels, while E_1, η_1 and η_n values were smaller in the opaque gels.   These results occured the transparent gels have a small instaneous deformation but a larger retarded deformation with time up to 300 sec. than the opaque gels.   The results suggested that the Su-ovalbumin gels were more elastic, while the opaque gels were more viscous.

Laser Light Scattering   It is well known that the gelling properties may be affected by the cross linking of unfolded molecules as a result of hydrogen bonding and ionic and hydrophobic interactions (*10*).   The turbidity of the ovalbumin gels depends on the pH and ionic strength of the heated protein solution (*11*).   In the past, low-angle laser light scattering was used to estimate the molecular weight distribution of heat-induced ovalbumin and dry-heated egg white protein aggregates (*4*).   However, this system has been hampered severely by solvent noise due to impurities.   The MALLS system dramatically reduces the effect of noisy background, resulting in cleaner data and highly reproducible results (*5*).   The molecular weight and RMS of heat-induced soluble aggregates were analyzed by the MALLS system.   Fig. 2 shows the relationship between retention volume (HPLC) and molecular weight of heat-induced protein aggregates.   Ovalbumin aggregates have molecular weights between 1.7×10^6 to 1.6×10^8, while the range for Su-ovalbumin was lower (2.1×10^5 to 4.2×10^7).   In general, the retention volume by HPLC gel filtration reflects the size of and conformation of proteins eluted at a particular retention volume.   Figure 3 shows the relationship between the RMS radius and the retention volume (HPLC) of the heat-induced aggregates of the proteins.   Interestingly, there was little difference in the RMS radius for both the aggregates except large retention volume values in spite of the considerable differences of absolute molecular weight.

Table 2 summarizes the laser light scattering characteristics of heat-induced ovalbumin and Su-ovalbumin aggregates.   The data indicated that the molecular weight distribution of the heat-induced ovalbumin aggregates was large and heterogeneous.   By contrast, Su-ovalbumin was smaller and less heterogeneous. Interestingly, no significant differences on the RMS of the aggregates between ovalbumin and Su-ovalbumin were shown.   These results suggest that succinylation of ovalbumin contributes to the formation of lower molecular weight and narrower molecular size distribution of the aggregates upon heating.   The formation of extended structures of heat-induced Su-ovalbumin polymer was also predicted from the RMS

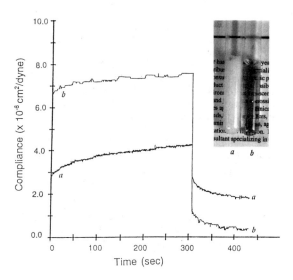

Figure 1. Creep and creep-recovery curves of (a) ovalbumin and (b) Su-ovalbumin gels.

Table 1. Viscoelastic Parameters of Ovalbumin and Su-ovalbumin Gels

parameters	ovalbumin	Su-ovalbumin
$P (\times 10^4)$	4.99	2.50
$E_0 (\times 10^5)$	3.57	1.52
$E_1 (\times 10^6)$	1.46	2.05
$\eta_1 (\times 10^7)$	4.86	3.31
$\eta_N (\times 10^8)$	4.00	6.08
τ_1	33.19	16.13

P, stress; E_0, instantaneous modulus; E_1, retarded elastic modulus;
η_1, retarded viscous modulus; η_N, Newtonian modulus

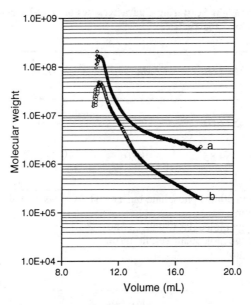

Figure 2. Relationship between absolute molecular weight and retention volume (HPLC) of (a) ovalbumin and (b) Su-ovalbumin soluble aggregates.

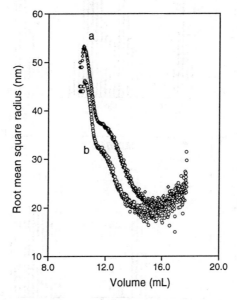

Figure 3. Relationship between RMS radius and retention volume (HPLC) of (a) ovalbumin and (b) Su-ovalbumin soluble aggregates.

Table 2. Laser Light Scattering Characteristics of Heat-Induced Ovalbumin and Su-ovalbumin Aggregates[a]

samples	Mw (x10⁴)[b]	Mn (x10⁴)[c]	Mw/Mn	RMS (w)[d]
ovalbumin	322.60	9.24	34.91	25.2
Su-ovalbumin	59.35	10.01	5.93	23.9

[a] 0.1% protein in 50mM phosphate buffer (pH 7.0) was heated at 80° C for 20 min.
[b] Weight-average molecular weight (g/mol).
[c] Number-average molecular weight (g/mol).
[d] RMS radius (nm).

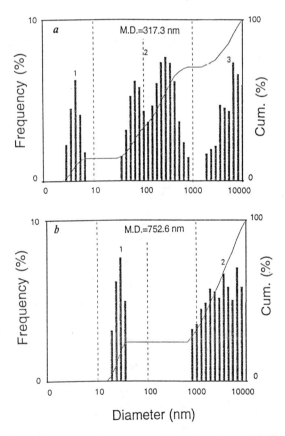

Figure 4. Particle diameter distributions for (a) ovalbumin and (b)Su-ovalbumin gels. M.D.; mean diameter, Cum.; accumulation.

data for the aggregates.   Similar results were observed in the case of dry-heated egg white protein which showed excellent gel strength (*4, 13*).   Thus, the aggregation behaviour of denatured proteins appears to play a crucial role in the subsequent gelling process of proteins.

The mean diameter and distribution of the diameters of the gel matrix obtained by a dynamic light scattering system is shown in Fig. 4.   As shown in Fig. 4, the diameter distribution of the gel matrix in the histogram was spread widely.   Ovalbumin gel matrix gave three spread distribution peaks and that of Su-ovalbumin gave two. This indicates that the theoretical approach of the *"continuum model"* shows ideal conditions for a gel matrix, however, the actual gel matrix of protein polymers is more complicated.   The smaller Peak 1 for ovalbumin gel (Fig. 4a) may be caused by dangling chain which are not linked completely the gel matrix.   The hydrodynamic radius of the Su-ovalbumin polymer was larger than that in the ovalbumin gel.   These data indicate that an elastic gel is composed of a firm polymer linkages and a wide gel matrix.   An opaque (less elastic) gels is made from a heterogeneous aggregate gel network.   The number of elastically active chains may increase with increasing transparency of the gel. Hydrophobic interactions and repulsive electrostatic interactions affect the physical characteristics of protein gels (*10*).   The aggregation behaviour of Su-ovalbumin may be caused by increasing the electrostatic repulsion through chemical modification and subsequent formation of elastic gels.

Structural changes in protein molecules between elastic and less elastic gels are also of great interest in food systems.   However, fundamental information on this issue is limited because of the difficulty in analyzing solid-like structures at high protein concentration.   Raman spectral analysis indicated an increase of β-sheet structure with a simultaneous decrease of helical structure in heated transparent gels of α-lactalbumin (*14*).   However, Ozaki *et al*. (*15*) reported that most α-helix structure changed into unordered structure in egg white proteins upon heating using near-infrared FT-Raman spectroscopy.   Combination of the laser light scattering technique and protein structural analysis using FT-Raman spectroscopy *in situ* would be desirable for further investigation.

Conclusion

This study demonstrated that multi-angle laser light scattering is a useful tool for analyzing the heat-induced aggregation behaviour of proteins and network structure of protein polymers *in situ*, which may be avaluable in evaluating protein functionality in food systems.   In this work, it was shown that elastic gels consisted of lower molecular weight and more extended protein polymers, were found during subsequent heat gelation with widely spaced network and transparency.   Less elastic gels, however, were composed of higher molecular weight aggregates.   The aggregation behaviour of denatured proteins appears to play a crucial role in the gelling process of proteins.

Acknowledgments
This research was supported by a grant from the Ontario Egg Producer's Marketing Board (Ontario, Canada).   The author is also grateful to Dr. Y. Matsumura (The Research Institute for Food Science, Kyoto University, Japan) for helping the creep analysis of the gels.

Literature Cited
1.   Clark, A.K. In *Physical Chemistry of Foods*;   Schwartzberg, H.G., Hartel, R.W. Eds.; Marcel Dekker Inc.; New York, N.Y., 1992.
2.   Nakamura, R.; Sugiyama, H.; Sato, Y. *Agric. Biol. Chem.* **1978**, *42*, 819-824.

3.   Kitabatake, N.; Tani, Y.; Doi, E. *J. Food Sci.* **1989**, *54*, 1632-1638.
4.   Kato, A.; Ibrahim, H.; Takagi, T.; Kobayashi, K. *J. Agric. Food Chem.* **1990**, *38*, 1868-1872.
5.   Wyatt, P.J. *Anal. Chim. Acta.* **1993**, *272*, 1-40.
6.   Groninger, H.S. Jr. *J. Agric. Food Chem.* **1973**, *21*, 978-981.
7.   Concon, J.M. *Anal. Biochem.* **1975**, *66*, 460-480.
8.   Kamata, Y.; Rector, D.; Kinsella, J.E. *J. Food Sci.* **1988**, *53*, 589-591.
9.   Tanaka, T. In *Dynamic Light Scattering. Applications of Photon Correlation Spectroscopy;* Pecora, R. Ed.; Plenum; New York, N.Y., 1985, pp345-362.
10.   Damodaran, S. In *Protein Functionality in Food Systems*; Hettiarachchy, N.S., Ziegler, G.R. Eds; Marcel Dekker, Inc., New York, N.Y., 1994, pp1-38.
11.   Hatta, H.; Kitabatake, N.; Doi, E. *Agric. Biol. Chem.* **1986**, *50*, 2083-2089.
12.   Kato, A.; Takagi, T. *J. Agric. Food Chem.* **1987**, *35*, 633-637.
13.   Mine, Y. *Food Res. Int.* **1996**, in press.
14.   Nonaka, M.; Li-Chan, E.; Nakai, S. *J. Agric. Food Chem.* **1993**, *41*, 1176-1181.
15.   Ozaki, Y.; Cho, R.; Ikegaya, K.; Muraishi, S.;   Kawauchi, K. *Appl. Spec.* **1992**, *46*, 1503-1507.

Chapter 9

Aggregation and Gelation of Bovine β-Lactoglobulin, α-Lactalbumin, and Serum Albumin

Jacquiline Gezimati, Harjinder Singh, and Lawrence K. Creamer

Department of Food Technology, Massey University, Palmerston North, New Zealand and New Zealand Dairy Research Institute, Private Bag 11029, Palmerston North, New Zealand

Solutions (10% w/v) of bovine β-lactoglobulin, α-lactalbumin, and serum albumin (BSA), in a buffer simulating the whey protein concentrate environment, were heated to 75°C and held for 60 min. Aggregation and gelation were measured using native- and sodium dodecyl sulfate polyacrylamide gel electrophoresis on natural and reduced samples, and dynamic oscillation rheology, respectively. BSA formed stiffer gels than β-lactoglobulin whereas α-lactalbumin did not gel under these conditions. Heating 10% mixtures of β-lactoglobulin and BSA showed that the gelling behavior of the mixture was similar to that of BSA when BSA was the major protein in the mixture and vice versa. In contrast, the gelling behavior of a mixture β-lactoglobulin and α-lactalbumin was dominated by β-lactoglobulin. Both β-lactoglobulin and BSA formed hydrophobically bonded aggregates and disulfide-bonded polymers, but α-lactalbumin did not form aggregates under the conditions used. However, α-lactalbumin aggregated rapidly when heated in the presence of β-lactoglobulin. These results indicate that the protein aggregates forming the gel network in heated BSA and β-lactoglobulin solutions are different than those in mixtures of β-lactoglobulin and α-lactalbumin.

Whey protein isolates and concentrates are currently being used for a variety of applications in the food industry. Heat-induced gelation of whey proteins is one of the most important functional properties for their use as ingredients. Whey proteins are a heterogeneous group of proteins, consisting mainly of β-lactoglobulin, α-lactalbumin, bovine serum albumin (BSA) and immunoglobulins. Although the denaturation and gelation behavior of β-lactoglobulin, the major protein component of whey, have been studied extensively (*1-8*), the roles of the constituent proteins of whey in the denaturation and gelation processes have not been fully explained (*9-11*).

McSwiney et al. (*5-6*) studied the thermal aggregation and gelation of β-lactoglobulin by dynamic oscillation rheometry and polyacrylamide gel electrophoresis (PAGE) under native conditions and in sodium dodecyl sulfate (SDS) buffer without and with reducing agents. It was concluded that the development of

0097–6156/96/0650–0113$15.00/0

rheologically significant structure was probably preceded by the formation of hydrophobically associated aggregates and disulfide-bonded aggregates. Gezimati et al. (*12*) extended the study by McSwiney et al. (*5-6*) to examine the protein aggregation and gelation of β-lactoglobulin in admixture with BSA. In the present study, we investigated the aggregation and gelation behavior of β-lactoglobulin, α-lactalbumin and BSA when they were heated separately and in admixture and their interactive contribution to gel formation in an ionic environment similar to that existing in whey protein concentrate solutions.

Materials and Methods

β-Lactoglobulin (product No. L-2506), BSA (product No. A-4378) and α-lactalbumin (product No. 5385) were purchased from Sigma Chemical Co. (St Louis, MO). All other chemicals used were of analytical reagent grade.

Measurement of Protein Gelation. The gelation properties of β-lactoglobulin, α-lactalbumin, BSA, and various protein mixtures dispersed in a buffer designed to simulate the mineral and lactose contents of a 12% (w/v) whey protein concentrate powder solution (2.03 g K_2HPO_4, 0.63 g NaCl, 2.06 g *tri*-sodium citrate dihydrate, 1.31 g $CaCl_2.2H_2O$ and 8 g lactose made up to 1 liter) were determined using a Bohlin VOR rheometer (Bohlin Reologi AB, Lund, Sweden), as described by Gezimati et al. (*12*). The protein solution was placed in a measuring system consisting of a cup and a bob. A torsion bar of 4.4×10^{-3} Nm was used; the frequency was set to 1 Hz and the shear strain was at 0.0206 ± 0005. The protein solution was heated from 25°C to 75°C at a rate of 1°C/min and the temperature was maintained for 60 min. G', G'', and the phase shift angle were recorded every 30 s for 1 h from the attainment of the heating temperature.

Protein Aggregate Formation. The protein solutions were heated to 75°C in the Bohlin rheometer cup at 1°C/min under the same conditions as in the gelation studies. Aliquots were removed at 0 (when the heating temperature was just reached), 1, 2, 4, 6, 10 and 14 min (where possible) and mixed with chilled sample buffers for subsequent analysis using native- and SDS-PAGE followed by densitometry using the methods of McSwiney et al. (*5-6*).

Results

Gelation of Protein Mixtures. The formation of a gel upon heating protein solutions at 75°C was observed by recording the continuous increase in G' with time. After heating 10% β-lactoglobulin solution at 75°C for up to 60 min, G' began to increase after about 20 min heating; in contrast, 10% BSA solution had gelled during the heat-up period (Figure 1A). Examination of protein mixtures with a total protein concentration of 10% showed that the rate of G' development increased as the relative amount of BSA in the mixture was increased. However, the G' value (after 60 min heating) attained for 2% β-lactoglobulin and 8% BSA was slightly higher than that of 10% BSA alone.

When 10% α-lactalbumin was heated at 75°C for 60 min there was no definite change in G' indicating that a gel network was not formed (Figure 1B). G' values at 75°C for 10% β-lactoglobulin solution began to increase after 12 min and continued to do so with further heating. When mixtures of β-lactoglobulin and α-lactalbumin were heated at 75°C, the rate of development of G' was slower than that for 10% β-lactoglobulin; G' decreased as the relative concentration of α-lactalbumin was increased in the mixture (Figure 1B).

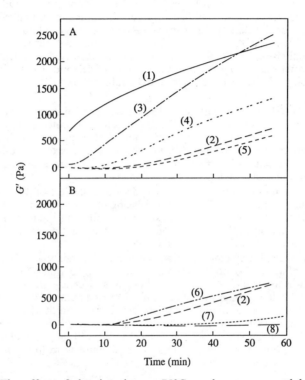

Figure 1. The effect of heating time at 75°C on the storage modulus (G') of: A, mixtures of β-lactoglobulin and BSA; B, mixtures of β-lactoglobulin and α-lactalbumin.   See text for further details.
(1), 10% BSA; (2), 10% β-lactoglobulin; (3), 8% BSA and 2% β-lactoglobulin; (4), 5% BSA and 5% β-lactoglobulin; (5), 2% BSA and 8% β-lactoglobulin; (6), 2% α-lactalbumin and 8% β-lactoglobulin; (7), 5% α-lactalbumin and 5% β-lactoglobulin; (8), 8% α-lactalbumin and 2 % β-lactoglobulin.

In all cases, it was clear that gelation was not complete after 60 min heating, suggesting that there were continual rearrangements of protein molecules within the gels.

Protein Aggregate Formation Prior to Gel Formation.   Solutions of protein mixtures containing either β-lactoglobulin and BSA or β-lactoglobulin and α-lactalbumin, and individual proteins were heated in the Bohlin rheometer cup, and analyzed using native- and SDS-PAGE. Three forms of proteins were identified in the heated samples. The protein band that showed as monomeric on native-PAGE (pH ~ 8.3) was operationally classified as 'native-like' under PAGE conditions. In SDS-PAGE (pH ~ 8.3, 0.1% SDS), the non-covalently linked aggregates that were dispersed into monomers could be called 'SDS-monomeric'. Finally the analysis of heated protein by SDS-PAGE after reduction, which converts the disulfide bonds to free thiols, gave the 'total reducible and dispersible protein in SDS solution' or 'total' protein.

Native-like Protein Concentrations.   The results for a mixture of 5% BSA and 5% β-lactoglobulin together with comparable solutions of pure proteins analyzed by native-PAGE are shown in Figure 2A. It is clear that at 75°C native-like BSA was lost from the system faster than native-like β-lactoglobulin. The loss of native-like BSA was faster from the mixture than from the 5% solution of pure protein. There was a substantial loss of native-like BSA during the heat-up time.

When the protein solutions were heated at 75°C, the loss of β-lactoglobulin was faster from the mixture than from a 5% β-lactoglobulin solution although this was comparable to the loss from a 10% β-lactoglobulin solution (Figure 2A). The amount of native-like β-lactoglobulin present at the end of the heat-up period was less for the mixture than either the 5% control or the 10% control.

When 8% α-lactalbumin solution was heated separately at 75°C, there was no apparent loss of native-like protein throughout the heating period (Figure 2B). Heating 5 or 10% β-lactoglobulin solutions resulted in a decrease in native-like protein concentrations, with a greater rate of decrease occurring in the 10% solution. The loss of β-lactoglobulin from a mixture of 5% α-lactalbumin and 5% β-lactoglobulin was substantially faster than from 5 or 10% solutions of β-lactoglobulin. The loss of α-lactalbumin from the mixture was even faster than that of β-lactoglobulin. About 78% of total β-lactoglobulin and 84% of total α-lactalbumin were lost from the mixture during the heat-up time at 75°C.

SDS-monomeric Protein Concentrations.   The results from SDS-PAGE analysis of the heated β-lactoglobulin and BSA solutions used for the results shown in Figure 2A are shown in Figure 3A.   The results for the loss of SDS-monomeric BSA and β-lactoglobulin (Figure 3A) were qualitatively similar to those for the loss of native-like protein (Figure 2A), but in all cases the loss of SDS-monomeric protein was less than that of native-like protein.

Analysis of solutions of β-lactoglobulin and α-lactalbumin heated at 75°C (Figure 3B) showed that the loss of SDS-monomeric β-lactoglobulin was faster from a 5% β-lactoglobulin and 5% α-lactalbumin mixture than from 5 or 10% β-lacto-globulin controls; the loss of α-lactalbumin was even greater than that of β-lactoglobulin. The loss of β-lactoglobuin from the 5% control was comparable to that from the 10% control. In all cases, the loss of SDS-monomeric protein was less than that of native-like protein.

SDS-PAGE analysis of samples reduced by 2-mercaptoethanol showed no changes in the concentrations of β-lactoglobulin, BSA, or α-lactalbumin under all conditions, indicating that disulfide bonds were the only form of covalent cross-linking.

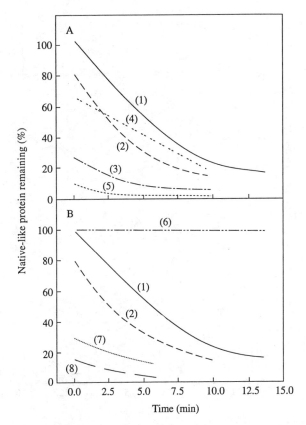

Figure 2. The effect of heating time at 75°C on the loss of native-like proteins from mixtures of: A, β-lactoglobulin and BSA; B, β-lactoglobulin and α-lactalbumin. See text for further details.
(1), 5% β-lactoglobulin;   (2), 10% β-lactoglobulin;   (3), 5% BSA;
(4), β-lactoglobulin in a mixture of 5% BSA and 5% β-lactoglobulin;
(5), BSA in a mixture of 5% BSA and 5% β-lactoglobulin;
(6), 8% α-lactalbumin;
(7), β-lactoglobulin in a mixture of 5% α-lactalbumin and 5% β-lactoglobulin;
(8), α-lactalbumin in a mixture of 5% α-lactalbumin and 5% β-lactoglobulin.

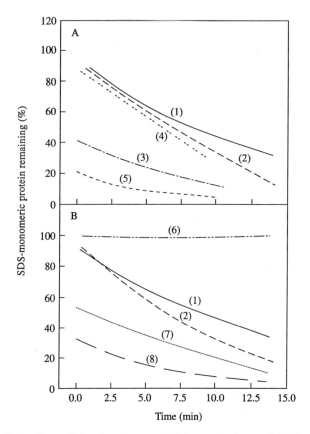

Figure 3. The effect of heating time at 75°C on the loss of SDS-monomeric proteins from mixtures of: A, β-lactoglobulin and BSA; B, β-lactoglobulin and α-lactalbumin. See text for further details.

(1), 5% β-lactoglobulin; (2), 10% β-lactoglobulin; (3), 5% BSA;

(4), β-lactoglobulin in a mixture of 5% BSA and 5% β-lactoglobulin;

(5), BSA in a mixture of 5% BSA and 5% β-lactoglobulin;

(6), 8% α-lactalbumin;

(7), β-lactoglobulin in a mixture of 5% α-lactalbumin and 5% β-lactoglobulin;

(8), α-lactalbumin in a mixture of 5% α-lactalbumin and 5% β-lactoglobulin.

Relationship between G' and Protein Aggregate Formation. Figure 4A shows the relationship between the PAGE results and the development of G' during heating at 75°C for 5% β-lactoglobulin and 5% BSA mixtures. It can be seen that G' did not increase until essentially all of the native-like and SDS-monomeric BSA and about 75% of total β-lactoglobulin had been polymerized (Figure 4A).

The relationship between the loss of native-like and SDS-monomeric proteins and the changes in G' on heating the 5% β-lactoglobulin and 5% α-lactalbumin mixture at 75°C is shown in Figure 4B. It is clear that the concentrations of both the native-like and and the SDS-monomeric α-lactalbumin and β-lactoglobulin decreased to less than 10% of the original concentration prior to any marked increase in G'.

Discussion

The electrophoresis results (Figures 2A and 3A) clearly indicate that, when 10% β-lactoglobulin or BSA solution is heated at 75°C, three types of protein are present in the reaction mixture: native, disulfide-bonded and a third type that is monomeric in SDS but is not native. McSwiney et al. (*6*) and Gezimati et al. (*12*) speculated that this third type of protein is formed by the aggregation of protein via hydrophobic interactions and that the protein molecules within these aggregates may have structures that are comparable in some fashion to the molten globule state (*13*). In contrast, when 10% α-lactalbumin solution is heated alone, it does not form aggregates that continue to exist at low temperature. This result supports earlier findings (*9, 11, 14*), although differential scanning calorimetry has shown that α-lactalbumin undergoes a reversible thermal transition at about 64°C (*15*).

The general mechanism of globular protein gelation that has been proposed by various researchers (*16*) involves a partial reorganization of the surface structures of the protein molecules that allows an increased exposure of hydrophobic groups, but without the loss of the native secondary structures. This is followed by the formation of spherical and bead-like soluble protein aggregates, which subsequently interact to form the strands of protein that constitute the gel network (*17-18*). McSwiney et al. (*5-6*) showed that during gelation of a 10% β-lactoglobulin solution hydrophobic aggregates that are stable at room temperature are formed initially. These aggregates are further strengthened by the formation of inter- and intramolecular disulfide bonds via sulfhydryl-disulfide interchange or sulfhydryl oxidation reactions which also determines the size of the aggregates and the type of gel network formed. On the basis of the present (Figures 2A, 3A, and 4A) and other results (*12*), it seems likely that the gelation mechanism for BSA follows the same general principles outlined for β-lactoglobulin (*6, 12*).

When the mixtures of β-lactoglobulin and BSA were heated at 75°C, BSA aggregates were formed at a much faster rate than those of β-lactoglobulin, and the aggregation rate for BSA appeared to increase slightly in the presence of β-lactoglobulin (Figure 2A). Both BSA and β-lactoglobulin form aggregates that are linked by non-covalent interactions and disulfide cross-links. However, it is not clear from this study whether BSA and β-lactoglobulin in behave independently in the mixture; that is, each protein forms hydrophobic aggregates and disulfide-bonded polymers with proteins of the same type, or they interact to form mixed aggregates. Limited PAGE analysis of the mixtures heated at 70 and 75°C (*12*) did not provide any evidence for interaction between the two proteins at these temperatures.

Hines and Foegeding (*11*) examined aggregation in a 1% (w/v) protein solution with a 1:1 molar ratio of BSA and β-lactoglobulin in TES (N-tris[hydroxymethyl]-methyl-2-aminoethanesulfonic acid) buffer heated at 80°C, and found that the aggregation of β-lactoglobulin was faster in the presence of BSA, but that the rate of BSA aggregation did not appear to be altered. Matsudomi et al. (*10*) observed that

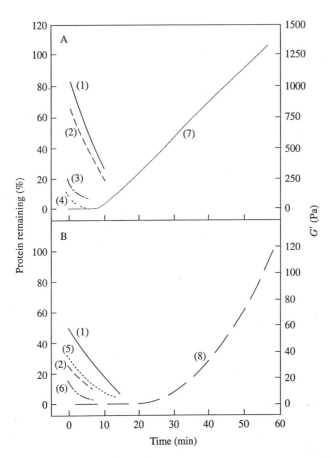

Figure 4. The effect of heating mixtures of proteins at 75°C on G' and the quantities of native-like and SDS-monomeric proteins. See text for further details.
(1), SDS-monomeric β-lactoglobulin; (2), native-like β-lactoglobulin;
(3), SDS-monomeric BSA; (4), native-like BSA;
(5), SDS-monomeric α-lactalbumin; (6), native-like α-lactalbumin;
(7), G' of the mixture of 5% BSA and 5% β-lactoglobulin;
(8), G' of the mixture of 5% α-lactalbumin and 5% β-lactoglobulin.

β-lactoglobulin interacted with BSA to form soluble aggregates through thiol-disulfide interchange reaction after heating 0.2% (w/v) protein solutions at 80°C for 30 min. It is not possible to make a direct comparison with our own results because of the differences in experimental conditions, i.e. heating temperatures, methods of heating, protein concentrations and buffer systems.

Clearly, the aggregation behavior of mixtures of β-lactoglobulin and α-lactalbumin is different than that of mixtures of β-lactoglobulin and BSA. The formation of aggregates of α-lactalbumin occurred only in the presence of β-lacto-globulin, possibly due to interaction between the two proteins. These results support the findings of Hines and Foegeding (*11*) and Matsudomi et al. (*9*). It appears from our studies (Figures 2B and 3B), that hydrophobic aggregates involving α-lactalbumin are formed as a consequence of heating at 75°C, a temperature above the thermal transition of both proteins (*15*).

The rheological experiments on β-lactoglobulin and BSA mixtures heated at 70°C showed that G' increased with an increase in the proportion of BSA in the mixture, suggesting that at this temperature the strands in the gel network are likely to consist predominantly of BSA aggregates (*12*). The role of β-lactoglobulin, under these conditions could be to act as 'filler' in the spaces between the BSA aggregates. However, the possibility of some partially modified β-lactoglobulin molecules interacting with the gel matrix via hydrophobic interactions cannot be ruled out completely. At 75°C, the G' values for the gels are dependent on the proportions of β-lactoglobulin and BSA in the mixture (Figure 1A). At high β-lactoglobulin to BSA concentration ratios in the mixture, the overall gelation behavior is somewhat similar to that of pure β-lactoglobulin; at high BSA to β-lactoglobulin ratios, the overall behavior is similar to that of pure BSA. Because of the differences in thermal stability and aggregation rates, BSA will aggregate and form gel strands and a network before β-lactoglobulin aggregation and gelation has proceeded, particularly at 70°C (*12*). Therefore, it can be concluded that BSA and β-lactoglobulin may form interpenetrating gel networks, provided sufficient quantities of each protein are present at the time of heating. A schematic presentation of events that may occur during the heating of BSA and β-lactoglobulin at 75°C is shown in Figure 5A.

When α-lactalbumin and β-lactoglobulin mixtures were heated at 75°C, the G' values after 60 min increased with increases in the relative proportion of β-lacto-globulin in the mixture, suggesting that β-lactoglobulin is the main protein supporting the gel network. However the G' values were considerably greater for some of the mixtures than for the individual proteins heated at the same protein concentration. This suggests that inclusion of α-lactalbumin in the mixture increased the rigidity of the gels. Matsudomi et al. (*9*) also found that, when mixtures of α-lactalbumin and β-lactoglobulin were heated at 80°C for 30 min and examined at room temperature, the gels from the mixtures were firmer than those composed of β-lactoglobulin alone. It is not immediately apparent why the addition of α-lactalbumin to β-lactoglobulin enhances the gel strength. Studies on ovalbumin gelation (*19-20*) showed that the gelation was affected by the extent of formation of soluble aggregates and subsequent polymerization. It is possible that the aggregates formed by the interaction between β-lactoglobulin and α-lactalbumin are more extensively crosslinked by disulfide bonds than the aggregates formed by β-lactoglobulin alone. In addition, there may be a greater number of disulphide bridges binding the aggregates together in β-lactoglobulin and α-lactalbumin mixtures. The gel network formed from a mixture of β-lactoglob-ulin and α-lactalbumin can be visualized as a single heterogeneous network, consisting of co-polymers of β-lactoglobulin and α-lactalbumin, as depicted in Figure 5B.

The development of gel stiffness with time in the two different types of mixtures (see the 5% mixtures in Figure 1) probably depends on the different ways the gel strands develop in the two systems to give the final gels. It can be speculated that

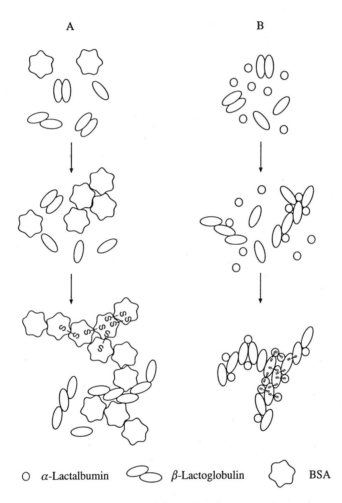

O α-Lactalbumin β-Lactoglobulin   BSA

Figure 5. Schematic presentation of the molecular events that may occur during heating and gelation 75°C of: A, a mixture of β-lactoglobulin and BSA; B, a mixture of α-lactalbumin and β-lactoglobulin.

BSA forms stiffer strands (possibly because of a large number of inter-protein disulfide bonds) or that BSA can link to more than two other protein molecules and thus there are more cross-links between the strands. Consequently a gel containing a higher proportion of BSA will be stiffer. Conversely a gel containing a higher proportion of α-lactalbumin may not be able to form a strand of comparable stiffness or to form cross-links as readily.

Acknowledgements

We are grateful to the Ministry of Foreign Affairs for a scholarship to Jacquiline Gezimati, to E. A. Foegeding, M. J. Boland, and J. P. Hill for assistance with the manuscript, and to the Foundation for Research, Science and Technology and the New Zealand Dairy Board for research grants.

Literature Cited

1. Paulsson, M.; Hegg, P.-O.; Castberg, H. B. *J. Food Sci.*, **1986**, *51*, 87-90.
2. Mulvihill, D. M.; Kinsella, J. E. *Food Technol.* **1987**, *41* (9), 102-111.
3. Matsudomi, N.: Rector, D.; Kinsella, J. E. *Food Chem.* **1991**, *40*, 55-69.
4. Foegeding, E. A.; Kuhn, P. R.; Hardin, C. C. *J. Agric. Food Chem.* **1992**, *40*, 2092-2097.
5. McSwiney, M.; Singh, H.; Campanella, O. *Food Hydrocolloids* **1994**, *8*, 441-453.
6. McSwiney, M.; Singh, H.; Campanella, O.; Creamer, L. K. *J. Dairy Res.* **1994**, *61*, 221-232.
7. Roefs, S. P. F. M.; de Kruif, K. G. *Eur. J. Biochem.* **1994**, *226*, 883-889.
8. Griffin, W. B.; Griffin, M. C. A.; Martin, S. R.; Price, J. *J. Chem. Soc. Faraday Trans.* **1993**, *89*, 3395-3406.
9. Matsudomi, N.; Oshita, T.; Sasaki, E.; Kobayashi, K. *Biosci. Biotechnol. Biochem.* **1992**, *56*, 1697-1700.
10. Matsudomi, N.; Oshita, T.; Kobayashi, K. *J. Dairy Sci.* **1994**, *77*, 1487-1493.
11. Hines, M. E.; Foegeding, E. A. *J. Agric. Food Chem.* **1993**, *41*, 341-346.
12. Gezimati, J.; Singh, H.; Creamer, L. K. *J. Agric. Food Chem.* **1996**, *44*, 804-810.
13. Kuwajima, K. *Proteins: Struct., Funct., Genet.* **1989**, *6*, 87-103.
14. Calvo, M. M.; Leaver, J.; Banks, J. M. *Int. Dairy J.* **1993**, *3*, 719-727.
15. Ruegg, M.; Moor, U.; Blanc, B. *J. Dairy Res.* **1977**, *44*, 509-520.
16. Clark, A. H.; Lee-Tuffnell, C. D. In *Functional Properties of Food Macromolecules;* Mitchell, J. R.; Ledward, D. A., Eds.; Elsevier Applied Science: London, **1986**; pp 203-272.
17. Nakamura, T.; Utsumi, S.; Mori, T. *J. Agric. Food Chem.* **1985**, *33*, 1201-1203.
18. Mori, T.; Nakamura, T.; Utsumi, S. *J. Agric. Food Chem.* **1986**, *34*, 33-36.
19. Kitabatake, N.; Tani, Y.; Doi, E. *J. Food Sci.*, **1989**, *54*, 1632-1638.
20. Matsudomi, N.; Yamamoto, N.; Fujimoto, E.; Kobayashi, K. *Biosci. Biotechnol. Biochem.* **1993**, *57*, 134-135.

Chapter 10

Gelation Properties of Myosin: Role of Subfragments and Actin

S. F. Wang, A. B. Smyth[1], and D. M. Smith[2]

Department of Food Science and Human Nutrition, Michigan State University, East Lansing, MI 48824

The muscle protein myosin is most responsible for gel formation in processed meat products. The objectives of this paper were to illustrate (1) how myosin and myosin subfragments are involved in denaturation, aggregation and matrix formation during heating; and (2) how the gelation of myosin is modulated by the presence of actin. Chicken breast muscle myosin unfolded at 36°C with transitions (Tm) at 50, 57 and 67°C. By comparing the enthalpy profile of several purified myosin subfragments against the Tm of 10 theoretically determined myosin domains, many of the transitions were assigned to discrete regions of the myosin molecule. This information was used to interpret myosin aggregation and rheological profiles during heat-induced gelation. When bound to myosin, actin stabilized subfragment-1 and light chains. This effect was not observed when actin was dissociated from myosin. These studies suggest that specific changes in environmental conditions and/or heating rates can be used to alter the mechanisms of myosin gelation to alter properties of the final heat-induced gel.

Gelation is one of the most important functional properties in foods. Gels exhibit diverse microstructural and mechanical properties, contributing both quality and sensory attributes to various processed food products. Muscle foods are a good source of protein, vitamins and minerals; however, processors are trying to improve the nutritional value of their products by decreasing the fat and salt content. Reductions in fat and salt usually cause undesirable changes in product quality due to decreased yields and undesirable textural changes. Processors are also continually trying to reduce the cost of their products while maintaining quality by substitution of lower cost ingredients. The heat-induced gelation of proteins is essential for producing meat products with desired properties. An understanding of the mechanism of muscle protein gelation will enable processors to manipulate proteins to achieve desired functionality in a variety of product types.

[1]Current address: Department of Food Chemistry, University College Cork, Ireland
[2]Corresponding author

The muscle protein myosin is most responsible for gel formation in processed meat products. Myosin, a 521 kDa protein, is composed of two heavy chains (MHC) and four light chains (1) (Figure 1). With limited digestion by trypsin, myosin is cleaved into two subfragments: heavy meromyosin (HMM) and light meromyosin (LMM) (2). Similarly, myosin can be hydrolyzed by papain into subfragment-1 (S-1), which contains the globular "head" of myosin, and rod, which comprises the α-helical "tail" of the molecule. The HMM subfragment can also be hydrolyzed by papain into two subfragments: S-1 and subfragment-2 (S-2), which is the rod-like part of HMM (2).

Several models have been generated to explain the thermal gelation process (3, 4). Gelation is usually initiated by denaturation or unfolding of proteins, followed by aggregation into progressively more numerous and bigger aggregates, that at the gel point cross-link throughout the matrix to form a viscoelastic network of protein. Protein unfolding and orientation of the unfolded molecules during aggregation influence the development of a gel network and subsequent rheological properties of the final gel (5, 6). Differential scanning calorimetry (DSC) can be used to study the thermodynamics of structural/conformational changes occurring in proteins on heating. Changes in turbidity are often used to follow protein aggregation, whereas small strain dynamic testing can be used to follow development of a viscoelastic cross-linked gel matrix (7).

Myosin is a multi-domain protein that exhibits several endothermic transitions during heating. A domain is defined as an independent, cooperative unit in a folded protein (8, 9). Several myosin subfragments have also been reported to contain more than one domain. For example, it has been reported that myosin rod, LMM and S-2 may contain six, five and three domains, respectively (10-12). The transition (Tm) or unfolding temperature of each domain and the order in which the domains within a given molecule may unfold can change with environmental conditions, such as pH or salt to alter properties of the final gel (13).

The objectives of this paper are to illustrate (1) how myosin and myosin subfragments are involved in denaturation, aggregation and matrix formation during heating; and (2) how the gelation of myosin is modulated by the presence of actin.

Myosin Denaturation

When heated at 1°C/min, chicken breast muscle myosin in 0.6 M NaCl, 50 mM Na phosphate buffer, pH 6.5, started to unfold at 36°C with transitions at 50, 57 and 67°C and a shoulder at 49°C when measured by DSC (7, 14, 15) (Figure 2). Myosin has been deconvoluted into at least 10 domains (7, 14). By comparing the enthalpy profile of several purified myosin subfragments against the Tm of the ten theoretically determined domains, Smyth et al. (14) were able to assign many of the transitions to discrete regions of the myosin molecule (Table I). When the enthalpy profile of myosin was deconvoluted it was assumed that each transition state represents a two-state unfolding process and that only a single domain unfolds at each temperature. Many of the deconvoluted transitions matched the unfolding temperatures of individual myosin subfragments. However, some of the theoretical transitions were assigned to two myosin regions and some transitions did not match those of any purified subfragment. For example, the transition with Tm of 50.9°C was assigned to both LMM and S-2. Different proteases are known to produce myosin subfragments of different sizes. It is also possible that proteolytic digestion during preparation of the subfragments may cleave some domains into two parts. It has been suggested that the least stable domain of

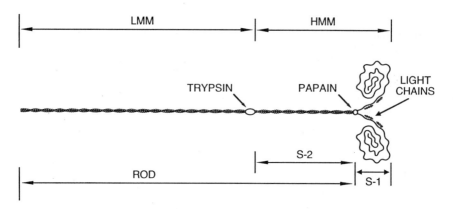

Figure 1. Schematic representation of the myosin molecule: HMM, heavy meromyosin; LMM, light meromyosin; HMM-S$_1$, subfragment-1; S$_2$, subfragment-2; DTNB, 5,5-dithiobis-(2-nitrobenzoate) light chains (Adapted from ref. 1).

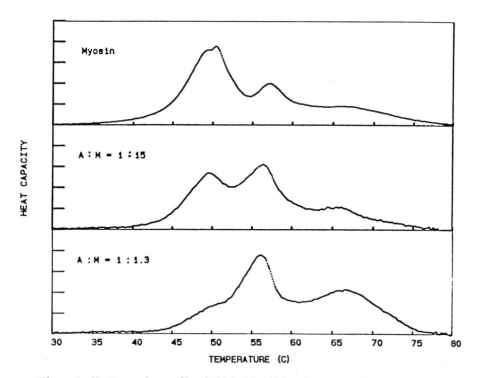

Figure 2. Heat capacity profile of 1% (w/v) chicken breast myosin and actomyosin in 0.6 M NaCl, 50 mM Na phosphate buffer, pH 6.5, heated from 25 to 85°C at 1°C/min. Actomyosin used at actin to myosin weight ratios of 1:15 and 1.1.3. (Reproduced with permission from ref. 15. Copyright 1995 American Chemical Society).

LMM and rod, at the LMM/S-2 junction, may be cleaved during tryptic digestion of the rod (16).

Table I. Transition temperatures and assigned myosin
subfragments of chicken breast muscle in 0.6 M NaCl, 50 mM
Na phosphate buffer, pH 6.5[a]

Peak	Transition Temperature (°C)	Myosin subfragment[b]
1	39.6±0.7	LMM
2	43.7±0.3	Rod
3	46.2±0.2	S-2
4	47.5±0.2	S-1, alkali LC
5	50.9±0.2	S-2, LMM
6	54.0±0.3	S-2
7	57.4±0.3	DTNB LC
8	61.5±0.4	Unknown
9	63.1±0.1	Rod and HMM
10	67.2±1.3	Unknown

[a] Adapted from ref. 14.
[b] Abbreviations: LMM, light meromyosin; S-2, subfragment 2;
S-1, subfragment 1; LC, light chain; DTNB, 5,5-dithiobis-
(2-nitrobenzoate); HMM, heavy meromyosin.

Smyth et al. (14) assigned the first theoretical transition at 39.6°C to a domain of LMM. The second Tm of 43.7°C in the theoretical myosin profile was attributed to the rod region, but did not match any transitions of LMM or S-2. This transition may have been from a domain near the junction of LMM and S-2 (hinge region) that was cleaved during preparation of the subfragments. The transitions at 46.2 and 54.0°C were assigned to S-2. The transition at 50.9°C was assigned to both LMM and S-2 and may arise from a domain near the hinge region between these subfragments that was present in both fragments due to differential hydrolysis by the proteases used during preparation. The transition at 47.5°C was assigned to both S-1 and the alkali light chains. The 5,5-dithiobis-2-nitrobenzoate (DTNB) light chains unfolded at 57.4°C. The theoretical peak at 63.1°C corresponded to a transition in MHC, rod and HMM, but S-2 showed no evidence of this peak suggesting this domain may be near the junction of rod and S-1. The theoretical transitions at 61.5 and 67.2°C were undefined. More work is needed to positively identify all domains, but the melting temperatures and the sequence that the myosin domains unfolded can be used to help interpret myosin aggregation and rheological profiles during gelation.

Myosin Aggregation and Viscoelasticity Development

Myosin undergoes a two-stage aggregation process during heating at 1°C/min; a rapid increase in absorbance between 50 and 54°C was followed by a plateau from 55 to 60°C (Figure 3). Another slower increase in aggregation was observed between 60 and 70°C (14). The changes in myosin leading to this relatively complicated aggregation profile upon heating are not clear; however, similar changes in viscoelasticity were seen using small strain rheological testing. Using small strain dynamic testing, Wang and Smith (7) found that the storage modulus (G') or the elastic-like properties of chicken breast myosin in 0.6 M NaCl, pH 6.5 increased sharply at 54°C, followed by a plateau between 58 and 60°C, and then increased gradually until 80°C (Figure 4).

Although purified myosin light chains did not aggregate when heated from 25 to 85°C, they influenced the aggregation and viscoelastic profile of myosin as compared to that of MHC. Removal of myosin light chains decreased the stability of MHC by decreasing the aggregation temperature by 5°C (45 vs 50°C). Also, when the light chains were removed, no plateau was observed in the aggregation or rheological profile of MHC (Figure 3). These results suggest that the plateau from 55 to 60°C in the aggregation profile and from 59 to 62°C in the rheological profile of myosin may be due to the denaturation of DTNB light chains with a Tm of 57.4°C. Denaturation of DTNB light chains seem to momentarily disrupt aggregation and matrix formation during the heat-induced gelation of myosin. Morita and Ogata (17) also suggested that the DTNB light chains might influence myosin gelation even though purified light chains do not gel.

To further understand the role of domains in gelation, Smyth et al. (14) compared the aggregation profiles of several myosin fragments (Figure 5). LMM was responsible for aggregation below 55°C, while the turbidity of S-2 increased from 60 to 75°C. The authors suggested that the 5°C gap between the aggregation of LMM and S-2 might be due to the hinge region between those two subfragments, as the rod aggregated continuously between 25 and 80°C. The turbidity of S-1 also increased between 45 and 55°C and was thought to contribute to the aggregation of myosin in that temperature range.

Myosin does not need to be completely denatured for a gel matrix to form. Only the first four domains were unfolded prior to the development of gel elasticity as determined by small strain rheological testing (7). The largest increase in G' occurred immediately after S-1 and the alkali light chains were unfolded. Thus, DSC, aggregation and rheological results suggest that LMM and S-1 play a critical role in myosin gelation below 55°C. Unfolding of other domains at higher temperatures, such as those within S-2, appear to modify the structure of the network formed by LMM and S-1 (Smyth, A.B., Smith, D.M., Vega-Warner, V.A., O'Neill, E. *Lebensmittel Wissenschaft Technol.*, in press). Wu et al. (18) have suggested that the unfolding and interactions of myosin domains below 55°C are critical to gel properties based on isothermal heating experiments.

Intact myosin aggregates at a slower rate and at a higher temperature than purified LMM and rod (14). The rate of aggregation of LMM and rod between 30 to 48°C was two and four times greater, respectively, than that of myosin. Although purified LMM and rod began to aggregate at 25°C, myosin did not significantly aggregate until 50°C. Thus, the rate of aggregation was slower and the onset temperature of aggregation was higher for these regions within the myosin molecule as compared to the purified fragments.

Hermansson (6) has suggested that more ordered gels with a finer network are formed when the rate of protein aggregation is slower than the rate of protein unfolding. When the rate of protein aggregation is high relative to the rate of unfolding, coarser and less organized gel networks are formed. Light meromyosin appeared to undergo both denaturation and aggregation processes almost simultaneously, whereas the aggregation of S-1 and S-2 fragments was not observed until after denaturation was complete. These findings indicate that the rate of denaturation and aggregation vary within domains of the same molecule.

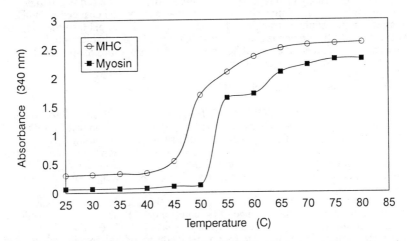

Figure 3. Aggregation profile of 1% (w/v) chicken breast muscle myosin and myosin heavy chain (MHC) in 0.6M NaCl, 50 mM Na phosphate buffer, pH 6.5 heated from 25 to 80°C at 1°C/min (Reproduced with permission from ref. 14. Copyright 1996 American Chemical Society).

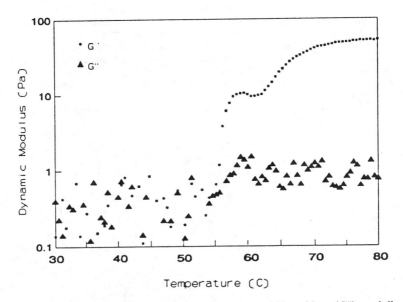

Figure 4. Representative rheogram showing storage (G') and loss (G") moduli of myosin (10 mg/ml) heated at 1°C/min in 0.6 M NaCl, 50 mM Na phosphate buffer, pH 6.5 (Reproduced with permission from ref. 7. Copyright 1994 American Chemical Society).

Bertazzon and Tsong (16) reported that the endotherm of the rod was highly sensitive to pH. The relative sizes of two domains in the rod changed with pH, suggesting migration of structure. S-1 was less sensitive to pH than the rod, but was more sensitive to heating rate. These findings suggest that it might be possible to manipulate the unfolding or aggregation properties of specific domains to achieve desired gel properties.

Actin Influences the Gelation of Myosin

In post-rigor meat, actin binds to the S-1 region of myosin to form actomyosin. Sano et al. (19) found that changes in the filamentous or F-actin:myosin ratio affected the rheogram of storage modulus of fish actomyosin during heating. Although actin does not form a gel (7, 19), addition of F-actin has been reported to increase myosin rod cross-linking in some cases. Asghar et al. (20) found that optimal gel strength was obtained with myosin:F-actin or weight ratio of 15:1 (molar ratio of 2.7:1).

Wang and Smith (15) found that the enthalpy profile of myosin was significantly altered in the presence of F-actin (Figure 2). The enthalpy profile of both myosin and actomyosin contained three peaks (50, 57, 67°C). When compared to myosin, the initial unfolding temperature of actomyosin was increased by 2 and 4°C at an actin:myosin weight ratio of 1:15 (AM 1:15) and 1:1.3 (AM 1:1.3), respectively. Also, the heat capacity of the peak at 50°C was decreased, but was increased at both 57 and 67°C as the concentration of F-actin was increased. Based on the work of Smyth et al. (14), the thermal transition at 50°C was due primarily to unfolding of S-1 and alkali light chains. The authors suggested that the stability of S-1 and the alkali light chains might be increased when actin binds the S-1 region of myosin head (21) and interacts with residues 1-41 of the light chain (22, 23). Therefore, it is possible that S-1 was stabilized by 7°C such that the broad peak at 50°C disappeared and became superimposed on the peak at 57°C. When pyrophosphate was added to actomyosin solutions to dissociate actin from myosin, the peaks in the endotherms at 50 and 57°C were similar to those of free myosin, further confirming that binding to actin increased the stability of S-1.

The peak in the endotherm of actomyosin at 67°C increased as actin concentration was increased in both the presence and absence of pyrophosphate (15). No transition peak was seen at 75.5°C in any actomyosin endotherm. Although purified F-actin has a Tm of 75.5°C and monomeric or G-actin has a Tm of 53°C (0.6 M NaCl, 50 mM Na phosphate buffer, pH 6.5), the authors attributed the peak at 67°C to the denaturation of actin and suggested that F-actin was somehow destabilized in the presence of myosin.

The rheograms of myosin and actomyosin were different when heated from 30 to 80°C at 1°C/min (15). The storage moduli of AM 1:15 (29% of protein as actomyosin) showed a gelation onset temperature of 51°C; 3°C below that of myosin. The storage modulus of AM 1:15 exhibited a broad peak between 55 and 61°C, and decreased abruptly between 61 and 64°C (Figure 6). Only a small increase in G' was observed when actomyosin was heated from 65 to 80°C. In contrast, the rheogram of myosin showed a plateau from 58 to 60°C, followed by an increase in G' as myosin was heated from 61 to 80°C. The G' and phase angle of myosin and actomyosin 1:15 gels were the same at 80°C. When pyrophosphate was added, G' increased rapidly from 51 to 57°C, peaked at 60°C, then decreased rapidly from 60 to 65°C. The G' then increased slightly as the temperature was increased to 80°C. The differences in the actomyosin and myosin rheograms can be explained by looking at changes in their respective endotherms.

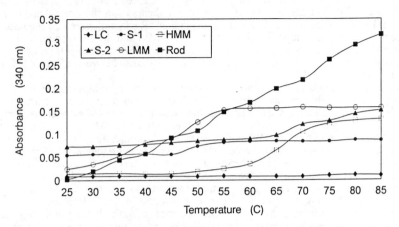

Figure 5. Aggregation profile of myosin subfragments in 0.6 M NaCl, pH 6.5. heated from 25 to 80°C at 1°C/min. Subfragment-1 (S-1) and heavy meromyosin (HMM) 0.4% protein (w/v); light chains (LC), rod, light meromyosin (LMM), and subfragment-2 (S-2) 0.2% protein (w/v) (Reproduced with permission from ref. 14. Copyright 1996 American Chemical Society).

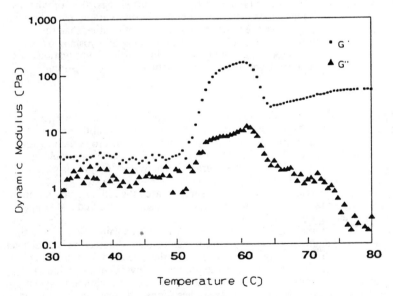

Figure 6. Representative rheogram of storage (G') and loss (G") moduli of 10 mg/mL actomyosin at actin to myosin weight ratio of 1:15, heated at 1°C/min in 0.6 M NaCl, 50 mM Na phosphate buffer, pH 6.5 (Reproduced with permission from ref. 15. Copyright 1995 American Chemical Society).

The initial increase in storage modulus of myosin resulted from the aggregation of LMM and S-1. The endotherms of actomyosin indicated that the thermal stability of S-1 was increased by $7\,^{\circ}$C due to actin binding. We hypothesize that the initial increase in G' of actomyosin solutions is due to LMM and the broadening of the plateau near $57\,^{\circ}$C is due to the unfolding of stabilized S-1. When actin is not present or S-1 is dissociated from actin, the broad plateau between 55 and $61\,^{\circ}$C is not observed.

Using electron microscopy, Yamamoto (24) observed head-to-head aggregation of rabbit myosin, forming "daisy-wheel-shaped oligomers with myosin tails extending radially", when heated isothermally at $40\,^{\circ}$C. We suspect that head-to-head aggregation may actually hinder aggregation of LMM and increase the gelation onset temperature. When S-1 is stabilized through actin binding, head-to-head aggregation should be delayed allowing LMM to aggregate at a lower temperature. This may explain the lower gelation onset temperature of actomyosin as compared to myosin. In the presence of pyrophosphate the gelation onset temperature was the same for actomyosin and myosin. If S-1 is destabilized, the aggregation of S-1 occurs before the "needed" domains of LMM are completely unfolded. Although more evidence is required, our results suggest that gel quality may be altered and controlled by modulating the temperature of S-1 aggregation relative to that of LMM.

An abrupt decrease in storage and loss moduli of actomyosin was observed between 62 to $65\,^{\circ}$C that was not seen in myosin solutions. Sano et al. (19) reported a similar effect for F-actin in fish muscle myosin and proposed that these changes resulted from the dissociation of actin from myosin, fragmentation of the actin filament, and subsequent breakdown of gel matrices. According to our observations, this decrease in G' in actomyosin occurred regardless of whether actin was free (with pyrophosphate) or bound to myosin (without pyrophosphate), suggesting this change was due to the presence of F-actin rather than the dissociation of actomyosin. It may be that the unfolding of myosin at $67\,^{\circ}$C interferes with the ordered cross-linking of myosin. Further investigations are needed to fully understand the influence of F-actin on myosin gelation.

Conclusion

Myosin gelation was defined by denaturation, aggregation and gel network formation. Using DSC enthalpy profiles of myosin and its subfragments, most of the theoretical deconvoluted peaks of myosin were ascribed to certain regions of the myosin molecule. It was found that LMM and S-1 were responsible for aggregation below $55\,^{\circ}$C, while the aggregation above $60\,^{\circ}$C was assigned to S-2. S-1 and LMM were responsible for the initial increase in the storage modulus of myosin due to crosslinking. Intact myosin decreased the rate of aggregation of LMM at lower temperatures, suggesting that myosin and purified LMM might have different aggregation mechanisms.

These domain assignments were also used to explain differences in the unfolding and aggregation mechanisms of actomyosin. Interaction between actin and myosin stabilized S-1 and light chains, and thus, delayed their unfolding. This stabilization effect was decreased in the presence of pyrophosphate due to dissociation of actomyosin. The stabilized myosin domains altered the gelation mechanism and gel properties. These studies suggest that specific changes in environmental conditions and/or heating rates can be used to alter the mechanisms of myosin unfolding, aggregation or crosslinking to alter properties of the final heat-induced gel. Therefore, information on the gelation mechanism of myosin will help us to manipulate protein functionality, and thus, efficiently develop new meat products or improve existing products.

Literature Cited

1. *Principles of Biochemistry: Mammalian Biochemistry;* Smith, E.L.; Hill, R.L.; Lehman, I.R.; Lefkowitz, R.J.; Handler, P.; White, A., Eds.; 7th ed.; McGraw-Hill: Singapore, **1983**.
2. Lowey, S.; Slayter, H.S.; Weeds, A.; Baker, H. *J. Mol. Biol.* **1969**, *42*, 1-29.
3. Clark, A.H. 1992. In *Physical Chemistry of Foods;* Schwartzberg, H.G.; Hartel, R.W., Ed.; Marcel Dekker: NY, **1992**, 263-305.
4. Foegeding, E.A.; Hamann, D.D. In *Physical Chemistry of Foods;* Schwartzberg H.G.; Hartel R.W., Ed.; Marcel Dekker: NY, **1992**, 423-441.
5. Ferry, J.D. *Adv. Protein Chem.* **1948**, *4*,1-78.
6. Hermansson, A.-M. *J. Texture Studies* **1978**, *9*, 33-58.
7. Wang, S.F.; Smith, D.M. *J. Agric. Food Chem.* **1994**, *42*, 2665-2670.
8. Privalov, P.L. *Adv. Protein Chem.* **1982**, *35*, 1-104.
9. Shriver, J.W.; Kamath, U. *Biochemistry,* **1990**, *29*, 2556-2564.
10. Potekhin, S.A.; Trapkov, V.A.; Privalov, P.L. *Biofizika* **1979**, *24*, 46-50.
11. Lopez-Lacomba, J.L.; Guzman, M.; Cortijo, M.; Mateo, P.L.; Aguirre, R.; Harvey, S.C.; Cheung, H.C. *Biopolymers* **1989**, *28*, 2143-2159.199
12. Bertazzon, A.; Tsong, T.Y. *Biochemistry* **1990**, *29*, 6453-6459.
13. Potekin, S.A.; Privalov, P.L. *J. Mol. Biol.* **1982**, *159*,519-535.
14. Smyth, A.B.; Smith, D.M.; Vega-Warner, V.; O'Neill, E. *J. Agric. Food Chem* **1996**, *44*, 1005-1010.
15. Wang, S.F.; Smith, D.M. *J. Agric. Food Chem.* **1995**, *43*, 331-336.
16. Bertazzon, A.; Tsong, T.Y. *Biochemistry* **1989**, *28*, 9784-9790.
17. Morita, J.-I.; Ogata, T. *J. Food Sci.* **1991**, *56*, 855-856.199
18. Wu, J.Q.; Hamann, D.D.; Foegeding, E.A. *J. Agric. Food Chem.* **1991**, *39*, 229-236.
19. Sano, T.; Nogushi, S.F.; Matsumoto, J.J.; Tsuchiya, T. *J. Food Sci.* **1989**, *54*, 800-804.
20. Asghar, A.; Morita, J.I.; Samejima, K.; Yasui, T. *Agric. Biol. Chem.* **1984**, *48*, 2217-2224.
21. Mornet, D.; Pantel, P.; Audemard, E.; Kassab, R. *Biochem. Biophys. Res. Commun.* **1979**, *89*, 925-932.
22. Sutoh, K. *Biochemistry* **1982**, *21*, 4800-4804.
23. Sutoh, K. *Biochemistry* **1983**, *22*, 1579-1585.
24. Yamamoto, K. *J. Biochem.* **1990**, *108*, 896-898.

Chapter 11

Effects of Macromolecular Interactions on the Permeability of Composite Edible Films

Tara Habig McHugh

Western Regional Research Center, Agricultural Research Service, U.S. Department of Agriculture, 800 Buchanan Street, Albany, CA 94710

Polysaccharides, proteins and lipids represent the three main categories of macromolecules found in edible films. Polysaccharide and protein films are good gas barriers, but poor moisture barriers. Pure lipid films are good moisture barriers, yet poor gas barriers. Considerable interest exists in the development of composite edible films that utilize the favorable functional characteristics of each class of macromolecule. A considerable amount of research has focused on polysaccharide-lipid edible films. Polysaccharides commonly do not form true emulsions with lipids; therefore, the resultant films exhibit bilayer structures with drastically improved permeability. Lipid particle size distributions significantly affect the water vapor permeability of protein-lipid emulsion films. Decreases in particle diameters result in improved water barriers, due to increasing interfacial interactions. Effects of protein-polysaccharide coacervate interactions are also reviewed.

Effective edible films can retard degradative mass transfer in food systems to prolong food product shelf life and improve product quality. By replacing or supplementing synthetic packaging materials with edible films, packaging waste can be reduced while increasing the recyclability of synthetic packaging materials.

The three main classes of edible film forming components are proteins, polysaccharides and lipids. Each of these components imparts both favorable and unfavorable characteristics on the final product. Tables 1 and 2 display characteristic water vapor and oxygen permeability values for each of the various classes of edible film forming components (1). Plasticized protein and polysaccharide films can be formed from aqueous solutions at room temperature and exhibit good mechanical properties. Proteins and polysaccharides act as cohesive structural matrixes in multicomponent film systems. They are good oxygen barriers, but are poor moisture barriers. Pure lipid films are good moisture barriers; however they often require solvent or high-temperature casting and exhibit poor mechanical and oxygen barrier properties. By combining proteins or polysaccharides with lipids, the advantages of each can be exploited to improve the properties of the film system.

Table 1: Water Vapor Permeability[a] of Various Edible Films

Proteins	
Whey Protein/Sorbitol (1.6:1)	62.0
Gluten/Glycerin (3.1:1)	54.4
Polysaccharides	
Hydroxypropylmethylcellulose (HPMC)	9.1
HPMC/Polyethylene Glycol (9:1)	6.5
Lipids	
Shellac	0.60
Beeswax	0.05
Paraffin Wax	0.02
Protein/Lipid	
Whey Protein/Beeswax/Sorbitol (3.5:1.8:1)	20.4
Polysaccharide/Lipid	
HPMC/Stearic Acid (1:1.1)	0.11

[a]   Water vapor permeability units are $g \cdot mm/m^2 \cdot d \cdot kPa$ measured between 25°C and 30°C with water on one side of the film and desiccant on the other.
(Adapted from ref. 1.)

Table 2: Oxygen Permeability[a] of Various Edible Films

Proteins	
Whey Protein/Sorbitol (3.5:1) (70[b])	42.3
Whey Protein/Sorbitol (3.5:1) (46[b])	1.5
Gluten/Glycerin (2.5:1) (0[b])	3.8
Polysaccharides	
Hydroxypropylcellulose/Polyethylene Glycol (9:1) (0[b])	308
Methylcellulose (MC)/Polyethylene Glycol (9:1) (0[b])	150
Lipids	
Beeswax (0[b])	932
Shellac (0[b])	75
Protein/Lipid	
Whey Protein/Beeswax/Sorbitol (3.5:1.8:1) (70[b])	101.1
Whey Protein/Beeswax/Sorbitol (3.5:1.8:1) (46[b])	8.6
Polysaccharide/Lipid	
Palmitic Acid/MC/Polyethylene Glycol (3:9:1) (79[b])	785

[a]   Oxygen permeability units are $(cm^3 \cdot \mu m/m^2 \cdot d \cdot kPa)$ measured between 23°C and 30°C.
[b]   Percentage relative humidity conditions employed during test.
(Adapted from ref. 1.)

Various physical, chemical and enzymatic treatments have been employed in the past to enhance macromolecular attractions between molecules in the same class. Examples of treatments that have been employed in the past include: 1) heat treatment of proteins to allow for the formation of intermolecular disulfide bonds by thiol-disulfide interchange and thiol oxidation reactions, 2) ionic crosslinking and pH adjustment as means to noncovalently link macromolecules and 3) transglutaminase and peroxidase enzymatic crosslinking of macromolecules. Interactions between macromolecules in the same class will not be reviewed in this chapter. This chapter will review the different types of interactions prevalent between protein, polysaccharide and lipid macromolecules and their effects on the permeability properties of composite edible film systems. Common methods for the manufacture of these films will be described and effects of chemical structure and physical characteristics on their permeability properties will be examined. Protein-lipid, polysaccharide-lipid, and protein-polysaccharide systems will be discussed.

Protein - Lipid Interactions

Emulsions. Emulsions are colloidal systems containing two partially miscible fluids and, almost invariably, a surfactant. The formation of emulsions involves an increase in the interfacial area between the two phases, as well as an increase in free energy. Mechanical work is necessary to overcome this free energy barrier. These systems are generally unstable; therefore, emulsifying agents or surfactants are often necessary to improve their stability. Emulsifying agents collect at the naked interface and lower the interfacial tension between the dispersed and continuous phases. Emulsion particle diameters are ordinarily greater than 0.2 μm. (2,3)

Protein-lipid emulsion coatings were first identified as effective mass transfer barriers by Ukai (4), who patented the use of coatings containing an aqueous water soluble high polymer and a hydrophobic substance in an aqueous emulsion or suspension for coating food products. By taking advantage of the good emulsification properties of proteins, emulsion film systems can be formed.

Guilbert investigated emulsified films made from ethanolic solutions of stearic acid, palmitic acid and carnuaba wax in aqueous solutions of caseins or gelatin (5). These films formed bilayers during drying, as did the films produced by Kamper and Fennema (6) using the "emulsion" technique, which will be discussed in more detail later in this chapter.

Krochta et al. developed emulsion films by combining lipids and caseinate (7). Avena-Bustillos and Krochta (8) continued research on caseinate-based emulsion films, finding that beeswax films exhibited lower WVPs than other lipid films.

Tables 1 and 2 illustrate the effect lipid addition has on the permeability properties of whey protein-based emulsion films. Beeswax addition to whey protein emulsion films results in water vapor permeability (WVP) reductions as well as increases in oxygen permeability values.

More recently, the effect of lipid addition on the relationship between relative humidity and WVP was determined in whey protein (WPI) emulsion film systems (Figure 1). The percent reduction in WVP through the addition of beeswax was greater at higher relative humidities than at low relative humidities. Lipid addition significantly lowered film WVP at all relative humidities. (9)

Hauser and McLaren (10) developed a method to predict hydrophilic film WVPs with variable relative humidity on the wet side using curves such as those shown in Figure 1. These curves are extremely useful for food applications, enabling prediction of film permeability properties for food products stored under any relative humidity conditions.

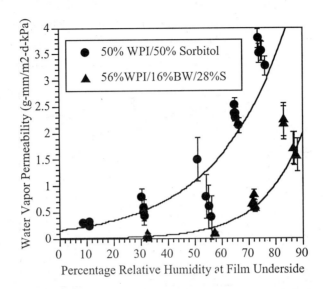

Figure 1: Effect of Lipid Addition on the Relationship between Relative Humidity Test Conditions and Water Vapor Permeability at 25°C. (Reproduced with permission from the ref. 9. Copyright 1994   Amer. Oil Chem. Soc.)

Increasing chain length of both fatty acids and alcohols resulted in significant reductions in WVP (*9*). The relative wt. fraction of the apolar portion of the molecule was responsible for this (*11*). Solubility is increased when solute molecules are attracted more strongly to the solvent than to themselves. As chain length increases, the length of the apolar section of the molecule increases, resulting in reduced aqueous solubility coefficients and WVPs (*9*).

The effect of lipid concentration on the WVP of beeswax, palmitic acid and stearyl alcohol emulsion films is shown in Figure 2 (*9*). Film WVP decreased as lipid concentration increased. This effect was more apparent in beeswax and palmitic acid films than in stearyl alcohol films. The downward, convex shape of these curves indicates that the dispersed phase was the barrier polymer and the continuous phase was the nonbarrier polymer. Additional increases in lipid concentration would result in a shift from the protein matrix being the continuous polymer to the lipid being the continuous polymer and a consequent shift in the curve to an upward, concave shape (*12*). Curves such as these can be used to tailor film WVP properties to specific food applications. By adding different amounts and/or types of lipid, the WVP characteristics of protein-lipid emulsion film systems can be regulated for application to a variety of food products.

Lipid type also has a dramatic effect on the WVP properties of whey protein emulsion films (Figure 2) (*9*). Beeswax emulsion films exhibited very low WVPs, followed by fatty acid films. Fatty alcohol emulsion films displayed significantly higher WVP values. This effect can be attributed to the lipid's solubility properties. Fatty alcohols are soluble in most organic solvents, whereas, fatty acids have extremely limited solubility, due to their strong tendency to dimerize through the formation of hydrogen bonded networks (*11*). Beeswax is composed of long ester molecules. The long hydrocarbon chains in beeswax result in its low solubility in most solvents (*9*). The solubility of lipids in various solvents can be used as an index for their relative WVP properties.

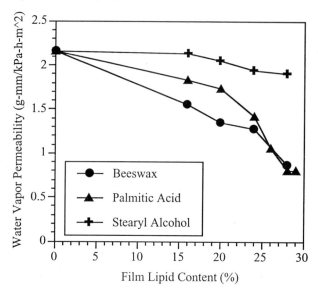

Figure 2: Effect of Lipid Type and Concentration on the Water Vapor Permeability of 56% Whey Protein/ X% Lipid/(44%-X%) Sorbitol Emulsion Films at 25°C. (Reproduced with permission from the ref. 9. Copyright 1994   Amer. Oil Chem. Soc.)

Particle size also affects the water vapor permeability properties of protein-lipid emulsion film systems.   The water barrier properties of 56% whey protein(WPI)/28% beeswax(BW)/16% sorbitol(S) emulsion films improved as particle size decreased (Figure 3) (*13*). Mean particle diameters were calculated from laser light scattering measurements. Decreased mean particle diameters resulted in linear decreases in WVP values.

Decreases in WVP values with decreased mean particle diameters originate due to increased interfacial areas provided by small particles at constant lipid content. Microfracture was not considered a valid explanation for this effect, since microfracture would have resulted in increases in WVP values with decreased mean particle diameters. The exact nature of protein/lipid reactions at interfaces remains unknown and is the subject of much controversy. LeMeste et al. (*14*) used electron spin resonance spectroscopy to examine the interaction between milk proteins and fatty acids. The milk protein and lipid were found to interact through their polar groups. The milk protein was thought to sit on the surface of the lipid droplets. No evidence of protein or lipid reorganization at the interface was found. Phillips (*15*), on the other hand, found that flexible ß-casein molecules were present at the water/oil interface in emulsion systems with protein chains, loops and tails protruding into both phases. Globular lysozyme behaved in a similar manner. Barford et al. (*16*) developed a similar model of lipid-protein interactions in an oil-water emulsion interface, concluding that protein molecule chain, loops and tails interacted at the interface to stabilize the lipid particles.

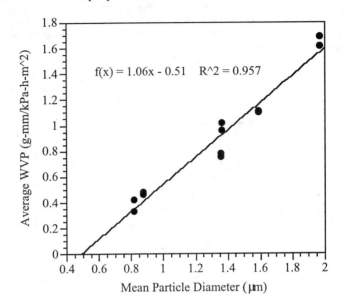

Figure 3: Effect of Mean Particle Diameter on the Water Vapor Permeability of 56% Whey Protein/28% Beeswax/16% Sorbitol Emulsion Films at 23°C. (Reproduced with permission from ref. 13. Copyright 1994 Food & Nutrition Press.)

McHugh and Krochta (*13*) hypothesized that whey protein-beeswax interfacial interactions in emulsion films resulted in reduced protein phase interfacial mobility and decreased WVP values. The Maxwell model (*17*), the tortuosity model (*18*), and the interaction model (*19,20*) were used to explain and quantify the effects of particle size on WVP (*13*). Both the Maxwell and the tortuosity models assume that the continuous phase properties are unaffected by the dispersed phase; whereas the interaction model accounts for interactions between the continuous and dispersed phases.

Based on a WVP of 2.16 g-mm/kPa-h-m^2 for 78% WPI/22% sorbitol films, Maxwell's model predicted that the WVP for the 56% WPI/28% BW/16% sorbitol emulsion films would be a constant value of 0.98 g-mm/kPa-h-m^2, independent of particle size. Experimentally, water vapor permeability was found to vary with average particle diameter (Figure 3); therefore, the Maxwell model proved insufficient to describe protein-lipid edible film WVP properties.

Similarly, the tortuosity model predicted a constant WVP of 1.36 g-mm/kPa-h-m^2, demonstrating the model's inability to account for particle size effects.

The interaction model was employed to calculate an interaction factor, B, using Equation 1. X_A and P_A are, respectively, the volume fraction and the WVP of the amorphous continuous phase. P_C is the predicted WVP of the composite film and T is the average distance a molecule must travel to get through a film divided by the thickness of the film. Specific tortuosity factors can be calculated based on the shape of the particles as well as their distribution in the film (*13*). The interaction factor, B, increased as particle size decreased (13). According to Rogers (19,20), the logarithm of the interaction factor should correlate linearly with the particle diameter squared. This was verified for the 56% WPI/28% BW/16% sorbitol edible emulsion films (Figure 4). It was concluded that immobilization of proteins at the lipid interface in emulsion films was responsible for the decrease in WVP values observed with decreasing particle diameters (13).

$$P_C = \frac{X_A P_A}{TB}$$

Equation 1

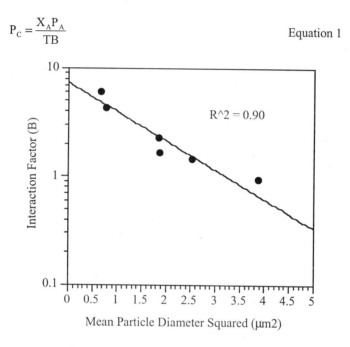

Figure 4: Relationship between Mean Particle Diameter and Interaction Factor. (Reproduced with permission from ref. 13. Copyright 1994 Food & Nutrition Press.)

By controlling lipid particle size and distribution in emulsion films and coatings, moisture loss, respiration and oxidation in foods can be tuned. Emulsion film lipid and aroma barrier properties may be similarly controlled, as could their mechanical properties.

Microemulsions. Unlike macroemulsions, microemulsions are generally thermodynamically stable under given conditions, such as formulation and temperature. Microemulsion particle diameters range between 0.01 μm and 0.2 μm. They ordinarily include a fourth component, often an alcohol, called a cosurfactant (2,3). Many floor polishes are microemulsions. Flavor oils are often dispersed as microemulsions for the soft drink industry. Mouthwash flavors are similarly frequently prepared as microemulsions.

The mechanism for forming microemulsions differs from that used in making macroemulsions. Microemulsion systems are dependent on formulation, no amount of work input nor excess emulsifier will produce the desired product if the formulation is not correct. When the formulation is correct, microemulsification occurs almost spontaneously. (*2*)

To date, no one has published on the potential of microemulsion systems for the formation of effective water barrier edible films and coatings. As noted in the previous section, recent research has shown that as macroemulsion particle sizes decrease, water barrier properties improve, indicating microemulsion systems may act as extremely effective edible water barriers.

Bilayers/Laminants. Edible films formed from emulsions tend to separate during drying. Emulsion separation and creaming during film dehydration are responsible for the nonisotropic nature of emulsion films. After drying, the film side facing the plate was shiny, indicating the presence of elevated protein concentrations. The side facing the air was dull. Emulsion instability results from the limited ability of whey proteins as emulsifiers (*21*). The low density lipids creamed to the film surface during film dehydration. Although emulsion separation occurred within these film systems, true bilayer films were not formed. Discrete lipid particles remained within the protein matrix, as shown by scanning and transmission electron microscopy (*9*).

This separation is apparent from the effect of film orientation in the test cup has on the WVP of emulsion films (*9*). Emulsion films oriented with the dull, lipid enriched side facing down (towards the inner, high relative humidity environment of the cup) exhibited significantly lower WVPs than films oriented with the lipid side up. This was presumably due to the exponential effect that relative humidity has on the WVP of hydrophilic whey protein-based films (Figure 1). When the lipid rich side of the film faced the high relative humidity environment, the protein rich side of the film experienced a lower relative humidity environment, lowering its contribution to the overall film WVP (*9*).

In the future it may be advantageous to promote the separation process once the emulsion has been applied to the food surface in order to produce a true bilayer film. Basic mass transfer principles indicate that the greatest permeability advantages will be achieved in true bilayer or laminant film systems. Cohesive laminant films are often difficult to form and delamination often occurs over time. Furthermore, application of the lipid layer in laminant films often requires the use of solvents or high temperatures, making production more costly and less safe than emulsion film systems which can be formed from aqueous solutions and applied to foods at room temperature. De-emulsification techniques have been developed extensively in the oil industry. Agents capable of displacing the stabilizing surfactant have been developed, leaving a non-stabilizing surfactant (*2*). These technologies may be applied to the formation of effective edible films and coatings in the future. This is another largely unexplored edible film area which awaits future research.

Lipophilization. Chemical modification of proteins and lipids under mild conditions offers the potential to improve the functional properties of these mixtures in edible film systems. One example of a chemical modification technique which may be beneficial is called lipophilization. Lipophilization is a technique used to increase the surface hydrophobicity of a protein molecule by covalent attachment of hydrophobic ligands, such as fatty acids (*22*). In the future this and other covalent attachment techniques, chemical or enzymatic, may be beneficial in forming highly effective protein-lipid based edible films and coatings.

Polysaccharide - Lipid Interactions

Bilayers. Much of the original research on bilayer polysaccharide-lipid edible films was performed by Fennema's research group at the University of Wisconsin (6,23). Tables 1 and 2 illustrate the effect of lipid addition on the water vapor and oxygen permeability properties of polysaccharide-based films. The addition of stearic acid to hydroxypropyl methylcellulose (HPMC) resulted in a significant reduction in WVP. At the same time fatty acid addition resulted in an increase in oxygen permeability.

Bilayer film systems composed of HPMC and various kinds of lipids have been formed using two techniques. The "coating technique" involves casting a lipid layer onto a dried edible film, whereas the "emulsion technique" involves adding the lipid to the film forming solution prior to film casting (6). HPMC is not a very effective emulsifier; therefore, upon drying the "emulsion" films separate into bilayer films. The stearic/palmitic acid blend prepared by the "coating" technique applied to a HPMC matrix exhibited a water vapor transmission rate of 43.9 $g/m^2 \cdot d$, whereas a similar composition film prepared by the "emulsion" technique had a water transmission rate of 3.1 $g/m^2 \cdot d$. The lower water vapor transmission rate for the "emulsion" film was attributed to the favorable orientation of fatty acid molecules at the air/water interface in the second film (6). This result was unexpected.

Lipid application techniques were also compared (23). Solvent, ethanol-soluble application versus a hot molten application of beeswax was compared. Films prepared using the solvent method exhibited greater water vapor permeability values, but reduced oxygen permeability values (23).

Researchers in France have recently expanded on Fennema's research on the properties of polysaccharide-lipid films. They performed a second study investigating the effects of film preparation technique on final film properties (24). Three techniques were investigated: 1) "emulsion", 2) "emulsion" and "coating" and 3) dipping. Films prepared using the "emulsion" technique exhibited the highest water vapor permeabilities. The dipped films were better water barriers than the "emulsion" and "coated" films (24).

A later study by Martin-Polo et al. (25) found that films produced by the "coating" method exhibited significantly better water barrier properties than films produced using the "emulsion" technique, due to the more homogeneous distribution of lipid in the coated film. This effect was modeled using equations for resistances in parallel and equations for resistances in series. Modeled and experimental results correlated well.

Chitosan was recently explored as an edible film forming ingredient, although the material is not approved by the FDA. Fatty acids were added to chitosan films in an attempt to reduce their water vapor permeabilities (26). Water vapor permeability increases did not increase with chain length increases as previously noted by numerous edible film researchers. In this case, lauric acid was the only fatty acid which successfully decreased film water vapor permeability. The report concluded that the morphological arrangement of the fatty acid chains with respect to the chitosan polymers resulted in microstructures that regulated the permeability of the final film (26).

Protein - Polysaccharide Interactions

The least studied type of macromolecular interactions are between proteins and polysaccharides. In recent years, several researchers have begun to explore the effects of protein/polysaccharide interactions on the permeability of edible films.

Shih (*27*) mixed soy isolate with propylene glycol alginate. The two polymers interacted to form electrostatic complexes and/or covalent complexes based on the treatment conditions. Reductive alkylation was accomplished through the addition of sodium cyanoborohydride, a reducing agent. Films treated with KOH had better moisture resistance than films treated with pH 8 buffer. Lauric acid addition reduced water vapor permeability.

Parris et al. (*28*) found that the addition of milk protein to alginate and pectin films resulted in a reduction in WVP values by 35%. Whey and nonfat dry milk addition resulted in higher WVP values than films containing casein or whole milk. Film forming solutions were made by first dissolving the polysaccharide in water then adding appropriate amounts of plasticizer and protein to yield a 1% final (w/w) mixture. Covalent linkages were not formed between the polysaccharides and the proteins. Film solutions were then dried and their permeability properties analyzed.

In the future alternative means for ionic crosslinking and/or covalent crosslinking proteins and polysaccharides should be explored and their effect on composite film permeability properties measured.

Conclusions

The advantages and disadvantages are known for each class of macromolecules used in the formation of edible films and coatings. Effects of macromolecular interactions on the permeability of composite edible films and coatings, containing different classes of macromolecules, are only beginning to be understood. Researchers must follow the lead of polymer scientists and look toward more complex film systems to solve the challenges presented to edible films and coatings. Multicomponent edible films must be made to resemble synthetic packages where moisture-sensitive packaging materials that are very effective oxygen barriers, such as ethylene vinyl alcohol and polyamides, are sandwiched between layers of moisture resistant polymers, such as polyvinyl chloride. Chemical and enzymatic means for covalently crosslinking the various types of macromolecules in composite edible film systems should also be explored.

Literature Cited

1. McHugh, T.H. and Krochta, J.M. In *Edible Films and Coatings to Improve Food Quality*; Krochta, J.M., Baldwin, E.A. and Nisperos-Carriedo, M., Eds.; Technomic Publishing Company: Lancaster, PA, 1994; pp 177-182.
2. *Basic Principles of Colloid Science*; Everett, D.H., Ed.; Royal Society of Chemistry: Piccadilly, London, 1988.
3. *Physical Chemistry of Surfaces*; Adamson, A.W., Ed.; John Wiley & Sons: New York, NY, 1990.
4. Ukai, N., Shingo, Y., Toshio Tsutsumi, I., and Kyoichi Marakami, K. *U.S. Patent 3,997,674.* **1976.**
5. Guilbert, S. In *Food Packaging and Preservation*, Mathlouthi, M., Ed.; Elsevier Publishing: Essex, England, 1986; pp 371-394.
6. Kamper, S.L. and Fennema, O. *J. Food Sci.* **1984**, *49*, pp. 1478-1481, 1485.
7. Krochta, J.M., Pavlath, A.E. and Goodman, N. In *Engineering and Food, Preservation Processes and Related Techniques*; Spiess, W.E.L. and Schubert, H., Eds.; Elsevier Publishing: Essex, England, 1990, Vol. 2; pp 329-340.
8. Avena, Bustillos, R.J. and Krochta, J.M. *J. Food Sci.* **1993**, *58*, pp. 904.

9. McHugh, T.H. and Krochta, J.M. *J. Amer. Oil Chem. Soc.* **1994**, *71*, pp. 307-312.
10. Hauser, P.M. and McLaren, A.D. *Ind. Eng. Chem.* **1948**, *40*, pp. 112.
11. Mead, J.F., Alfin-Slater, R.B., Howton, D.R. and Popjak, G. In *Lipids: Chemistry, Biochemistry and Nutrition*; Mead, J.F., Alfin-Slater, R.B., Howton, D.R. and Popjak, G., Eds.; Plenum Press: New York, NY, 1986.
12. DeLassus, P. In *Kirk-Othmer Encylcopedia of Chemical Technology*; John Wiley, 1992; pp. 931-962.
13. McHugh, T.H. and Krochta, J.M. *J. Food Proc. Pres.* **1994**, *18*, pp. 173-188.
14. Le Meste, M., Closs, B., Courthaudon, J.L. and Colas, B. In *Interactions of Food Proteins*; Parris, N. and Barford, R., Eds.; Amer. Chem. Soc.: Washington, D.C., 1991.
15. Phillips, M.C. *Food Tech.* **1981**, *Jan.*, pp. 50-57.
16. Barford, N.M., Krog, N. and Buchheim, W. In *Food Proteins*; Kinsella, J.M. and Socie, W.G., Eds.; Amer. Chem. Soc.: Washington, D.C., 1989.
17. *Treatise on Electricity and Magnitism*; Maxwell, C., Ed.; Oxford University Press: London, England, 1873.
18. Nielsen, L.E. *J. Macromol. Sci.* **1967**, *A1*, pp. 929-942.
19. Rogers, C.E. In *Physics and Chemistry of the Organic Solid State,* Fox, D., Labes, M.M. and Weissberger, A., Eds.; Interscience Publishers: New York, N.Y., 1965.
20. Rogers, C.E. In *Polymer Permeability*, Comyn, J., Ed.; Elsevier Applied Sci. Publishers: New York, N.Y., 1985.
21. Kinsella, J.E. and Whitehead, D.M. *Adv. Food & Nutrit. Res.* **1989**, *33*, pp. 343.
22. Haque, Z. and Kito, M. *J. Agric. Food Chem.* **1983**, *31*, pp. 1225-1230.
23. Greener, I.K. and Fennema, O. *J. Food Sci.* **1989**, *54*, pp. 1393-1399.
24. Martin-Polo, M., Mauguin, C. and Voilley, A. *J. Agric. Food Chem.* **1992**, *40*, pp. 407-412.
25. Debeaufort, F., Martin-Polo, M. and Voilley, A. *J. Food Sci.* **1993**, *58*, pp. 426-429, 434.
26. Wong, D.W.S., Gastineau, F.A., Gregorski, K.S., Tillin, S.J. and Pavlath, A.E. *J. Agric. Food Chem.* **1992**, *40*, pp. 540-544.
27. Shih, F.F. *J. Amer. Oil Chem. Soc.* **1994**, *71*, pp. 1281-1285.
28. Parris, N., Coffin, D., Joubran, R. and Pessen, H. *J. Food Sci.* **1995**, *43*, pp. 1432-1435.

Chapter 12

Films from Pectin, Chitosan, and Starch

Eastern Regional Research Center, Agricultural Research Service, U.S. Department of Agriculture, 600 East Mermaid Lane, Wyndmoor, PA 19038

Clear films from aqueous blends of chitosan HCl and high methoxy pectin, with glycerol added as plasticizer, had good dynamic mechanical properties. The pectin stabilized chitosan HCl films and prevented shrinkage and opacity, which presumably was the result of slow crystallization of chitosan HCl. Starch, added as a composite to chitosan HCl films, also stabilized the chitosan films with some reduction in dynamic mechanical properties.

Biodegradable films of pectin and starch have been made from renewable, stable agricultural sources (1, 2). Laminated films from two such readily available polysaccharides, pectin and chitosan have also been made with glycerol or lactic acid as a plasticizer (3). Applications are being sought for these films in order to replace some petroleum based films or to develop new specialized films in response to increasing environmental concerns. Pectin is the second most abundant plant polysaccharide (after cellulose), and is widely available from underutilized agricultural waste material. Pectin has a complex structure that can be characteristic of the plant source and the region in a given plant cell wall. It is here that pectin is found and has its biological function. The main backbone chain of pectin is composed of galacturonan (α-(1-4) galacturanopyranosyl) segments linked α-(1-2) to a rhamnopyranosyl unit which in turn is linked by as yet unknown α/β conformation (1-4) to the next galacturonan segment. The galacturonan segment length as well as the rhamnose content can also be variable. In addition, some of the rhamnose units may have attached short neutral sugar side chains consisting mainly of galactose and arabinose. Such substituted rhamnosyl units are often clustered into 'hairy' regions. The carboxyl groups of the galacturonan segments can be variably methyl esterified. Pectin, therefore, can be readily modified, through demethylation. High methoxy pectins can form excellent films (1, 2).

The scope of films made with pectin combined with other polysaccharides was widened to include chitosan for several reasons. Firstly, chitosan is derived from chitin, the second most abundant polysaccharide on the earth and which is commercially available from stable, renewable sources, such as waste from shellfish industries. Secondly, chitosan forms good films and membranes (4-9). Thirdly, the cationic properties of chitosan potentially enable electrostatic interactions with anionic, partially demethylated pectins (10 , 11). Lastly, chitosan films and membranes are biocompatible with a range of human tissues (12). Chitosan is poly-β-(1-4) 2-amino-2-deoxy-glucose with a content of N-acetyl groups that can vary from 0 to above 50%. Highly acetylated chitosan is chitin. The degree of acetylation can be modified to produce a polycationic polysaccharide less basic in nature than polylysine. Composites of chitosan with microcrystalline cellulose have suitable properties for biodegradable films (4). In many instances films made from chitosan solubilized with an organic acid have been treated with base to insolubilize the chitosan in order to form membranes. Some chitosan membranes have been used in permeation applications, such as the separation of water from ethanol. A composite membrane of poly(vinyl alcohol) and chitosan has been used to dehydrate alcohol by pervaporation (13). Membranes from blended chitosan and poly(ethylene oxide) have been investigated with application to haemodialysis (14). The propensity of chitosan to form ionic complexes with polyanions, such as proteins, has been exploited in the food industry to clarify juices and to treat waste water from poultry processing. Chitosan can react with polyanionic polysaccharides to form gels or insoluble gummy precipitates.

Films from chitosan and pectin in the past relied upon lamination, whereby a chitosan film was cast upon a preformed pectin film (3). Chitosan films were unsuitable foundation films, due to significant swelling. Chen et al., (7) reported that various chitosan films can absorb 40 to 80% of their weight in water. Strips of a chitosan film laminated to a pectin film curled up to form tubes, presumably due to differential swelling (15) . At no time during lamination of chitosan films to pectin films was there seen any indication of ionic interactions at the film-film interface (15). These laminated films were probably free of any significant degree of interpenetration interactions.

Experimental

Materials. Lime pectin was obtained from Grinsted Products, Inc. (Kansas City, KS) and was 65% methylated (Type 1400). Chitosan was obtained from Pronova Biopolymer, Inc. (Raymond, WA) and was 85% deacetylated (Seacure 343). Hydrochloric acid (HCl), nitric acid (HNO_3) and glycerol were ACS reagent grade. Distilled water was used to prepare all solutions.

Films. Solutions of 2.0 g of chitosan , 1.0 to 2.5 g of glycerol and 200 mL of 0.1 N HCl or HNO_3 were prepared by stirring for 1 h at room temperature. Solutions of 0.5 to 2.0 g of lime pectin and 50 mL of water were prepared by stirring at room temperature for 1 h. The two solutions were mixed and each resulting clear solution was filtered through glass wool to remove traces of residual shell particles. Films were cast with 30 mL portions of filtrate poured without bubble formation into 100 mm dia. plastic Petri dishes. In some instances, gelatinized starch was added to the filtrate (0.5 g. Amylose VII, 10 mL water, microwaved for 60 sec in a pressure safe microwave vessel (1,2)). Films from chitosan HCl or HNO_3 were also prepared without added pectin. All films were allowed to form during exposure to the atmosphere for 72 h.

Mechanical Testing. Dynamic tests were performed with a Rheometrics RSA II solids analyzer (Piscataway, NJ). Liquid nitrogen was used for temperature sweeps that started below room temperature. A nominal strain of 0.1% and an applied frequency of 10 rad/s (1.59 Hz) were routinely used. Films strips, 7.0 mm x 38.1 mm were excised from the center of the circular films with a razor blade. Sample thickness was measured with a Model No. 3 Dial Comparator micrometer, B. C. Ames Co. (Waltham, MA). Data were analyzed with Rheometrics RHIOS v3.0.1 software run on an IBM PC MSDOS platform. Tests were carried out on strips and a representative data set was selected for the comparisons that are presented in the Figures.

Results

Films prepared from chitosan HCl with or without added glycerol up to levels of 50% of total solids were initially clear. During the final stage of drying, the films shrank and turned opaque. Added starch (20% of weight of total polysaccharide) stabilized the films only when glycerol was present at a 44% total solids level. Pectin added to chitosan HCl solutions at total polysaccharide levels of from 28 to 50% resulted in clear, strong films that were stable for at least 6 months. Glycerol added to these films at levels up to 50% of total solids provided enough flexibility to prevent easy breakage from brittleness. These films readily dissolved in water (ca. 15 minutes) to give clear solutions when no starch was present. Films with composite starch gave turbid solutions characteristic of gelatinized starch. Films prepared with HNO_3 instead of HCl gave similar results. Chitosan HNO_3 films shrank and turned opaque upon standing at room temperature for 48 h. Addition of pectin stabilized chitosan HNO_3 films that were clear. When treated with 2% NaOH for 15 minutes and then washed with water, the chitosan HNO_3 film became insoluble in water and retained both its shrunken size and opacity to give a membrane. In like manner, a chitosan HCl-pectin-glycerol film was made water insoluble by treatment with 2% NaOH for 15 minutes, followed by thorough washing with water. The film retained its general structure and remained clear.

Storage and loss moduli plots against temperature for blends of high methoxy pectin and chitosan are shown in Figure 1. Markers

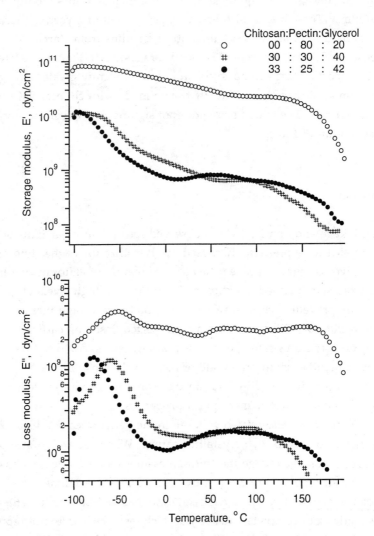

Figure 1. Storage and loss moduli of films made from blends of chitosan and pectin with added glycerol as plasticizer.

represent a typical profile and the extent of observed variation between 6 data sets was within 4x of the width of the markers. The blended films of pectin and chitosan had significantly lower storage and loss moduli than pectin films. These blends also had a higher content of glycerol. Storage and loss moduli plots against temperature for films from starch composites of pectin-chitosan HCl blends are shown in Figure 2. Here again, the pectin-starch composite film had significantly greater storage and loss moduli than those films that were made with chitosan HCl or chitosan acetate. The added starch did not significantly lower the storage or loss moduli of films.

Discussion

The shrinkage and opacity (white) observed for initially clear films of chitosan HCl were presumably caused by crystallization of the chitosan. Added glycerol did not prevent this transformation of chitosan HCl films. Gelatinized starch added to chitosan HCl films with high levels of glycerol apparently prevented crystallization. The solubility of high methoxy pectin in chitosan HCl solutions was unexpected based on earlier observations that this same pectin, even at low levels, could not be dissolved in either chitosan acetate or lactate solutions. Moreover, high methoxy pectin dissolved in chitosan HCl solutions produced clear stable films with good dynamic mechanical properties.

Completely deacetylated chitosan in the dehydrated solid state has been shown to exist in two crystalline forms. When the amino groups have each a positive charge the chitosan assumes an extended, rod-like crystalline structure. Without positive charges, the crystalline structure of chitosan is more compact and has been described as a spherulite (16). Films made from completely deacetylated chitosan HCl were not reported to shrink or turn opaque (16). That this occurred with both HCl and HNO$_3$ salts of 85% deacetylated crab chitosan can at this time possibly be best explained by the presence of the acetamido groups. Presumably, these groups allowed the chitosan in the film to convert from an extended to a compact, possibly less hydrated, structure. The high methoxy pectin in films of chitosan HCl or HNO$_3$ appears to have constrained the chitosan

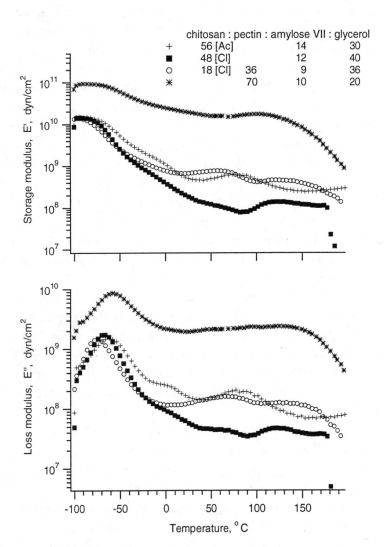

Figure 2. Storage and loss moduli of films made from blends of chitosan and pectin with added composite starch and glycerol as plasticizer.

to an extended, possibly hydrated, structure. When these latter blended films were treated with NaOH to deprotonate amino groups, the films retained their general shape and clarity. Below pH ca. 3 the unesterified carboxyl groups of high methoxy pectin (65% DE) would be negatively charged. This would tend to maintain an extended structure for pectin. Crystallinity of pectin is suppressed by rhamnose-induced kinks between galacturonan segments that constitute the pectin backbone and possibly by neutral sugar side chains clustered in 'hairy' regions. These structural features undoubtedly account, in part, for the remarkably strong films that have been prepared from pectin (1,2).

Films prepared from chitosan HCl or HNO3 were similar in their appearance (shrinkage and opacity). This would not be unusual save for the observation of Saitô et al. (17) that powders of completely deacetylated chitosan HCl and HNO3 had different solid state C_{13} spectra. From this and complementary x-ray diffraction studies they have proposed two helical structures for chitosan inorganic salts. Type 1 (HNO3) is a 2-fold helical structure that excludes water in its crystalline state. Type 2 (HCl) is a more open 4-fold or 8-fold helical structure that can accommodate some water. The shrinkage of the chitosan HCl or HNO3 films suggests that type 1 crystallinity prevails with a tighter structure that excludes water. The overall structural influence of acetamido groups (15%) in the chitosan HCl or HNO3 films probably overrides the notable structural differences between completely deacetylated chitosan HCl and HNO3 powders.

In earlier work we observed that pectin films had greater storage and loss moduli than pure chitosan acetate or lactate films containing comparable levels of glycerol (3). To date, films made from pectin have had the highest storage and loss moduli of all films made from any polysaccharide (1, 2). The finding that starch added as a composite to pectin-chitosan HCl films did not significantly lower storage and loss moduli is in agreement with observations by Coffin and Fishman (1, 2) for films made from composites of pectin and starch with various levels of glycerol added as plasticizer. A plasticizer, such as glycerol, was required to increase flexibility of the chitosan HCl films, with or without added pectin and/or starch. Hosokawa et al. (4) obtained good flexibility in films made from chitosan-cellulose composites without added glycerol.

Coffin and Fishman (1, 2), however, found that glycerol added to films made from pectin-starch composites improved flexibility. In the case of chitosan, satisfactory flexibility was obtained with films made from chitosan and lactic acid (3). Here the lactic acid served both to solubilize the chitosan and to plasticize. We have also found that films made with citric acid and chitosan and/or pectin are prone to brittleness even with added glycerol (15).

Chitosan HCl or HNO_3 films made with pectin are stable over time and readily dissolve in water. When acetic acid is used to prepare chitosan films, the acetic acid can slowly leave the film by evaporation to pull the hydrolysis of the weak amine acetate salt towards the free amine state (18). This process is encouraged by free water held in the film at room temperature. The solubility of the film is thereby reduced. Lactic acid does not leave chitosan films exposed to air, since it probably hydrogen bonds to the polysaccharide (3, 18).

Films from blends of chitosan inorganic salts of HCl or HNO_3 with high methoxy pectin can be prepared in both water soluble and water insoluble forms. Flexibility can be controlled by added glycerol and the films can be filled with starch. The films had good dynamic mechanical properties. These films can be prepared from renewable agriculture sources and are biodegradable.

Acknowledgments

I am grateful to Mr. Trung Ly for his technical assistance. Drs. David Coffin and Marshall Fishman are gratefully acknowledged for their helpful discussions. Mention of brand or firm names does not constitute an endorsement by U. S. Department of Agriculture above others of a similar nature not mentioned.

Literature Cited

1. Coffin, D. R.; Fishman, M. L. *J. Agric. Food Chem.* **1993**, *41*, 1192-1197.
2. Coffin, D. R.; Fishman, M. L. *J. Appl. Polym. Sci.* **1994**, *54*, 1311-1320.
3. Hoagland, P. D.; Parris, N. *J. Agric. Food Chem.* **1996**, *44*, 1915-1919.
4. Hosokawa, J., Nishiyama, M., Yoshihara, K., Kubo, T. *Ind. Eng. Chem. Res.* **1990**, *29*, 800-805.
5. Nishiyama, M. *Gekkan Fudo Kemikaru* **1993**, *9*, 98-105.

6. El Ghaouth, A.; Arul, J.; Ponnampalam, R.; Boulet, M. *J. Food Process. Preserv.* **1991**, *15*, 359-368.

7. Chen, R. H.; Lin, J. H.; Yang, M. H. *Carbohydr. Polym.* **1994**, *24*, 41-46.

8. Yang, T.; Zall, R. R. *J. Food Sci.* **1984**, *49*, 91-93.

9. Uragami, T.; Matsuda, T.; Okuno, H.; Miyata, T. *J. Membrane Sci.* **1994**, *88*, 243-251.

10. Mireles, C.; Martino, M.; Bouzas, J.; Torres, J. A. In Advances in Chitin and Chitosan; Brine, C. J.; Sanford, P. A.; Zikakis, J. P., Eds.; Elsevier: London, 1992, pp. 506-515.

11. Takahashi, T.; Takayama, K.; Machida, Y.; Nagai, T. *Int. J. Pharm.* **1990**, *61*, 35-41.

12. Muzzarelli, R. A. A. *Chitin*, Pergamon Press, New York, 1977, pp. 255-265.

13. Wu, L. G.; Zhu, C. L.; Liu, M. *J. Membrane Sci.* **1994**, *90*, 199-205.

14. Amiji, M. M. *Biomaterials* **1995**, *16*, 593-599.

15. Unpublished observation.

16. Samuels, R. J. *J. Polym. Sci. Polym. Phys. Ed.* **1981**, *19*, 1081-1105.

17. Saitô, H., Tabeta, R., Ogawa, K. *Macromolecules* **1987**, *20*, 2424-2430.

18. Demarger-Andre, S.; Domard, A. *Carbohydr. Polym.* **1994**, *23*, 211-219.

EMULSION

Chapter 13

Lipid–Protein Interaction at an Emulsified Oil Surface: Protein Structures and Their Roles in Lipid Binding

M. Shimizu[1] and M. Saito[2]

[1]Department of Applied Biological Chemistry, Division of Agricultural Life Sciences, University of Tokyo, Bunkyo-ku, Tokyo 113, Japan
[2]National Food Research Institute, Ministry of Agriculture, Forestry and Fisheries, Tsukuba, Ibaraki 305, Japan

Adsorption of proteins on emulsified oil surfaces is an important step in emulsification by proteins. Although the adsorption is mainly caused by hydrophobic interaction between protein molecules and oil, the molecular structure of the oil-binding site in adsorbed proteins is still obscure. We have applied various biochemical and immunochemical methods to reveal the protein structure at an emulsified oil surface. The results obtained by those studies using αs1-casein, β-lactoglobulin, serum albumin and lactophorin suggested that the amphiphilic helix structure present in these protein molecules is likely to play a role in lipid-protein interaction at the surface. The importance of the amphiphilic helix was also evaluated using synthetic peptides with different amphiphilic structure.

When oil is emulsified in a protein solution, protein molecules are adsorbed at the dispersed oil surface and stabilize the emulsion. During the adsorption process, protein molecules are considered to be unfolded, exposing their hydrophobic region to the oil surface. However, there is only limited information on the mode of hydrophobic interaction between proteins and emulsified oil surfaces, particularly at a molecular level. Many questions as follows, for example, remain to be answered; How does a water-soluble hydrophilic protein express the hydrophobic property that is necessary for strong interaction with lipids? Does the adsorptivity of a protein molecule depend only on the content of hydrophobic amino acids? Is the cluster of hydrophobic amino acids necessary for the interaction with lipids? The significance of the amphiphilic α-helix as a structural factor for the oil-binding site is discussed in this paper.

Lipid-protein Interaction in Biosystems

There are several examples of lipid-protein interaction in biosystems. Among them, lipid-protein interactions in cell membranes and serum lipoproteins have been particularly well-studied.

Cell membrane integral proteins are known to have one or more transmembrane domains. These domains are composed mainly of hydrophobic amino acids, forming highly hydrophobic α-helices, by which they are buried in the lipid bilayer (*1*). On the other hand, some of the apolipoproteins, such as apo-C-I in a high density lipoprotein, interact with lipids through the hydrophobic region exposed on the surface of the protein molecule. Such an apolipoprotein is known to have an amphiphilic α-helix on the surface of its molecule. The hydrophobic region exposed on one side of the α-helix is believed to play an important role in this lipid-protein interaction (*2*).

Although these types of lipid-protein interaction could be specific in biosystems, it would be possible that similar interactions also participate in the lipid-protein interaction in artificial emulsions. The protein adsorption on an emulsified oil surface through the amphiphilic α-helix as observed in apo-C-I is thought to be most feasible.

Presence of Amphiphilic α-Helices in the Proteins with Good Emulsifying Properties

We have studied many milk proteins in terms of the emulsifying and oil-binding properties. From the results obtained in these studies, some interesting aspects of the oil-binding sites in the protein molecules have appeared as follows.

αs1-Casein. αs1-Casein is known to have a unique amphiphilic structure in its sequence (*3*). The regions rich in hydrophobic amino acids are present in its sequence, which are thought to be important when caseins interact to form micelles in milk (*4*). In order to elucidate the contribution of these hydrophobic regions to the oil-binding, the cleavability of αs1-casein in an aqueous phase and that on the emulsified oil surface were compared using trypsin, chymotrypsin and pepsin as proteases (*5*). Some of the peptide bonds were observed to become unsusceptible to the proteases after emulsification, suggesting that the enzymes were not accessible to those peptide bonds when αs1-casein was adsorbed at the oil surface. The portions near these peptide bonds were therefore presumed to be oil-binding sites. From the experimental data, the presence of nine oil-binding sites in αs1-casein was suggested. Among these nine presumed oil binding sites, the residues 16-25 seemed to be the major ones, because the removal of the N-terminal 23 residues by limited cleavage of αs1-casein clearly diminished the adsorptivity of this protein, whereas the removal of other regions such as the C-terminal 54 residues did not diminish the adsorptivity (*6*). We also found that the N-terminal peptide of 23 residues, RPKHPIKHQGLPQEVLNENLLRF, was surface-active in itself (*7,8*). The surface-active property of this peptide is probably due to the amphiphilicity based on the positively-charged N-terminal and less-polar C-terminal regions in this peptide (*8*). In addition to this sequential amphiphilicity, we found that an amphiphilic structure appeared when the region corresponding to the residues 12-23 formed an α-helix (Figure 1a). It would be possible that this amphiphilic helix participated in the oil-binding properties of αs1-casein and its N-terminal peptide.

β-Lactoglobulin. β-lactoglobulin from bovine milk is known to have good emulsifying properties. Kaminogawa and his coworkers (*9*) have succeeded in eliciting several monoclonal antibodies (MAbs) recognizing the conformation of this protein. As shown in Table I, MAbs 21B3 and 31A4 reacted with denatured β-lactoglobulin with much higher affinity, while MAbs 61B4 and 62A6 reacted with the native β-lactoglobulin with higher affinity. These properties were observed to be highly useful for analyzing the conformational changes in this protein during the process of denaturation caused by heating or denaturants. Local

structural changes occurring in the β-lactoglobulin molecule during denaturation
(10) and renaturation processes (11) were successfully detected with high
sensitivity using these MAbs. Therefore, we applied these MAbs to the structural
analyses of β-lactoglobulin at an emulsified oil surface (12).

The reactivity of β-lactoglobulin with MAbs 21B3 and 31A4 increased about
100 times when β-lactoglobulin was adsorbed on an oil surface (Table II),
suggesting that the structure of the epitopes for these MAbs (the epitopes are
located around residues 15-29 for MAb 21B3 and 8-19 for 31A4) was drastically

**Table I. Reactivity of Monoclonal Antibodies (MAbs) with
Native and Denatured β-lactoglobulin (β-Lg)**

MAb	Epitopes	Binding Affinity
61B4	Thr^{125}-Lys^{135}	Native β-Lg >> RCM-β-Lg[a]
62A6	Close to the epitope for 61B4	Native β-Lg >> RCM-β-Lg
21B3	Val^{15}-Ile^{29}	Native β-Lg << RCM-β-Lg
31A4	Lys^{8}-Trp^{19}	Native β-Lg << RCM-β-Lg

[a]Denatured β-Lg was prepared by reduced carboxymethylation
(RCM).

**Table II. Binding Affinity (Kas) of MAbs with Native β-Lg
and β-Lg Adsorbed at an Emulsified Oil Surface**

| MAb | Kas with | | Conformation of the epitope |
	Native β-Lg	Adsorbed β-Lg	
61B4	1.5×10^{8}	1.2×10^{8}	unchanged
62A6	2.0×10^{8}	4.7×10^{7}	slightly changed
21B3	1.2×10^{6}	1.6×10^{8}	drastically changed
31A4	6.0×10^{5}	5.5×10^{7}	drastically changed

changed at the oil surface. On the other hand, the reactivity of β-lactoglobulin
with MAbs 61B4 and 62A6 did not change or only slightly changed on
emulsification (Table II), suggesting that the structure of the epitopes for these
MAbs (the epitopes are located around residues 125-135 for both of MAbs 61B4
and 62A6) was not changed very much. These results suggested that the surface
denaturation of β-lactoglobulin caused by emulsification was not accompanied by

the complete unfolding of the whole molecule, some portion such as the residues 125-135 and around retaining its native structure.

Interestingly, it was found that the portion including the residues 125-135 was able to form an amphiphilic helix (Figure 1b). Papiz et al. (*13*) have demonstrated using X-ray diffraction analysis that this portion was present as an α-helix in the native β-lactoglobulin molecule. This amphiphilic helix may have been kept intact even after the surface denaturation, although the possibility that the amphiphilic helix reformation from unfolded molecules was induced at the oil/water interface (*14*) cannot be ruled out.

Bovine Serum Albumin. Bovine serum albumin (BSA) is also known to express good emulsifying and oil-binding properties. We have found that the third domain of BSA played an important role in its adsorption on an emulsified oil surface, because the third domain (residues 377-582) produced by limited tryptic hydrolysis was adsorbed more preferentially on the emulsified oil surface compared with other peptides including the intact BSA (*15*). The hydrophobicity of this peptide estimated from the elution time in hydrophobic chromatography indicated that this peptide was the most hydrophobic among the peptides in the hydrolyzate (*15*).

In this domain, we found two sequences (residues 383-396 and 541-555) which could provide a typical amphiphilic α-helix structure as shown in Figures 1c and 1d. They may contribute to the strong oil-binding properties of this domain.

Lactophorin. The proteose-peptone is the heat-stable protein fraction in bovine milk whey and it shows good emulsifying properties. We found that the emulsifying activity of the proteose-peptone was much higher than that of the whole whey protein (*16*). The emulsion prepared using the proteose-peptone fraction was more stable than that using the whole whey protein over a wide range of pH (pH 3 - 9). The emulsion was also highly heat-stable, giving no oiling-off even after heating at 100°C. This suggests that the proteose-peptone components were strongly adsorbed on the oil surface covering the dispersed oil droplet surface. Formation of a stable adsorbed layer by the proteose-peptone components was also suggested by the fact that lipase-catalyzed hydrolysis of the emulsified lipid was strongly inhibited in the emulsion stabilized by the proteose-peptone (*16,17*).

In order to characterize the adsorbed proteins, the emulsified oil droplets (cream) were washed with aqueous buffer and the adsorbed materials were recovered from the oil surface. The reverse-phase high-performace liquid chromatography (HPLC) pattern for the adsorbed materials is shown in Figure 2. The major components of the proteose-peptone are known to be casein-derived peptide fragments, such as β-casein residues 1-28 and 29-105(or 107). However, these peptides (corresponding to peak C in Figure 2) were scarcely adsorbed on the oil surface. Instead, substances corresponding to peaks A and B in the HPLC chromatogram were shown to be concentrated at the emulsified oil surface (Figure 2). By sodium dodecylsulfate polyacrylamide gel electrophoresis (SDS-PAGE) followed by the periodic acid-Schiff staining for carbohydrate, peaks A and B were shown to contain major glycoproteins of the proteose-peptone fraction. Kanno (*18,19*) demonstrated that the major glycoprotein fraction of the proteose-peptone (called Component 3) consisted of polymorphic glycoproteins. He isolated and characterized the major glycoprotein of Component 3 and termed it lactophorin (*18,19*). Sorensen and Petersen (*20*) reported the primary structure of this glycoprotein. By N-terminal sequence analysis of peaks A and B on HPLC, we identified peak B as lactophorin. Although the other component, peak A, has not yet been characterized, the peak A glycoprotein would be a related molecule,

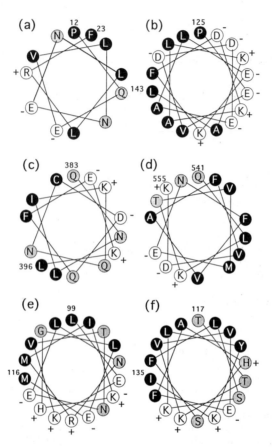

Figure 1. Helical Wheel Structures of αs1-casein residues 12-23 (a), β-lactoglobulin residues 125-143 (b), bovine serum albumin residues 383-396 (c) and 541-555 (d), lactophorin residues 99-116 (e) and 117-135 (f). Hydrophobic amino acids are shown by closed circles, hydrophilic charged amino acids by open circles, and hydrophilic uncharged amino acids by shaded circles.

considering its electrophoretic mobility which was very close to that of peak B
(*16*).

The primary structure of lactophorin reported by Sorensen and Petersen (*20*)
is shown in Figure 3. In spite of its high affinity for the oil surface, lactophorin
does not have a distinctive hydrophobic amino acid cluster in its sequence.
Hydrophobic amino acids are distributed rather homogeneously throughout the
whole sequence. Furthermore, there are three glycosylation and five
phosphorylation sites which make this molecule more hydrophilic. So, how does
this glycoprotein interact with lipids on the emulsified oil surface with such high
affinity? Interestingly, we found that the C-terminal region of this glycoprotein
had the potential to become highly hydrophobic. Figures 1e and 1f show the
helical wheel structures of the C-terminal regions of lactophorin, residues 99-116
and 117-135, respectively. By forming the α-helix, typical amphiphilicity is
shown to appear in these regions. The secondary structure prediction
demonstrated that the most probable conformation of this region was the α-helix.

In order to prove the importance of this C-terminal region in the oil binding,
we tried to prepare lactophorin lacking the C-terminal residues. Because
lactophorin has only two methionine residues located at residues 101 and 108, it
would be possible to remove the C-terminal portion from the lactophorin molecule
by CNBr treatment. Lactophorin was therefore chromatographically purified from
the proteose-peptone fraction and then treated with 1% CNBr in 70% formic acid.
The preliminary results indicated that the emulsifying activity of lactophorin was
markedly reduced by CNBr treatment. The adsorption of lactophorin on the oil
surface was also decreased (Table III). These results suggest that the C-terminal
region plays an important role in emulsification and oil-binding, although more
detailed analyses are needed.

**Table III. The Effect of CNBr Treatment on the Emulsifying
Properties of lactophorin**

	EAI[a] *(m2/g)*	*Amount of Adsorbed Protein (mg/g oil)*
Lactophorin	137.1	2.23
CNBr-treated lactophorin	84.4	1.67

[a]Emulsifying activity index.

Approaches Using Synthetic Peptides

Although there is no direct evidence that the peptide regions in Figure 1 participate
in the oil binding in emulsions and also these portions are present as α-helices at
the surface, the above examples suggest that the amphiphilic α-helix plays a role in
emulsification. In order to reveal the contribution of amphiphilic helices to protein
adsorption at an emulsified oil surface, we have started some model experiments
using synthetic peptides with simple sequences (*21*). Three peptides of 16
residues with the same amino acid composition, 8 leucine and 8 glutamic acids,
but with different sequences were designed and synthesized.

As shown in Figure 4, Peptide H was designed to form an amphiphilic α-helix.
The other peptides, S and R, were designed not to show such amphiphilicity when

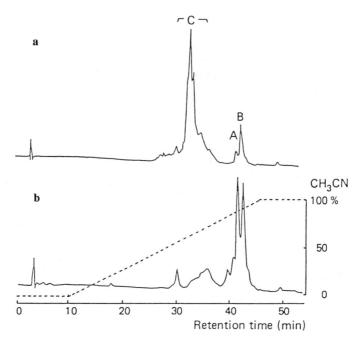

Figure 2. C₁₈-High Performance Liquid Chromatographic Patterns of the Proteose-peptone (a) and the Adsorbed Material Recovered from the Surface of the Emulsified Oil Droplets (b).

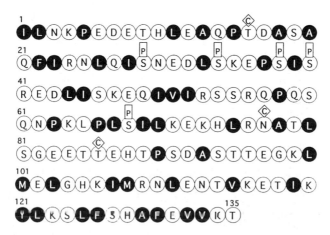

Figure 3. The Primary Structure of Lactophorin (20).
Hydrophobic amino acids are shown by closed circles. The phosphorylated amino acid residues are shown by "P" and glycosylated residues by "C".

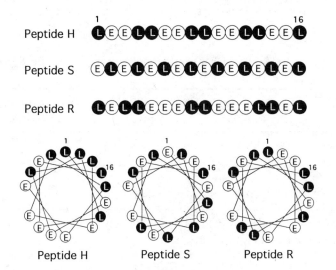

Figure 4. The Sequences and Helical Wheel Structures of the Three Synthetic Peptides.

they form an α-helix. By circular dichroism (CD) analysis, Peptide H was observed to form an α-helix in an aqueous solvent at pH 5.5 or in a helix solvent, trifluoroethanol (21). (In spite of these results, it must be emphasized that the results obtained by CD analysis in a homogeneous solution may not be very meaningful for estimating the protein structure at an oil/water interface, because the presence of a hydrophilic/hydrophobic interface could affect the conformation of proteins, sometimes inducing a new conformation (22)).

The surface pressure of the three peptides in a diluted aqueous solution at pH 5.5 is shown in Table IV. Although these peptides have the same hydrophobic amino acid content, their properties were markedly different. Peptide H showed the highest surface pressure, suggesting that this peptide was the most surface active among the three at pH 5.5. The emulsifying activity and adsorptive properties of the peptides at pH 5.5 were also the highest for Peptide H (Table IV). Although the real structure of the peptides at the oil/water interface was not revealed, these results strongly suggest that the amphiphilicity depending on the secondary structure is important in emulsification and adsorption on the oil surface.

Table IV. Surface and Emulsifying Properties of the Synthetic Peptides at pH 5.5

Peptide	Surface Pressure[a] (mN/m)	EAI[b] (m^2/g)	Amount of Adsorbed Peptide (mg/g oil)
H	29.6	455	0.43
S	15.2	263	0.18
R	10.0	47	0.05

[a]Measured at 1 min after the surface became still.
[b]Emulsifying activity index.

Concluding remarks.

From various experimental data, it is becoming clear that the surface denaturation of proteins is not always accompanied by the complete unfolding of the molecule. Some portion of the protein molecule may have a secondary structure such as an α-helix. In particular, the amphiphilic α-helix possibly plays an important role in lipid-protein interaction at the emulsified oil surface, as suggested from several examples described above.

In this text, we did not mention the β-sheet structure. Although the participation of the β-sheet structure in oil-binding has not yet been fully analyzed, the amphiphilic β-sheet structure, as well as the amphiphilic α-helix, may play some role in lipid-protein interaction. For example, the synthetic peptide S forms a β-sheet, one side of which is hydrophobic and the other side hydrophilic, in an aqueous solution at pH 7.0 (21). Under this condition, Peptide S showed a remarkably high emulsifying activity (data not shown), suggesting that the amphiphilic β-sheet was also important in emulsification.

Because of the lack of good analytical methods for estimating the protein conformation at an emulsified oil surface, it is so far difficult to provide direct evidence for the contribution of amphiphilic helices or sheets to protein-lipid interaction in an emulsion. Development of new analytical methods, including physicochemical and biochemical ones, and their application to this field of research will be necessary for the correct understanding of this interaction.

Literature Cited.

1. Jennings, M. L. *Ann. Rev. Biochem.* **1989**, *58,* 999-1028.
2. Jackson, R. J.; Morrisett, J. D.; Gotto, A. M. Jr. *Physiol. Rev.* **1976**, *56,* 259-376.
3. Mercier, J. -C.; Grosclaude, F.; Ribadeau-dumas, B. *Eur. J. Biochem.* **1971**, *23,* 41-51.
4. Kaminogawa, S.; Yamauchi, K.; Yoon, C. -H. *J. Dairy Sci.* **1980**, *63,* 223-227.
5. Shimizu, M.; Ametani, A.; Kaminogawa, S.; Yamauchi, K. *Biochim. Biophys. Acta* **1986**, *869,* 259-264.
6. Shimizu, M.; Takahashi, T.; Kaminogawa, S.; Yamauchi, K. *J. Agric. Food Chem.* **1983**, *31,* 1214-1218.
7. Shimizu, M.; Lee, S. W.; Kaminogawa, S.; Yamauchi, K. *J. Food Sci.* **1984**, *49,* 1117-1120.
8. Shimizu, M.; Lee, S. W.; Kaminogawa, S.; Yamauchi, K. *J. Food Sci.* **1986**, *51,* 1248-1252.
9. Kaminogawa, S.; Hattori, M.; Ando, O.; Kurisaki, J.; Yamauchi, K. *Agric. Biol. Chem.* **1987**, *51,* 797-802.
10. Kaminogawa, S.; Shimizu, M.; Ametani, A.; Hattori, M.; Ando, O.; Hachimira, S.; Nakamura, Y.; Totsuka, M.; Yamauchi, K. *Biochim. Biophys. Acta* **1989**, *998,* 50-56.
11. Hattori, M.; Ametani, A.; Shimizu, M.; Kaminogawa, S. *J. Biol. Chem.* **1993**, *268,* 22414-22419.
12. Shimizu, M.; Ametani, A.; Kaminogawa, S.; Yamauchi, K. *Protein and Fat Globule Modifications by Heat Treatment, Homogenization and other Technological Means for High Quality Food,* International Dairy Federation: Brussels, 1993; pp.362-367.
13. Papiz, M. Z.; Sawyer, L.; Eliopoulos, E. E.; North, A. C. T.; Finlay, J. B. C. *Nature.* **1986**, *324,* 383-385.
14. Brown, E. M. *J. Dairy Sci.* **1984**, *67,* 713-722.
15. Saito, M.; Monma, M.; Chikuni, K.; Shimizu, M. *Biosci. Biotech. Biochem.* **1993**, *57,* 952-956.
16. Shimizu, M.; Saito, M.; Yamauchi, K. *Milchwissenschaft* **1989**, *44,* 497-500.
17. Courthaudon, J. -L.; Girardet, J. -M.; Chapel, C.; Lorient, D.; Lindon, G. In *Food Macromolecules and Colloids*; Dickinson, E. and Lorient, D. Eds.; The Royal Society of Chemistry, Cambridge, 1995; pp.58-70.
18. Kanno, C. *J. Dairy Sci.* **1989**, *72,* 883-891.
19. Kanno, C. *J. Dairy Sci.* **1989**, *72,* 1732-1739.
20. Sorensen, E. S.; Petersen, T. E. *J. Dairy Res.* **1993**, *60,* 535-542.
21. Saito, M.; Ogasawara, M.; Chikuni, K.; Shimizu, M. *Biosci. Biotech. Biochem.* **1995**, *59,* 388-392.
22. DeGrado, W. F. *Nature.* **1993**, *365,* 488-489.

Chapter 14

Characteristics of the Products of Limited Proteolysis of β-Lactoglobulin

Harold E. Swaisgood, Xiaolin L. Huang, and George L. Catignani

Southeast Dairy Foods Research Center, Department of Food Science, North Carolina State University, Raleigh, NC 27695–7624

Application of limited proteolysis to β-lactoglobulin (β-Lg) using immobilized trypsin allows precise control of the extent of reaction resulting in the release of fragments of the central core β-barrel domain. These 4- or 5-stranded β-barrel fragments have β-structure that reversibly unfolds in a transition centered at 3.88 M urea. Conditions yielding 35-55% disappearance of intact protein produced 13-15% of the protein as domain fragments, the major fragment being β-Lg(f41-100 + 149-162) which is linked by a disulfide bond. Membrane fractionation of a limited proteolysate of β-Lg yielded a protein ingredient, consisting of domain fragments, in the fraction passing through a 30 kDa retention membrane and retained by a 3 kDa retention membrane. The Emulsifying Activity Index (EAI) of this fraction was two-fold larger than that of the intact protein throughout the pH range 3 to 9 and also was greater than that for egg white protein. Furthermore, the emulsified oil droplets were smaller when the domain fraction was used and the emulsions appeared to be more stable. Treatment of whey protein isolate (WPI) under conditions that should yield 10-15% of the domain fragments produced a protein ingredient that gave particulate gels, whereas the untreated WPI gave fine-stranded gels.

The functional properties of a protein are determined by its structure, structural stability and flexibility, its surface topology and the chemical characteristics of its surface. Therefore, any changes in the primary structure that affect these characteristics will also affect its functional properties. Limited proteolysis obviously alters the primary structure thus changing the stability and flexibility of the resulting oligopeptides, often exposing new protein surface, and altering the chemical properties of the surface. Many proteins are composites of multiple domains (1,2), where a domain is a region of tertiary structure with many interactions between residues within

0097–6156/96/0650–0166$15.00/0

the structure but few interactions with residues outside the domain. Hence, a domain possesses a compact, rather stable structural motif that is resistant to proteolysis; whereas, the protein chain linking domains is flexible and thus susceptible to proteolysis (2,3). Hence, proteinases are commonly used to probe the structure of proteins for the presence of multiple domains.

There are a number of important advantages derived by using immobilized proteinases rather than soluble enzyme for limited proteolysis, especially for the production of functional peptides (2-4). The most significant advantages are: 1) the extent of reaction is easily and precisely controlled so the proteolysis is reproducibly limited, 2) an enzyme denaturation step is not required to stop the reaction which would cause destruction of functional structure of the oligopeptides produced, and 3) enzyme autolysis does not occur so autolysis products are not present and the enzyme activity remains constant.

The x-ray crystal structure of the whey protein β-lactoglobulin (β-Lg) reveals a central structural motif consisting of an eight-stranded anti-parallel β-barrel with another β-strand and an α-helix lying on its surface (5,6). The β-barrel forms a calyx which is the most likely site of the high affinity binding of retinol (5,7,8) or the retinyl moiety (9,10). We have shown that limited proteolysis of β-Lg with immobilized trypsin removes the α-helix and some of the β-strands yielding, as major peptides, fractions of the central core calyx structure (7,11).

Limited Proteolysis of Whey Protein

Immobilization of Trypsin. The enzyme was covalently immobilized on succinamidopropyl-glass or succinamidopropyl-Celite using either the simultaneous or the sequential activation/immobilization procedure with water-soluble carbodiimide to activate carboxyl groups (12,13). Using *p*-tosyl-L-arginine methyl ester as substrate, either immobilization procedure typically yields a biocatalyst with a specific activity of 20-40 μmol/min/mL of beads. Titration of active sites with *p*-nitrophenyl-*p'*-quanidinobenzoate indicated that more than 20% of the immobilized trypsin molecules had competent active sites. For most of the studies described here, a 100-mL fixed-bed bioreactor of trypsin-Celite was used. This bioreactor has been used for more than one year without significant loss of activity.

Conditions for Limited Proteolysis. The rate of proteolysis depends upon the temperature, pH and the ratio of activity to substrate protein. To optimize the concentration of functional domain peptides, we have observed that conditions that result in 35-55% disappearance of the intact β-Lg appear to be optimal. We have used conditions ranging from 4°C to 24°C at pH 7.6 and 8.0 and activity/substrate ratios of 500 to 2400 U/g with various time periods of treatment to yield the desired degree of proteolysis.

Membrane Fractionation of the Limited Proteolysate. A membrane fractionation process was developed to obtain the oligopeptide fraction of the limited proteolysate for potential use as a value-added protein ingredient (14). The proteolysate was fractionated with a 30-kDa membrane (YM30, Amicon, Inc., Beverly, MA) and the resulting permeate was fractionated with a 3-kDa membrane (YM3) to yield the oligopeptides in the retentate (Figure 1). For these studies, a 100-mL bioreactor of trypsin-Celite was used to treat 2g of β-Lg at pH 7.6 for 60 min at 24°C. The β-Lg was prepared as described by Huang *et al.*(*12*). Because β-Lg exists as a 36-kDa dimer under these conditions, the intact protein is retained by the 30-kDa membrane. The functional oligopeptide fraction retained by the 3-kDa membrane contains 50% of an 8.4-kDa oligopeptide which is probably β-Lg(f41-100 + 149-162) (see Figure 2). This fraction also contains small amounts of a large oligopeptide, which may represent the dimer with a small peptide removed, and several other fragments. Most of the very small peptides are removed in the permeate of the 3-kDa membrane.

Characteristics of β-Lg(f41-100 + 149-162). This oligopeptide has native-like secondary and tertiary structure represented by 5 strands of the calyx motif of the intact protein and undergoes reversible unfolding in urea (*7*). However, the structure of this domain fragment is considerably less stable than that of the intact protein as indicated by measurements of their structural transitions in urea by CD spectroscopy and their thermal denaturation by DSC (Table I). The pH-dependent binding of the domain fragment to retinyl-Celite was similar to that of the intact protein, which also supports the conclusion that the oligopeptide exhibits the calyx motif (*7*).

Table I. Summary of Thermodynamic Data[1]

Protein	CD Spectra		DSC
	ΔG_D (kcal/mol)	[Urea]$_{1/2}$	T_m (°C)
β-Lg A & B	11.4	5.51	81.1
Domain fragments	6.82	3.88	56.9

[1] Taken from data presented by Chen *et al.*(*7*).

Emulsification Characteristics of β-Lg Domain Peptides

Emulsifying Properties of the Oligopeptide Fraction. The emulsifying activity index (EAI)(*15*) for the oligopeptide fraction is compared with that for the intact protein and for egg white protein at several pHs in the range of pH 3 to 9 in Figure 3. The emulsifying activity of both intact β-Lg and the oligopeptide fraction increased with increasing pH, possibly due to increasing flexibility of the structure; whereas, the activity of egg white protein decreased with increasing pH. More importantly, the activity for the oligopeptide fraction was two-fold greater than that of the intact protein throughout the pH range and became increasingly greater than that of egg white protein as the pH increased.

The stabilities of peanut oil emulsions prepared with intact β-Lg and the oligopeptide fraction were also compared. Using a turbidometric method based on turbidity measurements at 500 nm (*16*), similar stabilities were indicated by a similar dependence of the relative turbidity on time. However, after standing the two emulsions were visually distinct (Figure 4). The emulsion formed with intact protein

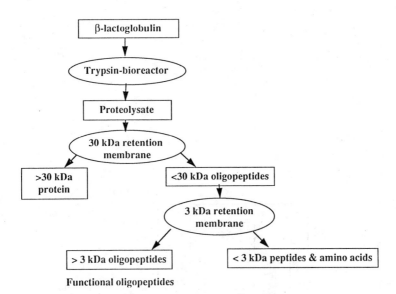

Figure 1. Schematic illustration of the bioprocess for obtaining functional oligopeptides by limited proteolysis.

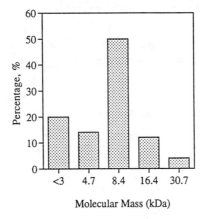

Figure 2. Peptide size distribution of the oligopeptide fraction as determined by size-exclusion chromatography in 20 mM Tris-HCl, pH 7.0, containing 15 mM NaCl and 0.02% NaN$_3$.

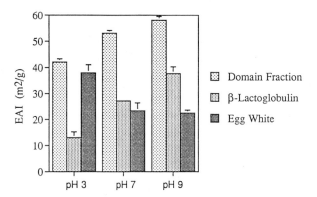

Figure 3. Comparison of the Emulsifying Activity Index (EAI) of the oligopeptide domain fraction, β-lactoglobulin and egg white protein in the pH range 3 to 9.

of about 500 U/g protein and resulted in depletion of 32% of the intact β-Lg during the 90-min treatment (Table II). Of course, limited proteolysis of the other whey proteins also occurred; however, as would be expected from the relative stabilities of β-Lg and α-lactalbumin at pH 7.6, the depletion of the latter protein was considerably less.

Table II. Composition of Whey Protein Isolate and Percent Depletion of Intact Protein Resulting from Limited Proteolysis[1]

Protein	Composition[2] (g/100g)	Depletion of intact protein (%)
β-Lactoglobulin	64.3	32.4
α-Lactalbumin	18.6	15.2
Bovine serum albumin	7.6	22.4
Immunoglobulin	9.5	39

[1]Limited proteolysis of WPI (4g) was at pH 7.6 in 10 mM ammonium acetate for 90 min at 24°C with the 100-mL immobilized trypsin bioreactor.
[2]Amounts of the individual proteins were determined from size-exclusion HPLC using a Superose column and 20 mM Tris-HCl, pH 7.0, containing 0.15 M NaCl as the elution buffer.

Comparison of the Gel Properties.   Thermal gels of WPI and enzyme-treated WPI were formed by heating 10% solutions of these proteins in 50 mM TES buffer containing 50 mM NaCl, pH 7.0, to 80°C, holding at that temperature, followed by cooling to 24°C. Some results of texture profile analysis are shown in Figure 5. The enzyme-treated WPI did not fracture in the analysis, exhibited substantially more cohesiveness, gumminess and chewiness. Furthermore, the enzyme-treated WPI gels were opaque, whereas WPI gels were more transparent. Thus, under these conditions of gelation, enzyme-treated WPI appeared to form particulate gels; whereas, WPI formed fine-stranded gels. Light microscopy of the gels also supported this conclusion as shown by the photographs shown in Figure 6.

Gel structure is primarily determined by the rate of association of the unfolding protein chains. This rate depends upon the balance of attractive and repulsive forces; for example, hydrophobic, ion-pair and H-bonding between unfolded portions of the protein promote association, while electrostatic repulsion promotes dissociation. Of course, the rate also depends upon the concentration of unfolded structure which is determined by the structural stability and the total protein concentration. Higher rates of association favor particulate type gel structures. If the protein has been previously denatured and therefore aggregated, it likely can not further unfold during thermal treatment and participate in formation of gel structure. Thus, many variables affect the gelling characteristics of WPI, including protein concentration and the amount of previous denaturation, pH, ionic strength and the type of ions present.

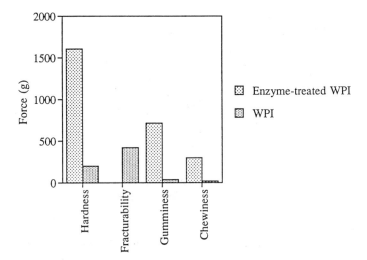

Figure 5. Comparison of the texture profile analysis data for protein gels prepared from WPI and enzyme-treated WPI. Gels were prepared with 10% protein in 50 mM TES buffer, pH 7.0, containing 50 mM NaCl by heating at 80°C for 30 min and cooling to 24°C in 30 min.

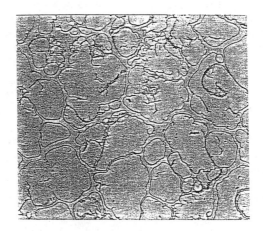

WPI: a typical fine-stranded gel

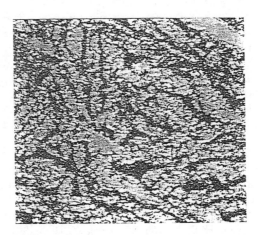

Enzyme-treated WPI: a typical particulate gel

Figure 6. Photographs of light microscopic patterns for 10% WPI and enzyme-treated WPI gels taken at 100-fold magnification.

Previous studies have shown that particulate gels can be formed between pH 4.5 and 6.0 (*28*) or by adding Ca^{2+} (*28,29*). At pH 7, fine-stranded transparent gels are formed with 10-15% WPI at low salt concentrations, but particulate gels can be formed by increasing NaCl or $CaCl_2$. Thus, ion concentrations have a dramatic effect on the gelling properties of whey proteins (*30*) and divalent cations in particular may have some special effects (*31*). Limited proteolysis transforms the gelling characteristics of WPI so that particulate gels are formed at pH 7 in low salt solutions rather than fine-stranded transparent gels. This effect may result from the decreased structural stability of the released oligopeptide central core domain of β-Lg (*7,32*). Ju *et al.* (*33*) did not observe gelation of trypsin-treated WPI at pH 7; however, their salt concentrations were very low because the WPI was dispersed in water and the degree of proteolysis was most likely greater.

Our results demonstrate that limited proteolysis can be used to change the gelling properties of WPI allowing particulate gels to be formed at pH 7 under low salt conditions. The importance of using immobilized trypsin for this purpose must be emphasized because heat treatment to inactivate soluble enzyme would denature the functional oligopeptides, thus destroying their gelling properties.

LITERATURE CITED

1. Neurath, H. *Chem. Scr.* **1986**, *26B*, 221-229.
2. Swaisgood, H. E.; Chen, S. X.; Oh, S.; Catignani, G. L. In *Protein Structure-Function Relationships in Foods*; Yada, R. Y.; Jackson, R. L.; Smith, J. L., Eds.; Blackie Academic & Professional Publishers: London, 1994, pp 43-61.
3. Swaisgood, H. E. and Catignani, G. L. *Methods Enzymol.* **1987**, *135*, 596-604.
4. Swaisgood, H. E. In *Food Enzymology*, Vol 2, Fox, P. F., Ed.; Elsevier Applied Science Publishers: London, 1991, pp 309-341.
5. Papiz, M. Z.; Sawyer, L.; Eliopoulos, E. E.; North, A. C. T.; Findlay, J. B. C.; Sivaprasadarao, R.; Jones, T. A.; Newcomer, M. E.; Kraulis, P. J. T *Nature* **1986**, *324*, 383-385.
6. Monaco, H. L.; Zanotti, G.; Spadon, P.; Bolognesi, M.; Sawyer, L.; Eliopoulos, E. E. *J. Mol. Biol.* **1987**, *197*, 695-706.
7. Chen, S. X.; Hardin, C. C.; Swaisgood, H. E. *J. Protein Chem.* **1993**, *12*, 613-625.
8. Cho, Y.; Batt, C. A.; Sawyer, L. *J. Biol. Chem.*, **1994**, *269*, 11102-11107.
9. Jang, H.-D.; Swaisgood, H. E. *J. Dairy Sci.*, **1990**, *73*, 2067-2074.
10. Wang, Q.; Swaisgood, H. E. *J. Dairy Sci.*, **1993**, *76*, 1895-1901.
11. Huang, X. L.; Catignani, G. L.; Swaisgood, H. E. *J. Agric. Food Chem.*, **1994**, *42*, 1281-1284.
12. Huang, X. L.; Catignani, G. L.; Swaisgood, H. E. *J. Agric. Food Chem.*, **1994**, *42*, 1276-1279.
13. Janolino, V. G.; Swaisgood, H. E. *Biotechnol. Bioengr.*, **1982**, *24*, 1069-1080.
14. Huang, X. L.; Wang, Q.; Catignani, G. L.; Swaisgood, H. E. *J. Dairy Sci.*, **1995**, *78(Suppl.1)*, Abstr. D149.
15. Pearce, K. N.; Kinsella, J. E. *J. Agric. Food Chem.* **1978**, *26*, 716-723.

16. Britten, M.; Giroux, H. J. *J. Food Sci.* **1991**, *56*, 792-795.
17. Jost, R.; Monti, J. C. *Le Lait* **1982**, *62*, 521-530.
18. Lee, S. W.; Shimizu, M.; Kaminogawa, S.; Yamauchi, K. *Agric. Biol. Chem.* **1987**, *51*, 161-166.
19. Chaplin, L. C.; Andrews, A. T. *J. Dairy Res.* **1989**, *56*, 544.
20. Turgeon, S. L.; Gauthier, S. F.; Paquin, P. *J. Agric. Food Chem.* **1991**, *39*, 673-676.
21. Turgeon, S. L.; Gauthier, S. F.; Mollé, D.; Léonil, J. *J. Agric. Food Chem.* **1992**, *40*, 669-675.
22. Waniska, R. D.; Shetty, J. K.; Kinsella, J. E. *J. Agric. Food Chem.* **1981**, *29*, 826-831.
23. Phillips, M. C. *Food Technol.* **1981**, *35*, 50-57.
24. Shimizu, M.; Kamiya, T.; Yamauchi, K. *Agric. Biol. Chem.* **1981**, *45*, 2491-2496.
25. Kato, A.; Komatsu, K.; Fujimoto, K.; Kobayashi, K. *J. Agric. Food Chem.* **1985**, *33*, 931-934.
26. Das, K. P.; Kinsella, J. E. *J. Dispersion Sci. Technol.* **1989**, *10*, 77-102
27. Swaisgood, H. E. In *Developments in Dairy Chemistry-1*; Fox, P. F., Ed.; Applied Science Publishers: London, 1982; pp 1-59.
28. Tang, Q.; McCarthy, O. J.; Munro, P. A. *J. Dairy Res.* **1995**, *62*, 469-477.
29. Zirbel, F.; Kinsella, J. E. *Milchwissenschaft* **1988**, *43*, 691-694.
30. Kuhn, P. R.; Foegeding, E. A. *J. Agric. Food Chem.* **1991**, *39*, 1013-1016.
31. Foegeding, E. A.; Kuhn, P. R.; Hardin, C. C. *J. Agric. Food Chem.* **1992**, *40*, 2092-2097.
32. Chen, S. X.; Swaisgood, H. E.; Foegeding, E. A. *J. Agric Food Chem.* **1994**, *42*, 234-239.
33. Ju, Z. Y.; Otte, J.; Madsen, J. S.; Qvist, K. B. *J. Dairy Sci.* **1995**, *78*, 2119-2128.

Chapter 15

Effects of High Pressure on Protein–Polysaccharide Interactions

V. B. Galazka and D. A. Ledward

Department of Food Science and Technology, University of Reading, Whiteknights, Reading RG6 6AP, United Kingdom

High pressure treatment (800 MPa for 20 min) at a pre-pressure pH of 7.0, causes bovine serum albumin (BSA) and β-lactoglobulin to undergo unfolding and aggregation. The aggregation following treatment is mainly thought to be due to the formation of disulphide bridges. Differential scanning calorimetry confirms that high pressure treatment has a major effect on the structure of the proteins. Pressure treatment of mixtures of β-lactoglobulin + dextran sulphate (DS) yielded structures with virtually zero enthalpy when heated from 10 - 100 °C, but, BSA + DS mixtures had structures with a significant enthalpy of denaturation. This suggests that DS either enables the pressure denatured protein to regain some secondary structure or protects the protein against unfolding. The effect of high pressure on the structure of β-lactoglobulin significantly affects its emulsification properties.

Over 100 years ago Bert Hite (1899) reported the potential of high pressure as a means of preserving and processing of food (*1*). This work was undertaken because of dissatisfaction and constraints with the methods of sterilization, cooling and pasteurization of milk then available. During the next 15 years, Hite and co-workers (*2*) extended their studies and demonstrated that high pressure could be used to sterilize and preserve a wide range of commercial fruit and vegetable products. An attempt was also made to study the kinetics of the death of several micro-organisms. A contemporary of Hite, Bridgman (*3*), observed that egg-white proteins could be denatured under certain conditions, demonstrating that high pressure treatment in addition to killing certain micro-organisms, can alter the three-dimensional structure of proteins.

This technique, though, was not developed commercially because of the technical problems and costs associated with high pressure processing units and routine handling of materials at such pressures. However, these limitations are now seen to be surmountable and a great deal of research effort is now being devoted to

0097–6156/96/0650–0178$15.00/0

understanding the effect of high pressure on food ingredients and food (*4-5*). With the consumer demand for safe, minimally processed and additive free foods it is clear that high pressure technology has a great deal to offer the food industry.

In the past 6 years Japanese researchers, under the leadership of Professor Hayashi, became involved in the applications of high pressure food technology. A special unit was set up in 1989 at Kyoto University to bring together the Japanese Ministry of Agriculture, Forestry and Fisheries, 21 food and engineering companies and academic research scientists to promote the commercialisation of high pressure technology for use in the food industry. In the last 4 to 5 years a number of products, mainly fruit and vegetable based, have become commercially available. As a result of these successes, the Japanese are undoubtedly the world leaders in this technology. Major research initiatives are also underway in mainland Europe, the USA, and the UK. These projects are being supported by both Government and Industrial monies and this suggests that high pressure processed products will soon be available to European and American consumers (*4-5*).

An important aspect of high pressure treatment is that the pressure is felt uniformly and rapidly throughout the food, so that all the food irrespective of its size is treated evenly throughout. This is, in contrast to thermal processing, where a thermal gradient exists between the exterior and interior as heat is transferred by conduction. This may lead to over-cooking the exterior, if the centre is to be adequately processed. Another important characteristic of high pressure processing is that it keeps covalent bonds intact, and affects only non-covalent ones (i.e., hydrogen, ionic and hydrophobic bonds). This results in the retention of pigments and flavour, with little or no loss of vitamins.

High pressure thus offers the potential for food product preservation and modification by an environmentally clean process. The destruction of micro-organisms and inactivation of enzymes are two main objectives in food preservation whereas protein denaturation and subsequent aggregation can lead to textural changes, which may be beneficial to product quality.

Effects of High Pressure on Biological Materials

At the molecular level, the effect of high pressure on protein stability and reactivity is best understood in terms of the Le Chatelier principle, which states 'that when a constraint is placed on a system, the system will change so as to minimise the effects of the constraint'. When applying pressure a constraint is placed on volume, and hence reactions with a negative volume change ($-\Delta V$) will be accelerated, whereas those involving a positive volume change ($+\Delta V$) will be inhibited. Pressure also affects the kinetics of reactions, and when applied to complex systems, many different reactions take place which can be affected differently depending on the size and sign of their respective ΔV or ΔV^* values. A negative activation volume ($-\Delta V^*$) will enhance reactions under pressure, while those with a positive activation volume ($+\Delta V^*$) will be retarded. Conventional treatment of food, on the other hand, will produce different responses depending on the free energy, ΔG, or activation free energy, ΔG^*, of the various reactions. This is the main difference between high pressure and thermal processing. A classic example of the differences between the two types of food processing can be demonstrated in egg (6,7). An egg which has been subjected to pressure treatment (800 MPa for 10 min) sets like a hard boiled egg, but the yolk of the pressure treated sample retains the colour (and flavour) of the raw egg, and the white though gelled has a glossier appearance than the heated sample (and is softer to touch). Furthermore, the ferrous sulphide ring located between the yolk and white (present in the boiled egg), is absent in the pressure treated egg i.e., H_2S is not liberated from cysteine in the pressure treated product, as it is in the heated system.

In general, the pressure range used in food processing is of the order of 100 MPa to 1GPa. Because of the time-scale over which research in the field has been carried out, a variety of units are found in the literature. Conversion units are shown in Table I.

Table I. Units used in High Pressure Food Processing

Unit	Conversion
100 MPa	986.9 atm
100 MPa	1.0 kbar
100 MPa	1019.7 kg f cm^{-2}
100 MPa	14503.8 lb f in^{-2}
100 MPa	6.475 ton f in^{-2}

As a comparison to this bench mark, some examples experienced in everyday life give some idea of the pressures used in the food industry. The pressure exerted at the bottom of the deepest ocean is in the order of 110 MPa, which can also be thought of the weight of a stack of five family cars bearing down on an area the size of a postage stamp. A lady weighing 60 kg, placing all her weight on one stiletto heal would exert a pressure of 8.8 MPa (8-9).

Intermolecular Interactions. Several studies (10-14) have reported changes in protein structure and function during high pressure treatment. Generally, it is assumed that, in protein molecules, pressure enhances the formation of hydrogen bonds, the separation of ion pairs and the rupture of hydrophobic interactions, since these reactions cause the overall volume of the system to decrease. However, hydrogen bonding is influenced to a lesser extent than ionic or hydrophobic interactions (6, 11-12,14). Pressure also favours the formation of charged species because the increased electrostriction of water around the charges produced decreases the total molar volume of the water. At higher pressures (300 MPa), and in the presence of oxygen, sulfhydryl groups may become oxidised to −S−S− bonds. However, opposite effects can also occur at higher pressure (300 - 400 MPa), e.g., enhancement of hydrophobic interactions, due to higher compressibility of free water as compared to that of bound water (15).

Buffers generally selected for high pressure treatment are largely based on their pressure-resistant characteristics (16-18). Table II gives some values of volume changes for the dissociation of some weak acids and bases.

Table II. Values of ΔV for some Weak Acids subjected to 600 MPa (*16,19*)

Weak Acid	ΔV (mL/mol)
Dihydrogen phosphate	-23.8
Acetic Acid	-11.2
Tris(hydroxymethyl)aminomethane	+1.0
Imidazole H$^+$	-2.0

Phosphate buffer undergoes a reversible decrease in pH during pressurization, but in contrast, Tris-HCl buffers maintain a fairly constant pH under pressure. Reported decreases in pH of buffers per 100 MPa are as follows: phosphate buffer 0.38, acetate buffer 0.19, Tris buffer 0.03 (*20*).

Protein Denaturation. Many studies have reported (*4-5, 8, 10-13*) that pressure treatment of proteins (i.e., up to 800 MPa) will cause most water soluble proteins to unfold. This pressure-induced denaturation appears to be dependent upon the native protein structure, magnitude of the applied pressure, solvent compostion, pH, ionic strength and temperature (*12*). Generally, reversible effects are observed at relatively low pressures (100 - 200 MPa). At these pressures biochemical studies indicate (*14*) that (i) oligomeric proteins dissociate into subunits, possibly by the weakening of hydrophobic interactions and (ii) monomeric structures partially unfold (denature), due mainly to the disruption of both hydrophobic bonds and intramolecular salt bridges. At pressures above *ca.* 200 MPa, many proteins tend to unfold, and reassociation of subunits from dissociated oligomers can occur. Several studies (*6,11-13,21-22*) have reported that, under certain conditions, pressure-induced dissociation and/or denaturation can be reversible; however, the renaturation process can be slow and both hysteresis behaviour and conformational drift can be seen. It is now well established that most proteins in dilute solution, will, upon release of pressure reform into a modified macromolecular conformation which is similar, but not necessarily the same as the native form (*21,22*). Recently (*23*) it has been demonstrated that pressure induces the dimerization of metmyoglobin, and the pressure-treated form of ovalbumin differs from the native form (*24*). Many globular proteins, at higher concentrations and under appropriate conditions, may denature irreversibly to form precipitates or gels – egg-white being a classic example (*3*). Beyond 300 MPa oxidation of sulphydryl groups (*25*) can occur leading to formation of disulphide-linked aggregates (*26,27*).

There is some correlation between the functional properties of biological molecules and their conformation and conformational changes (*28*). Pressure induced changes may therefore change the functional properties of the renatured protein, which could be advantageous. It has been reported (*29*) that the foaming characteristics of egg-white are reduced because of selective precipitation under certain temperature/pressure conditions. In contrast, pressure-treated ovalbumin and soy-protein have improved emulsifying properties (*30*), whereas pressure processing of whey protein concentrate leads to reduced emulsifying properties (*31*). Recent experimental data obtained at Reading shows the potential of high pressure treatment to inactivate enzymes (*32*), the enzyme activity of tyrosinase is affected by both the magnitude of applied pressure and treatment time.

Dextran sulphate forms an electrostatic complex at pH 7 with bovine serum albumin, although there is no evidence to suggest that β-lactoglobulin forms such a complex (*33,34*). It is not known how pressure affects such mixtures. During pressure treatment all carboxyl groups on proteins and polysaccharides ionise irrespective of the initial pH. This causes the pH to decrease in an unbuffered system, although on pressure release the degree of ionization will return to its pre-pressure value. The conformation of the protein may also be altered, which could also modify any protein–polysaccharide interactions.

This paper discusses some recent work on the effects of high pressure on the solution properties of two water soluble globular proteins (β-lactoglobulin and bovine serum albumin), and how these changes in conformation and charge modify electrostatic protein–polysaccharide interactions. The effect of pressurization on the emulsifying properties of β-lactoglobulin oil-in-water emulsions is also described.

Experimental

Proteins studied were β-lactoglobulin (3 x crystallized and lyophilized) and bovine serum albumin (99 % purity). These were purchased from Sigma Chemical Co., as were DL-Dithiothreitol (DTT), 5,5'-dithio-bis(2-nitrobenzoic acid) [DTNB], buffer salts, sodium azide, *n*-tetradecane. Dextran sulphate (~ 5 x 10^5 daltons, containing 0.5–2.0 % phosphate buffer salts, pH 6–8), made by treatment of dextran T-500 with chlorosulphonic acid in pyridine, was also obtained from Sigma Chemical Co., St. Louis, MO, U.S.A. Isoelectric focussing (IEF) gels and buffers were purchased from NOVEX™ (San Diego, CA, U.S.A.). Buffer solutions for the emulsion studies were prepared using Analar-grade reagents and deionized, double-distilled water.

Solutions of β-lactoglobulin and bovine serum albumin (0.1, 0.2 and 2.5 %) and mixtures of protein + polysaccharide, in sealed cryovac bags, were adjusted to the appropriate pH by addition of 0.05 M HCl or NaOH prior to treatment. The buffering in this system at pH 7 is primarily due to the histidine and α-amino groups on the protein and because ionization of these groups does not lead to a volume change (*18*), it seems likely that the pH will remain relatively constant on the application of pressure. These solutions were subjected to pressures of 800 MPa for 20 min at $23 \pm 2°C$ in a prototype high pressure rig as described previously (*35*).

The conformational changes of both proteins were compared using the reaction with 1-anilinonaphthalene-8-sulphonate [ANS] (*35*) to assess surface hydrophobicity, α-helical content by circular dichroism (*36*), assessment of the % sulfhydryl groups (*25*), gel permeation chromatography with and without of DL-Dithiothreitol (*25*), IEF to assess aggregate formation (*25*), and thermal stability by differential scanning calorimetry (*25*) in the presence and absence of dextran sulphate.

Pressure treatments (200, 600 or 800 MPa) for three different holding times (10, 20 or 40 min) were applied to β-lactoglobulin before emulsion preparation, and also to the emulsion prepared with the native protein. Emulsions (20 vol% *n*-

tetradecane, 0.4 wt% protein, 20 mM imidazole-HCl buffer, pH 7) were prepared using the Leeds small-scale 'jet homogenizer' (*37,38*) at an operating pressure of 400 bar. Sodium azide (0.1 wt%) was added following homogenization and high pressure treatment to retard microbial growth. Samples were stored in a water bath at 23 $\pm$ 1 °C and droplet size diameters (d_{43}) determined in pre-stirred samples immediately after emulsion preparation and as a function of time using a Malvern Mastersizer S2.01 with a presentation code 0405. Creaming behaviour was visually determined in samples of height 40 mm from time-dependent changes in the thicknesses of cream and serum layers.

Results

Protein aggregation. Recently it has been shown (*35,39-41*) that β-lactoglobulin and bovine serum albumin (BSA) undergo some limited aggregation and change in structure following pressure treatment. The effect of high pressure at 800 MPa for 20 min (pH 7.0) on these two globular proteins on various structural parameters are shown in Table III.

Circular dichroism measurements for BSA show a substantial reduction in the α-helical content (native = 69 %; treated = 44 %) after high pressure treatment, which compares well with Hayakawa's (*40*) results who found a 50 % decrease in their pressure-treated (1,000 MPa for 10 min) BSA samples. However, at lower pressures (600 MPa for 9 min) there was no significant change (*24*). Our circular dichroism data for β-lactoglobulin show a modest loss of α-helix (native = 28 %; treated = 25 %) following treatment. This suggests that there is possibly some refolding of the protein at low concentrations (0.2 %). It is known that β-lactoglobulin has a high degree of compressibility and flexibility (*41*) and due to this it is considered that a large proportion of unfolding is reversible, even though the exact mechanism is not understood. It seems that unfolding is extended by increasing the protein concentration – possibly due to reduced reversibility caused by aggregation (*25*).

Table III. Properties of Bovine Serum Albumin and β-Lactoglobulin solutions (0.1 or 0.2 %) following Pressurization at 800 MPa for 20 min at pH 7.0

Protein	Bovine Serum Albumin	β-Lactoglobulin
α-Helix	Native 69 %, Pressure 44 %	Native 28 %, Pressure 25 %
ANS Intensity (Surface Hydrophobicity)	Decrease 41 %	Increase 40 %
% Sulfhydryl Groups	Decrease 55 %	Decrease 42 %

Table III also shows the results of fluorometry studies of ANS bound to pressure-treated BSA at a pre-pressure of pH 7.0. There is a decrease (41 %) in fluorescence intensity, which suggests a significant decrease in protein surface hydrophobicity after pressure treatment. This could be due to the protein refolding into a slightly different conformation thus burying some of the hydrophobic groups or possibly to the lower number of hydrophobic groups binding to the ANS because of intermolecular interactions (24). On the other hand, spectrofluorometry data for β-lactoglobulin in aqueous solution indicates an increase in the protein surface hydrophobicity (40 %).

Each protein has a single free cysteine residue which is well shielded and under normal conditions is unavailable for reaction with other sulfhydryl or disulphide groups. As temperature is increased (over 60 °C) this group becomes more accessible leading to the formation of covalent disulphide bonds and dimerization, unfolding and eventual aggregation (42-45). Our DTNB data suggests that there is a concomitant loss of sulfhydryl groups (55 % for BSA; 42 % for β-lactoglobulin), indicating a decrease in free cysteine after high pressure treatment. Presumably the free cysteine forms disulphides or is irreversibly oxidised by some other mechanism.

In Figure 1 gel permeation chromatography data for both proteins is shown, and we see that BSA behaves similarly to β-lactoglobulin. The major peak for unpressurized native β-lactoglobulin (Figure 1a) elutes at approximately 20,000 which is in fair agreement with the reported molecular weight (18,200). After pressure treatment (Figure 1b) the apparent molecular size increases to approximately 40,000 which indicates the formation of dimers. These results suggest that dimers are produced on treatment and supports the view that aggregation occurs during pressurization (39). The addition of DTT (Figure 1c) to the pressure-treated sample reduces the level of dimerization which suggests that polymerization due to −S−S− bridging occurs during or after pressure treatment at 800 MPa for 20 min. These findings agree with those of Johnston and Murphy (46) who demonstrated that high pressure treatment reduces the number of −SH groups in milk. In the case of BSA, it is seen that the native sample (Figure 1d) has a molecular weight of approximately 70,000 with some indication of larger units. Pressurization at 800 MPa for 20 min (Figure 1e) induces extensive dimer, trimer and higher aggregate formation, which return to monomer status after DTT addition (Figure 1f). It appears that during high pressure treatment these units are stabilized by disulphide bonding and the presence of a single free −SH group coupled with this multimerisation suggests that some level of disulphide interchange occurs.

Studies on metmyoglobin (22,23) at pH 7 have shown that high pressure treatment induces dimerization, the dimers dissociating in the presence of SDS. This indicates that association is primarily due to hydrophobic interactions and not disulphide bonding as in the case of BSA and β-lactoglobulin.

Native IEF focussing (Figure 2) provides some evidence for aggregation in both BSA and β-lactoglobulin following high pressure treatment. To give some indication of the size of the aggregates formed samples were filtered (2μ millipore) and compared with unfiltered samples. The additional bands seen with BSA after high pressure processing correspond to the high molecular weight aggregates seen on the gel permeation chromatography (Figure 1e). The data for β-lactoglobulin is consistent with that of Nakamura et al., (47) who demonstrated extensive aggregation of β-lactoglobulin when whey protein concentrate is pressurized at 200 - 600 MPa. More recently several studies (31,35,39) have demonstrated that high pressure causes extensive aggregation and unfolding of this protein.

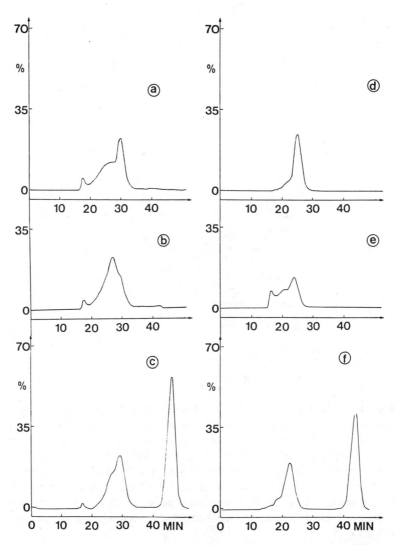

Figure 1. Gel permeation chromatography (Superose 12 column at pH 8.0) of protein constituents from native and pressure treated (800 MPa for 20 min) solutions; (a) native β-lactoglobulin; (b) pressure treated β-lactoglobulin; (c) pressure treated β-lactoglobulin with 5 mM DTT; (d) control BSA; (e) high pressure treated BSA; (f) treatment of pressure processed BSA with 5 mM DTT. Absorbance was measured at 280 nm. Note in the presence of DTT the pressure treated profile returns to that of the native material for both proteins. Reproduced with permission from reference 25.

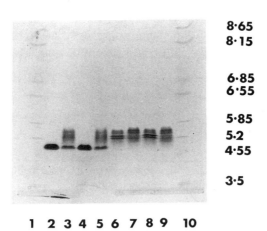

8·65
8·15

6·85
6·55

5·85
5·2
4·55

3·5

1 2 3 4 5 6 7 8 9   10

Figure 2. Effect of high pressure treatment (800 MPa for 20 min) on the aggregation properties of BSA and β-lactoglobulin. Isoelectric focussing of filtered (0.2 μ millipore) and unfiltered native and pressure treated protein solutions; (1) standards; (2) filtered native BSA; (3) filtered pressure processed BSA; (4) unfiltered control BSA; (5) unfiltered pressure processed BSA; (6) filtered native β-lactoglobulin; (7) filtered pressure treated β-lactoglobulin; (8) unfiltered native β-lactoglobulin; (9) unfiltered pressure processed β-lactoglobulin; (10) standards. Reproduced with permission from reference 25.

Protein denaturation. Differential scanning calorimetry (DSC) thermograms for solutions of native and pressure-treated β-lactoglobulin (2.5 %) and mixtures of protein + dextran sulphate (1:1 by weight) are shown in Figure 3. It is seen that the endothermal peak for the pressure-processed protein shifts to a lower temperature (native T_m = 73.32 °C; pressure treated T_m = 38.26 °C), with the enthalpy falling to virtually zero. This suggests that there is a major loss of tertiary structure during pressurization. The T_m (73.61 °C) does not change in the presence of dextran sulphate (not shown), however the mixture subjected to pressure treatment shows a T_m of 59.27 °C with a very low but discernable enthalpy. The presence of a small peak at 29 °C could well be due to impurities present within the sample or an artifact. Overall, the experimental data suggest that if complexes are formed between the protein and polysaccharide during pressurization, then the binding is likely to be reversible and weak.

Thermograms for solutions of native and pressure-treated (800 MPa for 20 min) BSA (2.5 %) and mixtures of BSA + dextran sulphate (1:1 by weight), show that the native BSA has an endothermal peak with a T_m of 59.24 °C (Figure 4) which shifts to a lower temperature (51.47 °C) after pressure treatment. The enthalpy falls to virtually zero, which indicates a significant loss of native structure, which is not recovered after pressure release. When dextran sulphate was added to the BSA there is a decrease in both T_m (53.94 °C) and enthalpy. High pressure treatment of the native mixture causes a decrease in calorimetric enthalpy with no significant change in T_m (54.18 °C) when compared with the untreated mixture. We see that the T_m and enthalpy for the pressure treated BSA + DS mixture is higher than for the BSA treated alone, which indicates that the presence of dextran sulphate during pressure treatment may involve complex formation between BSA and dextran sulphate at around neutral pH. This complex could help secondary structure formation from the pressure denatured form or inhibit its loss during pressure treatment.

β-Lactoglobulin emulsions. Emulsions of *n*-tetradecane-in-water (0.4 wt% protein, 20 vol% oil, 0.02M imidazole–HCl, pH 7.0) made with β-lactoglobulin are presented in Figure 5, where the average droplet diameter d_{43} is plotted as a function of maximum applied pressure for three different treatment times. The average diameter d_{43} is defined by $d_{43} = \Sigma\, n_i d_i^4 / \Sigma\, n_i d_i^3$, where n_i is the number of droplets of diameter d_i. This parameter, as determined by the Mastersizer, is a sensitive indicator of differences in protein emulsification efficiency and stability (*33*). The average droplet diameter for fresh emulsions made with native β-lactoglobulin at pH 7.0 and 23 °C is d_{43} = 0.86 ± 0.05 μm (mean of triplicates). Replacement of the untreated protein by a sample of β-lactoglobulin which had been subjected to high pressure treatment at 200, 600 or 800 MPa for 10, 20 or 40 minutes leads in each case to emulsions with larger average droplet size (d_{43}). Moreover, there was a general trend toward larger droplets with increase in the magnitude of applied pressure and treatment time. A modest treatment regime of 200 MPa for 10 min gives an emulsion with d_{43} = 1.12 ± 0.03 μm, whereas a treatment of 800 MPa for 40 min gives an emulsion with d_{43} = 1.58 ± 0.05 μm. Figure 5 shows, that at a fixed

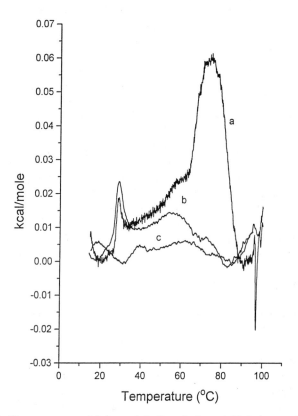

Figure 3. Thermograms of β-lactoglobulin solutions (pH 7.0) containing 2.5 % protein; (a) native β-lactoglobulin solution; (b) β-lactoglobulin + dextran sulphate (1:1 by weight) solution processed at 800 MPa for 20 min; (c) pressure treated β-lactoglobulin solution at 800 MPa for 20 min. Heating rate 1 °C min^{-1}. Reproduced with permission from reference 25.

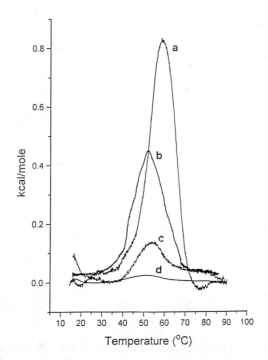

Figure 4. Thermograms of BSA solutions (pH 7.0) containing 2.5 % protein; (a) native BSA solution; (b) native BSA + dextran sulphate (1:1 by weight); (c) BSA + dextran sulphate (1:1 by weight) solution processed at 800 MPa for 20 min; (d) pressure treated BSA solution at 800 MPa for 20 min. Heating rate 1 °C min⁻¹. Reproduced with permission from reference 25.

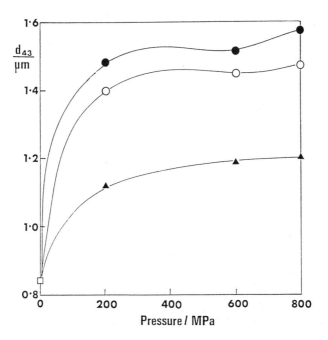

Figure 5. Effect of high pressure treatment on the emulsifying efficiency of β-lactoglobulin at pH 7.0. Oil-in-water emulsions (20 vol % oil, 0.4 wt % protein) were prepared at 400 bar and 23 °C using a 'jet' homogenizer. The average droplet size d_{43} is plotted as a function of applied pressure for three different treatment times: □, untreated; ▲, 10 min; ○, 20 min; ●, 40 min. Reproduced with permission from reference 35.

treatment time, there is relatively little effect of pressure on the emulsifying capacity in the range 200 - 800 MPa.

Figure 6 shows the creaming behaviour of 7 day old emulsions (which had not been subjected to agitation) as a function of pressure at constant treatment time. Cream separation was monitored visually, and the height, L, expressed as a % of the total sample height. We note that at 200 MPa for the different treatment times, and at 600 MPa for 20 min there is little change in creaming between the control sample and emulsions made with the pressure treated protein. However, the extent of creaming in emulsions made with pre-treated protein at 800 MPa was greater than for the control and other treatments. In particular, the protein pre-treated at 800 MPa for 40 min, has a value of L, twice that of the control.

The previous two paragraphs related to systems in which the β-lactoglobulin was treated <u>before</u> homogenization. We now look at the effect of high pressure treatment on the droplet size <u>after</u> homogenization. In Figure 7 we see the influence of pressure on the average droplet size (d_{43}) of 24 hour old emulsions prepared with native β-lactoglobulin. In this set of experiments, the average droplet diameter d_{43} increases only slightly with pressure treatment ($d_{43} = 0.9 \pm 0.05$ μm for the untreated sample to $d_{43} = 1.16$ μm for the pressure processed emulsion at 800 MPa for 40 min). Comparison with Figure 5, shows that the effect of high pressure processing on β-lactoglobulin stabilized oil-in-water emulsions is much less than when applied to the protein before homogenization. This may be because during emulsification the protein becomes partially unfolded and strongly interacting with its neighbours in a closely packed adsorbed monolayer, rendering it less susceptable to high pressure.

The significant change in the emulsifying capacity of β-lactoglobulin at neutral pH when subjected to pressures of 200 MPa is in agreement with other studies (*48-50*) of the effect of pressure on its solution properties. For example, the digestibility of this protein (in whey protein concentrate) by thermolysin is enhanced under pressure (200 MPa) – possibly as a direct result of conformational changes (*47*). We and others (*25,35,39,47*) have shown that β-lactoglobulin undergoes aggregation during high pressure treatment. Based on the results presented in this paper, the loss of emulsifying efficiency and stability can best be explained in terms of protein aggregation following exposure of hydrophobic groups during protein unfolding, followed by covalent –S–S– bond formation.

Conclusions

We find that high pressure processing of BSA and β-lactoglobulin causes these proteins to both unfold and aggregate. Protein aggregation after pressure treatment is probably due to disulphide bridging and exposed hydrophobic groups on the protein reacting to form oligomers. The addition of dextran sulphate to BSA seems to reduce the degree of unfolding, however this was not the case with β-lactoglobulin. Stronger complexes are formed between BSA + dextran sulphate than between β-lactoglobulin + dextran sulphate (*25*). Furthermore, the BSA in the pressure treated mixture still has some secondary structure, even at 2.5 % concentration. This may well be due to a large negative charge on the complexes

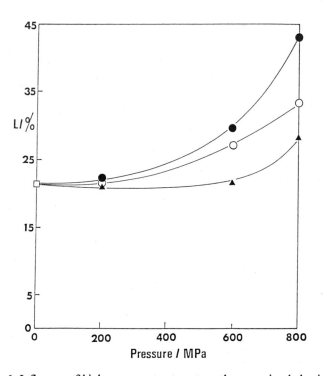

Figure 6. Influence of high pressure treatment on the creaming behaviour of emulsions (0.4 wt % protein, 20 vol % oil, pH 7.0, 23 °C) made with pressure treated β-lactoglobulin. Thickness, L, of the cream layer (percentage of total sample height) in 7 day old emulsions is plotted against applied pressure for three different treatment times: □, untreated; ▲, 10 min; ○, 20 min; ●, 40 min. Reproduced with permission from reference 35.

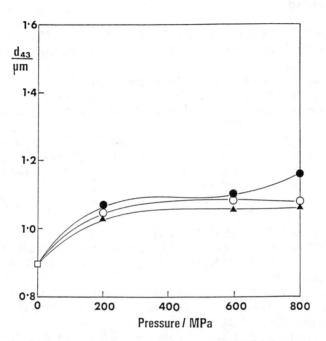

Figure 7. Effect of high pressure treatment on emulsions (0.4 wt % protein, 20 vol % *n*-tetradecane, pH 7.0, 23 °C) made with native β-lactoglobulin. Average droplet size d_{43} is plotted against the applied pressure for three different treatment times: □, untreated; ▲, 10 min; ○, 20 min; ●, 40 min. Reproduced with permission from reference 35.

which inhibits protein-protein interactions thus allowing some reformation of structure after high pressure treatment.

High pressure treatment of a β-lactoglobulin emulsified oil-in-water system appears to have relatively little effect on the emulsion stability; however, pressure treatment of β-lactoglobulin before homogenization has a detrimental effect. Thus, at ambient temperature, high pressure treatment does not appear to improve the emulsifying properties of β-lactoglobulin.

Literature Cited

1. Hite, B.H. *Bull. W. Va. Univ. Agric. Exp. Stn.* **1899**, *58*, pp. 15-35.
2. Hite, B.H.; Giddings, N.J.; Weakly, C.E. *Bull. W. Va. Univ. Agric. Exp. Stn.* **1914**, *146*, pp. 3-67.
3. Bridgman, P.W. *J. Biol. Chem.* **1914**, *19*, pp. 511-512.
4. Ledward, D.A. In *High Pressure Processing of Foods;* Ledward, D.A.; Johnston, D.E.; Earnshaw, R.G.; Hasting, A.P.M., Eds.; Nottingham University Press: Leicestershire, 1995, pp. 1-5.
5. Galazka, V.B.; Ledward, D.A. *Food Technol. Int. Eur.* **1995**, pp. 123-125.
6. Johnston, D.E. In *High Pressure Processing of Foods;* Ledward, D.A.; Johnston, D.E.; Earnshaw, R.G.; Hasting, A.P.M., Eds.; Nottingham University Press: Leicestershire, 1995, pp. 99-121.
7. Galazka, V.B.; Ledward, D.A. *Science and Technology Now.* **1995**, *Winter*, pp. 24-26.
8. Johnston, D.E. *Chem. Ind.* **1994**, *July*, pp. 499-501.
9. Johnston, D.E. In *The Chemistry of Muscle-Based Foods;* Ledward, D.A.; Johnston, D.E.; Knight, M.K., Eds.; Royal Society of Chemistry: London, 1992, pp. 298-307.
10. Heremans,K. In *High Pressure and Biotechnology;* Balny, C.; Hayashi, R.; Heremans, K.; Masson, P., Eds.; Colloque INSERM/John Libbey Eurotext: Montrouge, 1992, Vol. 224; pp 37-44.
11. Balny, C.; Masson, P.; Travers, F. *High Pressure Res.* **1989**, *2*, pp. 1-28.
12. Masson, P. In *High Pressure and Biotechnology;* Balny, C.; Hayashi, R.; Heremans, K.; Masson, P., Eds.; Colloque INSERM/John Libbey Eurotext: Montrouge, 1992, Vol. 224; pp 89-98.
13. Balny,C.; Masson, P. *Food Reviews Int.* **1993**, *9*, pp. 611-628.
14. Cheftel, J.-C. In *High Pressure and Biotechnology;* Balny, C.; Hayashi, R.; Heremans, K.; Masson, P., Eds.; Colloque INSERM/John Libbey Eurotext: Montrouge, 1992, Vol. 224; pp 195-208.
15. Ohmiya, K.; Kajino, T.; Shimizu, S.; Gekko, K. *Agric. Biol. Chem.* **1989**, *53*, pp. 1-7.
16. Neuman, R.C.; Kauzmann, W.; Zipp, A. *J. Phys. Chem.* **1973**, *77*, pp. 2687-2691.
17. Heremans, K. *Ann. Rev. Biophys. Bioeng.* **1982**, *11*, pp. 1-21.
18. Funtenberger, S.; Dumay, E.; Cheftel, J.-C. *Lebensm.-Wiss. u.-Technol.* **1995**, *28*, pp. 410-418.

19. Kunugi, S. *Prog Polym. Sci.* **1993**, *18*, pp. 805-838.
20. Kajiyama, N.; Isobe, S.; Vemura, K.; Nogucki, A. *Int. J. Food Sci. Technol.* **1995**, *30*, pp.147-158.
21. Zipp, A.; Kauzmann, W. *Biochemistry.* **1973**, *12*, pp. 4217-4227.
22. Defaye, A.B.; Ledward, D.A.; MacDougall, D.B.; Tester, R.F. *Food Chem.* **1995**, *52*, pp. 19-22.
23. Defaye, A.B.; Ledward, D.A. *J. Food Sci.* **1995**, *60*, pp. 262-265.
24. Hayakawa, I.; Kajihara, J.; Morikawa, K.; Oda, M.; Fujio, Y. *J. Food Sci.* **1992**, *57*, pp. 288-292.
25. Galazka, V.B.; Sumner, I.G.; Ledward, D.A. *Food. Chem.*, in press.
26. Schmid, G.; Lüdemann, H.D.; Jaenicke, R. *Eur. J. Biochem.* **1978**, *86*, pp. 219.
27. Aoki, K.; Hiramatsu, K.; Tanaka, M.; Kaneshina, S. *Biochim. Biophys. Acta.* **1968**, *160*, pp. 368.
28. Nakai, S.; Li-Chan, E. *Hydrophobic Interactions in Food Systems*; CRC Press: Boca Raton, Fl, 1988.
29. Knorr, D.; Böttcher, A.; Dörnenburg, H.; Eshtiaghi, M.; Oxen, P.; Richwin, A.; Seyderhelm, I. In *High Pressure and Biotechnology;* Balny, C.; Hayashi, R.; Heremans, K.; Masson, P., Eds.; Colloque INSERM/John Libbey Eurotext: Montrouge, 1992, Vol. 224; pp. 211-218.
30. Denda, A.; Hayashi, R. In *High Pressure and Biotechnology;* Balny, C.; Hayashi, R.; Heremans, K.; Masson, P., Eds.; Colloque INSERM/John Libbey Eurotext: Montrouge, 1992, Vol. 224; pp. 333-335.
31. Galazka, V.B.; Ledward, D.A.; Dickinson, E.; Langley, K.R. *J. Food Sci.* **1995**, *60*, pp. 1341-1343.
32. Gomes, M.R.A.; Ledward, D.A. *Food Chem.*, in press.
33. Dickinson, E.; Galazka, V.B. *Food Hydrocoll.* **1991**, *5*, pp. 281-296.
34. Dickinson, E.; Galazka, V.B. In *Gums and Stabilisers for the Food Industry;* Phillips, G.O.; Wedlock, D.J.; Williams, P.A., Eds; IRL Press: Oxford, 1992; Vol. 6, pp. 351-362.
35. Galazka, V.B.; Dickinson, E.; Ledward, D.A., *Food Hydrocoll.*, in press.
36. Galazka, V.B.; Ledward, D.A. In *Science, Engineering and Technology of Intensive Processing;* Akay, G.; Azzopardi, B.J., Eds; University of Nottingham: Nottingham, 1995, pp. 137-140.
37. Castle, J.; Dickinson, E.; Murray, A.; Stainsby, G. In *Gums and Stabilisers for the Food Industry;* Phillips, G.O.; Wedlock, D.J.; Williams, P.A., Eds; IRL Press: Oxford, 1988; Vol. 4, pp. 473-483.
38. Burgaud, I.; Dickinson, E.; Nelson, P.V. *Int. J. Food Sci. Technol.* **1990**, *25*, pp. 39-46.
39. Dumay, E.M.; Kalichevsky, M.T.; Cheftel, J.-C. *J. Agric. Food Chem.*, **1994**, *42*, pp. 1861-1868.
40. Hayakawa, I.; Kanno, T.; Tomita, M.; Fujio, Y. *J. Food Sci.*, **1994**, *59(1)*, pp. 159-163.
41. Gekko, K. In *Water Relationships in Food;* Levine, H.; Slade, L., Eds; Plenum Press: New York, 1991; pp. 753-771.
42. Katsuta, K.; Kinsella, J. E. *J. Food Sci.*, **1990**, *55*, pp. 1296-1302.

43. Katchalski, E.; Benjamin, G.S.; Gross, V. *J. Am. Chem. Soc.,* **1957,** *79,* pp. 4096-4099.
44. Wetzel, R.; Becker, M.; Behlke, J.; Billwitz, H.; Bohm, S.; Ebert, B.; Hamann, H.; Krumbiegel, J.; Lassmann, G. *Eur. J. Biochem.,* **1980,** *104,* pp. 469-478.
45. Peters, T. *Advances in Protein Chemistry,* **1985,** *37,* pp. 161-245.
46. Johnston, D.E.; Murphy, R. J. In *Food Macromolecules and Colloids;* Dickinson, E.; Lorient, D., Eds; The Royal Society of Chemistry: Cambridge, 1995; pp. 134-140.
47. Nakamura, T.; Sado, H.; Syukunobe, Y. *Milchwissenschaft,* **1993,** *48(3),* pp. 141-145.
48. Hayashi, R.; Kawamura, Y.; Kunugi, S. *J. Food Sci.,* **1987,** *52,* pp. 1107-1108.
49. Okamoto, M.; Hayashi, R.; Enomoto, A.; Kaminogawa, S.; Yamauchi, K. *Agric. Biol. Chem.,* **1991,** *55,* pp. 1253-1257.
50. Dufour, E.; Hervé, G.; Haertlé, T. In *High Pressure and Biotechnology;* Balny, C.; Hayashi, R.; Heremans, K.; Masson, P., Eds.; Colloque INSERM/John Libbey Eurotext: Montrouge, 1992, Vol. 224; pp. 147-150.

Chapter 16

Biopolymer Interactions in Emulsion Systems: Influences on Creaming, Flocculation, and Rheology

Eric Dickinson

Procter Department of Food Science, University of Leeds, Leeds LS2 9JT, United Kingdom

The stability and rheology of food oil-in-water emulsions are sensitive to the nature and strength of the biopolymer interactions at the surface of the droplets. Addition of polysaccharides to emulsions stabilized by food proteins or small-molecule emulsifiers may lead to a greater or poorer degree of stability with respect to creaming—depending on the type of polysaccharide and its concentration. Small-deformation shear rheological measurements can be very useful for predicting emulsion creaming behavior, for acting as a sensitive indicator of the character of droplet–polysaccharide interactions, and for distinguishing between the postulated bridging and depletion flocculation mechanisms. These general statements are illustrated in this article for model emulsions at neutral pH containing a number of different proteins (*i.e.*, casein(ate), bovine serum albumin and β-lactoglobulin) and polysaccharides (*i.e.*, dextran, xanthan, rhamsan, guar gum and dextran sulphate).

Biopolymer interactions have an important bearing on the shelf-life and texture of numerous food products. The primary macromolecular stabilizing agent in a typical food emulsion of the oil-in-water kind (homogenized milk, ice-cream, salad dressing, *etc.*) is a multicomponent mixture of adsorbed proteins. In addition, polysaccharides may be present as thickening or gelling agents, and there may be interactions between adsorbed proteins and polysaccharides which may have implications for the stability and rheology of the system (*1–6*).

To understand the general behavior of biopolymers in emulsions, it is convenient to distinguish between competitive and co-operative phenomena (*6*). Competitive adsorption involves the partial (or perhaps complete) displacement from the surface of one biopolymer by another. The rate and extent to which this occurs depends on a number of factors such as the concentrations and molecular characteristics of the biopolymers, the aqueous solution conditions (pH, temperature, *etc.*), the age of the

0097–6156/96/0650–0197$15.00/0

adsorbed protein layer, and the presence of other competing and/or interacting species (surfactants, calcium ions, *etc.*). Co-operative adsorption involves transient or permanent association of two (or more) biopolymers in one (or more) discrete layers at the interface. Whereas competition implies a net repulsion (or at least no strong attraction) between coadsorbing species, co-operation implies direct complexation between the biopolymers—due to covalent, electrostatic, hydrophobic or hydrogen bonding (or any combination of these). In many systems there is evidence for both competitive and co-operative effects within the same adsorbed layer.

The stability and rheological properties of a food colloid containing a mixture of dispersed droplets + biopolymer molecules are sensitive to the nature and strength of the droplet–droplet and droplet–biopolymer interactions (*1,2*). Whether these pair interactions are net attractive or net repulsive (weak or strong) is determined by the chemical nature of the biopolymer(s) present and the solution conditions (pH, temperature, *etc.*). The primary emulsifier layer around the emulsion droplets may consist of protein or low-molecular-weight surfactant. Where protein is the primary emulsifier, the overall structure and composition of the layer depends on the balance between the different kinds of biopolymer interactions, *i.e.* with the droplet surface, with the aqueous medium, or with other biopolymer molecules in the system. A binary aqueous solution of protein + polysaccharide may exhibit either complex coacervation or thermodynamic incompatibility, depending on whether the overall molecular pair-wise protein/polysaccharide interaction is, respectively, net attractive or net repulsive. In turn, the average strength and nature of the droplet–biopolymer interactions determines whether the emulsion is colloidally stable, or is flocculated, either by a bridging or a depletion mechanism (*4,5*).

This paper describes recent experimental results obtained at Leeds relating to the effect of added polysaccharide on the stability and rheology of oil-in-water emulsion systems. Two types of systems are mainly considered: (i) emulsions of small-molecule surfactant-coated droplets containing an added microbial polysaccharide, dextran or xanthan, and (ii) emulsions of milk protein-coated droplets with an added anionic polysaccharide dextran sulphate. In the former emulsions the droplet–polysaccharide interaction is non-associative or net repulsive, whereas in the latter systems it is net attractive.

Repulsive Droplet–Polysaccharide Interactions

Creaming and rheological behavior have been studied (*7,8*) for oil-in-water emulsions prepared with nonionic Tween 20 (*i.e.* polyoxyethylene sorbitan monolaurate) as emulsifier and with various concentrations of dextran (or some other polysaccharide) added after emulsification. Figure 1 shows the effect of the concentration of nonionic dextran (5×10^5 daltons) on the creaming stability at 25 °C of emulsions (30 wt % mineral oil, 1 wt % Tween 20, pH 7, ionic strength 0.05 M) having a fairly narrow droplet-size distribution and an average volume–surface diameter of $d_{32} = 0.55$ μm. Stability of each emulsion sample was visually assessed by observing the rate of appearance of a distinct serum layer at the bottom of an undisturbed flat-bottomed glass cylinder of height 60 mm. Whereas there was found to be no discernible serum separation over a storage period of 15 days for emulsion

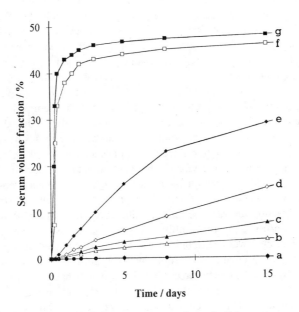

Figure 1. Influence of added dextran on creaming of oil-in-water emulsions (30 wt% oil, 1 wt% Tween 20, pH 7, ionic strength 0.05 M, d_{32} = 0.55 μm). Serum volume fraction (expressed as a percentage of total emulsion volume) is plotted against the storage time at 25 °C: (a), •, ≤0.1 wt%; (b), Δ, 10 wt%; (c), ▲, 5 wt %; (d), ◊, 2 wt%; (e), ♦, 1 wt%; (f), □, 0.5 wt%; (g), ■, 0.3 wt%. Reproduced with permission from reference 7.

samples without added polysaccharide (or containing just 0.1 wt % dextran), emulsions containing 0.3 or 0.5 wt % dextran were found to exhibit rather rapid serum separation. Further increases in dextran content to 1, 2, 5 and 10 wt % led to a gradual and systematic lowering of the serum separation rate, although even with 10 wt % added polymer the emulsion was still less stable than the polymer-free system.

Figure 2 shows the shear rheological behavior at 25 °C, as measured using both a Bohlin VOR controlled shear-rate rheometer and also a Bohlin CS controlled stress rheometer, for some of same dextran-containing emulsions. In the absence of added polysaccharide, the 30 wt % emulsion was found to be close to Newtonian (*i.e.* shear viscosity effectively independent of stress) with a relatively low apparent viscosity of *ca.* 4 mPa s. In the presence of 0.5 wt % dextran, however, the emulsion was found to become substantially pseudoplastic with a limiting zero-shear-rate viscosity (shear stress $\approx 10^{-2}$ Pa) which was more than 10^3 times larger than that of the polymer-free sample. With 1 or 10 wt % dextran, the limiting zero-shear-rate viscosity was found to increase to values more than 10^4 or 10^6 times that for the polymer-free sample.

The combined creaming and rheological results in Figures 1 and 2 can be interpreted in terms of a flocculation of the emulsion droplets induced by the added dextran. This destabilization of the colloidal system by the non-adsorbing polymer is explicable in terms of a depletion flocculation mechanism (*9–12*). In addition to the non-Newtonian rheological behavior, further experimental evidence for the depletion effect has been provided by complementary osmotic pressure measurements (*7*). The established theory of depletion flocculation predicts (*10*) that the strength of the interdroplet attraction increases with increasing concentration of the non-adsorbing polymer. As confirmed by the limiting zero-shear-rate viscosity in Figure 2, the emulsion containing 1 wt % dextran is more strongly flocculated than that containing 0.5 wt % dextran. Presumably, the droplet network in the former case is so strongly held together by the depletion forces that the structure cannot easily rearrange to expel serum phase quickly from the droplet network, and so the observed rate of creaming is reduced. This effect becomes increasingly evident with increasing polysaccharide content (> 1 wt %), although even at 10 wt % polymer the very high apparent viscosity of the emulsion is still not sufficient to prevent the rearrangement completely.

It is interesting to compare the above results with those obtained (*8,13*) for similar emulsions where dextran is replaced by different microbial polysaccharides, xanthan or rhamsan. Whereas dextran in aqueous solution at a relatively low concentration (say, 0.5 wt %) is a Newtonian solution of viscosity not much greater than that for water, an equivalent solution of xanthan or rhamsan is extremely shear-thinning with a limiting zero-shear-rate viscosity of the order of 10^6–10^7 times larger. This shear rheological behavior of the polysaccharide solution containing 0.5 wt % xanthan or rhamsan confers excellent creaming stability on emulsions prepared with nonionic surfactant, anionic surfactant, or protein as emulsifier (*8,13–17*). At much lower concentrations of xanthan or rhamsan (say, 0.05 wt %) when the shear rheological properties of the continuous phase are not so predominant, the emulsions exhibit evidence for the same extensive depletion flocculation and greatly enhanced serum separation as found with the dextran-containing systems. This behavior is illustrated by the data in Figure 3 for the effect of xanthan added after emulsification on the

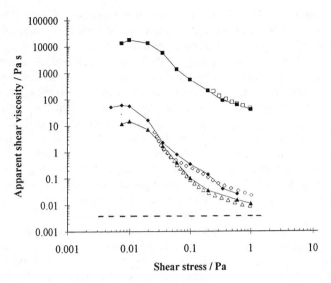

Figure 2. Influence of added dextran on the rheology of oil-in-water emulsions (30 wt % oil, 1 wt % Tween 20, pH 7, ionic strength 0.05 M, d_{32} = 0.55 μm). The apparent shear viscosity is plotted against shear stress at 25 °C: ▲, Δ, 0.5 wt %; ◆, ◊, 1 wt %; ■, □, 10 wt %. The solid symbols refer to the controlled stress experiments. The open symbols refer to the controlled strain-rate experiments. Dashed line refers to the system without dextran. Reproduced with permission from reference 7.

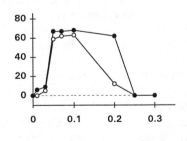

Figure 3. Influence of added xanthan on the creaming of oil-in-water emulsions (20 vol % oil, 2 wt % Tween 20, d_{32} = 0.54 μm). The serum volume fraction (expressed as a percentage of total emulsion volume) is plotted against the polysaccharide concentration in the aqueous phase following storage at 30 °C for 2 days (o) and 15 days (●).

creaming stability at 30 °C of sunflower oil-in-water emulsions (20 vol% oil, 2 wt% Tween 20, d_{32} = 0.54 μm) (*18*). The results show that, whereas serum separation is rather limited at very low xanthan concentrations (< 0.03 wt%), it is very rapid and extensive over the concentration range 0.05–0.2 wt%; however, at higher polymer contents (≥ 0.2 wt%) there is a slowing down again of the rate of serum separation, and creaming effectively stops due to rheological control at a xanthan content of *ca.* 0.3 wt%. Qualitatively similar behavior can be found when the small-molecule surfactant is replaced by a protein emulsifier such as sodium caseinate (*15,16*) or the microbial polysaccharide is replaced by some other type of non-adsorbing hydrocolloid. For instance, Figure 4 shows some data for the effect of guar gum added after emulsification on the creaming stability at 5 °C of a triglyceride oil-in-water emulsion (10 vol% oil, 1 wt% sodium caseinate, pH 6.6, d_{32} = 0.54 μm) (*19*). We can see that there is an optimum hydrocolloid content (~ 0.1 wt%) giving the greatest extent of destabilization of the emulsion. Substantially below this optimum value, the creaming stability is greater because the flocculation is weaker, and substantially above the optimum the creaming stability is greater because the flocculation is strong and the mechanical structure of the emulsion resists the gravitational settling (as with the higher concentration dextran systems in Figure 1).

Weakly Attractive Droplet–Polysaccharide Interactions

Small deformation rheological behavior has been investigated (*20*) for concentrated oil-in-water emulsions prepared with protein emulsifier (bovine serum albumin (BSA), β-lactoglobulin or sodium caseinate) and containing various concentrations of anionic dextran sulphate (DS) (5×10^5 daltons) added after emulsification. These particular mixed biopolymers systems were chosen for detailed rheological study following the observations (*5,21,22*) that the two globular proteins BSA and β-lactoglobulin behave differently with respect to their stability properties in emulsion systems containing dextran sulphate at neutral pH. Together with evidence from complementary surface shear viscosity experiments (*6,22,23*), the earlier results can be interpreted in terms of interfacial electrostatic complexation between BSA and DS at neutral pH, but not between β-lactoglobulin and DS under similar solution conditions.

Figure 5 shows the effect of DS concentration on the complex shear modulus G^* measured at 30 °C and 1 Hz for oil-in-water emulsions (40 wt% *n*-tetradecane, 2.6 wt% protein, pH 7, ionic strength 0.005 M) having a mean volume–surface diameter of $d_{32} \approx 0.6$ μm. Here, for convenience, the added polysaccharide concentration is expressed in terms of a (reduced) DS surface coverage Γ_{DS}, where $\Gamma_{DS} = 1$ corresponds to an added DS amount equal to that required to cover completely all the surface of the emulsion droplets to an arbitrary surface concentration of 2 mg m^{-2} (assuming all the DS is adsorbed). The results in Figure 5 indicate that the sensitivity of the emulsion rheology to the polysaccharide is very much dependent on the nature of the protein emulsifier present. With the BSA emulsion, addition of DS was found to produce (a) a large increase in G^* at very low Γ_{DS}, (b) a maximum complex modulus of $G^* \approx 150$ Pa at $\Gamma_{DS} \approx 0.1$, and (c) an approximately constant modulus ($G^* = 50 \pm 30$ Pa) for $\Gamma_{DS} \geq 0.3$. With the β-lactoglobulin emulsion, there was found

thickness (cm)

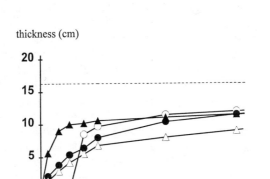

Figure 4. Influence of added guar gum on the creaming of oil-in-water emulsions (10 vol % oil, 1 wt % sodium caseinate, pH 6.6, d_{32} = 0.54 μm). The serum layer thickness is plotted against the storage time at 5 °C: o, 0.005 wt%; •, 0.01 wt%; ▲, 0.1 wt%; △, 0.2 wt%. Dashed line denotes total sample height.

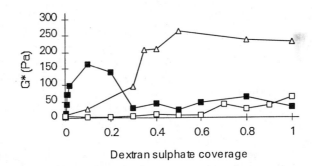

Dextran sulphate coverage

Figure 5. Influence of added dextran sulphate on shear rheology of oil-in-water emulsions (40 vol % oil, 2.7 wt % protein, pH 7.0, ionic strength 0.005 M, d_{32} ≈ 0.6 μm). The complex shear modulus G^* at 1 Hz is plotted against the (reduced) polysaccharide surface coverage Γ_{DS}: ■, BSA; □, β-lactoglobulin; △, caseinate.

to be no significant change in G^* for $\Gamma_{DS} \leq 0.2$, and just a gradual increase in modulus up to $G^* \approx 50$ Pa at $\Gamma_{DS} = 1$. And, with the sodium caseinate emulsion, there was found to be a steady increase in G^* from $\Gamma_{DS} = 0$ to $\Gamma_{DS} \approx 0.5$ with an approximately constant value of $G^* \approx 250$ Pa reached at the higher levels of polysaccharide addition.

The large increase in shear modulus of the BSA emulsion at very low added polymer concentrations—well below that required for full surface coverage—is consistent with droplet flocculation by a bridging mechanism (24). It has been noted that the addition of DS at $\Gamma_{DS} = 0.1$ to dilute BSA emulsions at pH 7 leads to a substantial increase in the measured d_{32} value, indicative of bridging flocculation of protein-coated droplets by the added biopolymer, whereas addition at $\Gamma_{DS} = 5$ (i.e. well above saturation coverage) gives no change in d_{32} and hence no flocculation. Restabilization at high DS concentrations can be explained in terms of a complete secondary stabilization layer of the highly charged polysaccharide adsorbed on top of the original protein layer. Additional evidence for an attractive interaction at pH 7 between DS and adsorbed BSA is available from separate measurements of surface viscosity (22,23), electrophoretic mobilities (5,6), emulsion stability (22), and foam stability (25). It has been suggested (6,22) that, even though both biopolymers carry a net negative charge at neutral pH, the origin of the attractive interaction is predominantly electrostatic between the highly charged anionic polysaccharide and small positive patches on the globular protein. This explanation is consistent with the observed reduction in the maximum value of G^* on increasing the ionic strength from 5 mM to 70 mM by addition of sodium chloride (20).

When experiments identical to those described above were repeated with β-lactoglobulin instead of BSA, no evidence was found (6,21,25) for any significant attractive interaction of DS with the adsorbed protein layer at pH 7. The absence of any bridging flocculation at low Γ_{DS} in the concentrated emulsions containing β-lactoglobulin + DS is confirmed by the small-deformation rheological data in Figure 5. There are various differences in molecular properties between BSA and β-lactoglobulin which may account for the difference in interaction with the anionic polysaccharide. It is known (26) that β-lactoglobulin monomer (18.4 kDa) has just one strong binding site for surfactants, whereas the larger BSA monomer (66.2 kDa) has up to 12 such sites (27). In addition, at pH 7, there are a much greater number of positively charged residues (e.g. 59 lysines) on the BSA molecule as compared with the β-lactoglobulin molecule, which has 15 lysines, of which some are buried in the interior of the molecule (28). Clustering of this smaller number of positive groups into patches on the surface of the β-lactoglobulin may be insufficient to generate complexation with the negatively charged DS under these conditions.

It is interesting to note that the effect of polysaccharide on the rheology of the sodium caseinate emulsion is different from that found with either of the globular proteins. Whilst the data in Figure 5 show no evidence for bridging flocculation of the casein emulsions at $\Gamma_{DS} \approx 0.1$, the relatively high value of G^* for $\Gamma_{DS} \geq 0.5$ is certainly indicative of some substantial protein–polysaccharide interaction. The 'lag phase' in the plot may be indicative of some sort of cooperativity in the interaction. The structure obtained here is rather insensitive to disturbance by high shear fields, in contrast to the bridging flocs which are irreversibly disrupted by high shear (20). The

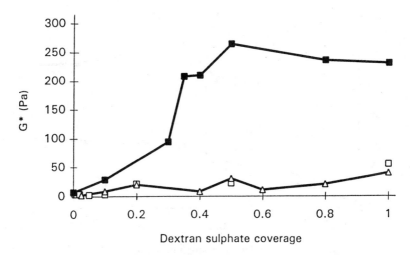

Figure 6. Effect of ionic strength on rheology of caseinate-stabilized emulsions (40 vol % oil, 2.7 wt % protein, pH 7.0, d_{32} = 0.58 μm) with added dextran sulphate. The complex shear modulus G^* at 1 Hz is plotted against the (reduced) polysaccharide surface coverage Γ_{DS}: ■, 5 mM imidazole; □, 5 mM imidazole adjusted with NaCl to ionic strength 70 mM; Δ, 50 mM phosphate buffer.

electrostatic origin of the casein–DS interaction at pH 7 is seemingly indicated by the large reduction in G^* on increasing the ionic strength from 5 mM to 50 or 70 mM as shown in Figure 6. As there is no evidence in this case for a specific attractive interaction, it could be that the origin of the high G^* in the DS + casein emulsion is due to structuring and possible thermodynamic incompatibility (*29*) in the aqueous phase under the influence of electrostatic repulsion between charged biopolymers present at high concentration. Such interactions would be expected to be reduced by increasing the ionic strength (Figure 6) or by lowering the pH towards the protein's isoelectric point (Figure 7).

The preceding set of results for emulsion systems containing mixed biopolymers with weak electrostatic protein–polysaccharide interactions shows that such systems are extremely sensitive to the nature of the protein adsorbed layer and to the aqueous solution conditions. It is clear that small-deformation rheology is an extremely sensitive and powerful technique for monitoring the biopolymer interactions and the mechanisms of flocculation in concentrated emulsions containing mixtures of proteins and polysaccharides. The reason for this sensitivity is that changes in small-deformation rheological behavior are highly responsive to changes in large-scale network formation in concentrated colloidal systems. Statistical mechanical theory shows (*30*) that the addition of a low concentration of weakly interacting polymer, modeled as small weakly adsorbing spheres that can fit into the gaps between larger spherical oil droplets, can produce a very substantial enhancement of the gel-like character of concentrated emulsions. Depending on whether the droplet–polymer pair

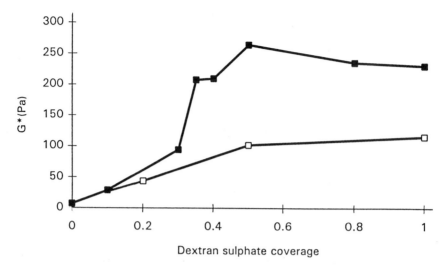

Figure 7. Effect of pH on shear rheology of caseinate-stabilized emulsions (40 vol % oil, 2.7 wt % protein, 5 mM imidazole, d_{32} = 0.58 μm) with added dextran sulphate. The complex shear modulus G^* at 1 Hz is plotted against the (reduced) polysaccharide surface coverage Γ_{DS}: ■, pH 7.0; □, pH 6.7.

interaction is weakly attractive or weakly repulsive, reversible (weak) gel formation may occur by either bridging flocculation or depletion flocculation. This article has presented several examples of these two types of flocculation behavior as induced by the presence of polysaccharides.

Acknowledgment

Support from the Biotechnology and Biological Sciences Research Council (U.K.) is gratefully acknowledged.

Literature Cited

1. Dickinson, E. In *Gums and Stabilisers for the Food Industry*; Phillips, G. O.; Wedlock, D. J.; Williams, P. A., Eds; IRL Press: Oxford, 1988; Vol. 4, pp. 249–263.
2. Dickinson, G.; Stainsby, G. In *Advances in Food Emulsions and Foams*; Dickinson, E.; Stainsby, G., Eds; Elsevier Applied Science: London, 1988; pp. 1–44.
3. Dickinson, E.; Euston, S. R. In *Food Polymers, Gels and Colloids*; Dickinson, E., Ed.; Royal Society of Chemistry: Cambridge, UK, 1991; pp. 132–146.
4. Dickinson, E. *J. Chem. Soc. Faraday Trans.* 1992, **88**, 2973–2983.
5. Dickinson, E.; McClements, D. J. *Advances in Food Colloids*; Blackie A & P: Glasgow, 1995; pp. 81–101.

6.   Dickinson, E. In *Biopolymer Mixtures;* Mitchell, J. R., Ed.; Nottingham University Press, Nottingham, 1995; pp. 349–372.
7.   Dickinson, E.; Goller, M. I.; Wedlock, D. J. *J. Colloid Interface Sci.* 1995, **172**, 192–202.
8.   Dickinson, E.; Ma, J.; Povey, M. J. W. *Food Hydrocoll.* 1994, **8**, 481–497.
9.   Vrij, A. *Pure Appl. Chem.* 1976, **48**, 471–483.
10.   Fleer, G. J.; Scheutjens, J. H. M. H.; Vincent, B. In *Polymer Adsorption and Dispersion Stability;* Goddard, E. D.; Vincent, B., Ed.; ACS Symposium Series No. 204; American Chemical Society: Washington, DC, 1984; p. 245.
11.   Vincent, B.; Edwards, J.; Emmett, S.; Croot, R. *Colloids Surf.* 1988, **31**, 267–298.
12.   Lekkerkerker, H. N. W.; Stroobants, A. *Physica A* 1993, **195**, 387–397.
13.   Dickinson, E.; Goller, M. I.; Wedlock, D. J. *Colloids Surf. A* 1993, **75**, 195–201.
14.   Gunning, P. A.; Hibberd, D. J.; Howe, A. M.; Robins, M. M. *Food Hydrocoll.* 1988, **2**, 119–129.
15.   Cao, Y.; Dickinson, E.; Wedlock, D. J. *Food Hydrocoll.* 1990, **4**, 185–195.
16.   Cao, Y.; Dickinson, E.; Wedlock, D. J. *Food Hydrocoll.* 1991, **5**, 443–454.
17.   Luyten, H.; Jonkman, M.; Kloek, W.; van Vliet, T. In *Food Colloids and Polymers: Stability and Mechanical Properties;* Dickinson, E.; Walstra, P., Eds; Royal Society of Chemistry: Cambridge, UK, 1993; pp. 224–234.
18.   Dickinson, E.; Ma, J.; Povey, M. J. W. *J. Chem. Soc. Faraday Trans.* 1996, **92**, in press.
19.   Heeney, L. M. Phil. Thesis, University of Leeds, 1994.
20.   Dickinson, E.; Pawlowsky, K. In *Gums and Stabilisers for the Food Industry;* Phillips, G. O.; Wedlock, D. J.; Williams, P. A., Eds; Oxford University Press: Oxford, 1996; Vol. 8, in press.
21.   Dickinson, E.; Galazka, V. B. *Food Hydrocoll.* 1991, **5**, 281–296.
22.   Dickinson, E.; Galazka, V. B. In *Gums and Stabilisers for the Food Industry;* Phillips, G. O.; Wedlock, D. J.; Williams, P. A., Eds; IRL Press: Oxford, 1992; Vol. 6, pp. 351–362.
23.   Larichev, N. A.; Gurov, A. N.; Tolstoguzov, V. B. *Colloids Surf.* 1983, **6**, 27–34.
24.   Dickinson, E.; Eriksson, L. *Adv. Colloid Interface Sci.* 1991, **34**, 1–29.
25.   Izgi, E.; Dickinson, E. In *Food Macromolecules and Colloids;* Dickinson, E.; Lorient, D., Eds; Royal Society of Chemistry, Cambridge, UK, 1995; pp. 312–315.
26.   Coke, M.; Wilde, P. J.; Russell, E. J.; Clark, D. C. *J. Colloid Interface Sci.* 1990, **138**, 489–504.
27.   Morr, C. V.; Ha, E. Y. W. *Crit. Rev. Food Sci. Nutr.* 1993, **33**, 431–476.
28.   Brown, E. M.; Pfeffer, P. E.; Kumosinski, T. F.; Greenberg, R. *Biochemistry* 1988, **27**, 5601–5610.
29.   Dickinson, E. In *Food Polysaccharides and their Applications;* Stephen, A. M., Ed.; Marcel Dekker, New York, 1995; pp. 501–515.
30.   Dickinson, E. *J. Chem. Soc. Faraday Trans.* 1995, **91**, 4413–4417.

CHEMICAL, ENZYMATIC, AND GENETIC MODIFICATION

Chapter 17

Phosphorylation of Proteins and Their Functional and Structural Properties

Fakhrieh Vojdani and John R. Whitaker

Department of Food Science and Technology, University of California, Davis, CA 95616

Phosphoproteins are very important biologically and have the potential to improve the functional properties of proteins as food ingredients. Food proteins can be phosphorylated by protein kinases, which require an encoded specificity site on the protein or by use of $POCl_3$, phosphorus pentoxide, other high energy phosphates or carbodiimides. In this paper we report the phosphorylation of α-lactalbumin, β-lactoglobulin and zein by $POCl_3$ and the effect on solubility, and functional and structural properties of the proteins. The paper also reviews published research (Chobert *et al. J. Am. Oil Chem. Soc.* **1987**, *64*, 1704-1711) in which tryptophan and lysine were covalently linked to zein when they were added to the solution along with the protein and $POCl_3$. The bound amino acids improved the nutritional value of zein from 4.5 to 48% relative to casein, when tested with *Tetrahymena thermophili*.

Chemical modification of proteins is important in biological systems, as well as *in vitro*. Modification occurs unintentionally during food processing, storage and cooking and intentionally in modification of proteins for improved functionalities (color, flavor, taste, texture, solubility). *In vivo*, there are at least 135 protein modifications known to occur during post-translational processing of proteins (1-3). All except one, hydrolysis of peptide bonds, occurs by the modification of side chains of amino acid residues of the protein. Intentional modifications of proteins are many; however, the most important are: phosphorylation, sulfhydryl/disulfide bond rearrangement, cross-linking reactions, proteolysis and deamidation. This paper will deal only with phosphorylation.

Phosphoproteins are very common in biological systems and are found in high levels in foods such as caseins in milk and phosvitin in egg yolk. Both are major sources of the essential phosphate during fetal, infant and adult lives of the animals. Phosvitin, for example, has a molecular weight of 35,500 and contains 126 serine residues (56% of total amino acid residues) of which 120 are phosphorylated (35% (w/w) of the protein). The -OH group serine residues are phosphorylated by protein kinases, which require ATP and Mg^{++}. Phosphate groups can be found in proteins

0097–6156/96/0650–0210$15.00/0

attached to the oxygen of seryl, threonyl, aspartyl (β-carboxyl) and tyrosyl residues, and via nitrogen to lysyl (ε-amino) and histidyl (1 and 3) residues. Phosphorylation is also important in regulation (such as the phosphorylation-dephosphorylation of α- and β-phosphorylases). There are thousands of publications on biological phosphorylation each year (4), indicating its key importance in biology. These naturally occurring phosphoproteins have important functional properties in foods (5) which can be modified by removal of some or all of the phosphate groups with alkaline phosphatases (6, 8). By contrast, there are relatively few publications each year on *in vitro* modification of proteins by phosphorylation.

Phosphate groups can be added *in vitro* to proteins by protein kinases and removed by alkaline phosphatases (7, 8). The number of added phosphate groups is small (5 to 7 for the caseins as example) because of the high specificity of the enzyme for an encoded amino acid sequence which must be on the surface of the folded protein. Therefore, under nondenaturing conditions, addition of one or two phosphate residues per mole is about the maximum achieved. Higher energy phosphate compounds react with groups on the side chains of proteins by non-enzymatic methods. These include phosphorylation of the -OH group of seryl, threonyl and tyrosyl residues and the N of the ε-amino and imidazole groups of lysyl and histidyl residues of proteins. Under some conditions the guanidino groups of arginine may be phosphorylated (above pH 11). Phosphorylation of O- and N- containing groups can be distinguished by nuclear magnetic resonance (NMR) and by lability in acidic or alkaline solvents (9, 10). At acidic pHs (pH 2-4) phosphorylation of O-groups is favored because of stability of the product, while phosphorylation of N-groups is favored at alkaline pHs where they are more stable and the ε-amino group of the lysyl residue is unprotonated.

Heidelberger *et al.* (11) reported the phosphorylation of egg albumin as early as 1941. Mayer and Heidelberger (12) reported the phosphorylation of crystalline horse serum albumin in 1946. Boursnell *et al.* (13) studied the chemical and immunological properties of several phosphorylated proteins in 1948, as did Ferrel *et al.* (14), using phosphorus pentoxide in phosphoric acid for phosphorylation. More recently, there have been a number of publications on phosphorylation of food proteins to modify their functional properties (15-28).

The major advantages of phosphorylation of food proteins are the increase in solubility and decrease in pI of the proteins, thereby changing the functional properties, especially near the pI of the original proteins. Two other advantages include increased ionic cross-linking of phosphoproteins via Ca^{2+} (in gel formation for example) and the potential simultaneous covalent incorporation of amino acids, peptides, carbohydrates and other compounds into proteins (23).

We report unpublished data on the phosphorylation of α-lactalbumin and β-lactoglobulin and review previously published research on zein (23), with emphasis on the changes in chemical and physical properties of the native and modified proteins.

Materials. α-Lactalbumin and β-lactoglobulin (genetic variant A) were purified from milk of a single cow of the University of California, Davis dairy herd. The milk was cooled immediately and the casein was precipitated at pH 4.6 (1 N HCl added dropwise) within 1 h of milking. After pH adjustment the samples were kept for 30-60 min at 4°C to complete precipitation of the casein. Following centrifugation at 12,000 g for 25 min, the fat was removed (top layer). The whey solution was filtered (Whatman #1 paper) and dialyzed against distilled water or deionized using a Sephadex G-25 column. The whey solution was frozen, lyophilized and stored at -20°C. All subsequent solutions contained 0.02% sodium azide. α-Lact-albumin and β-lactoglobulin were separated from each other on a Sephadex G-100 column (0.075 *M* Tris·HCl buffer, pH 7.5). The individual proteins were further purified by DEAE-cellulose anion column chromatography in 0.075 *M* Tris·HCl buffer, pH 7.5, with a linear gradient of 0.0 to 0.26 M NaCl in the same buffer, and by CM-cellulose cation column chromatography using 0.01 M sodium acetate buffer, pH 5.0, with a linear gradient to 0.5 N NaCl gradient in the same buffer. The β-lactoglobulin was further purified on a hydroxyapatite column (0.01 M sodium phosphate buffer, pH 7.0, initially, followed by a gradient of 0.01-0.3 M of the same buffer, and finally a 0.3 M buffer). The proteins were shown to be pure by polyacrylamide gel electrophoresis (PAGE) (29) and by sodium dodecyl sulfate polyacrylamide gel electrophoresis (SDS-PAGE) (30).

Zein was purchased from Sigma Chemical Co. and was used without further purification.

Methods. The following analytical and preparation methods were used.

Phosphorylation.   The phosphorylation of α-lactalbumin and β-lactoglobulin was performed as described by Matheis *et al.* (19). At the end of the reaction, the CCl4 layer was removed, as well as the precipitate formed between the CCl4 and aqueous phases. The aqueous phosphorylated protein solution was dialyzed against 0.5 M KCl and then against distilled water. The solution was lyophilized and the white powder stored at -20°C.

The total phosphate content of the sample after HClO4 digestion was determined (31). Inorganic phosphate was desorbed by treating the phosphorylated protein with an equal volume of 0.1 N NaOH for 30 sec, followed by precipitation of the protein with 10% trichloroacetic acid. The precipitate was removed by centrifugation and the soluble inorganic phosphate was determined (31) and subtracted from the total phosphate to give the organic bound phosphate.

Functional Properties.   The solubility, emulsifying activity and stability, and foaming capacity and stability were determined.

Solubilities.   Native and modified proteins were weighed and dispersed in distilled water. The pHs of different solutions were adjusted with HCl or NaOH to pH 2 to 10, with constant stirring. The volume was adjusted to give 0.1% protein (w/v). After 1 h equilibration at 23°C, a part of the solution was centrifuged at 18,500 x g for 15 min in a microfuge centrifuge. The supernatant was filtered using Microfilterfuge tubes containing 0.2 μm nylon-66 membrane filters. The protein content of the supernatants was determined by the Lowry method (32) using bovine serum albumin as standard.

Emulsifying Activities.   The emulsifying activities were determined by the spectroturbidity method of Pearce and Kinsella (33), with slight modifications. Calculation of E.A.I. (emulsifying activity index) was done according to Pearce and Kinsella (33) as shown by Eq. 1, where T = turbidity (= 2.3A/l, where A is absorbance

$$E.A.I. = 2T/\phi C \tag{1}$$

at 500 nm and l is light path in meters), ϕ is oil phase fractional volume (0.25 in all experiments reported here) and C is concentration of protein (0.1% in all cases) before the emulsion was formed.

The Pearce and Kinsella Equation was corrected recently by Cameron *et al.* (34) to E.A.I. = 2T/C(1-ϕ). According to Figure 1 of Cameron *et al.*, the "corrected" E.A.I. values at 0.25 used in our experiments in Figures 1C, 3C and 5B, D would be E.A.I./1.2 based on the particle size and E.A.I./3.0 for theoretical values.

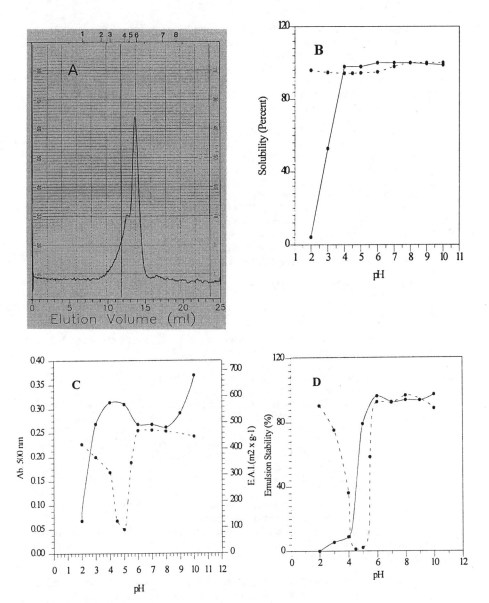

Figure 1.   Structural and functional properties of native (α-La) and phosphorylated α-lactalbumin (α-La-phos). A. Superose 12 gel filtration chromatogram of α-La-phos. Standard MW markers used: 1, Blue Dextran, 2000 kDa; 2, human transferrin, 80 kDa; 3, bovine serum albumin, 66 kDa; 4, ovalbumin, 43 kDa; 5, β-lactoglobulin, 18.3 kDa; 6, α-lactalbumin, 14.2 kDa; 7, bradykinin, 1.06 kDa; 8, prolylglycine, 0.172 kDa. B. Solubility-pH profile of α-La (--•--) and α-La-phos (—•—).   C. Emulsifying activity-pH profile of α-La (--•--) and α-La-phos (—•—). D. Emulsion stability-pH profile of α-La (--•--) and α-La-phos (—•—).

Emulsion Stabilities. Aliquots of the emulsions prepared for emulsifying activities were incubated at 23°C for 24 h. The tubes were rotated five times to obtain a homogeneous suspension and an aliquot removed and diluted 1000 x to determine turbidity (33). The remainder of the 24 h old emulsions were heated at 80°C for 30 min, cooled and mixed as above. An aliquot was diluted and the turbidity determined again. The emulsion stability was calculated as described by Pearce and Kinsella (33) according to Eq. 2.

$$\% \text{ Emuls. stab.} = \text{E.A.I.}\% = (\text{E.A.I.}_{max} - \text{E.A.I.}_{80°C}) \times 100/\text{E.A.I.}_{max} \qquad (2)$$

Foaming Capacity and Stability. The apparatus for measuring foaming capacity and stability, designed in our laboratory, was made of a calibrated glass column (41 cm length, 1.0 cm i.d.), surrounded by a water jacket connected to a pump and water bath at $25 \pm 0.5°C$. The bottom of the tube contained a fritted glass disc (1.0 cm diameter, 10 C Corning, Ace Glass Co. No. 31001). N_2 was bubbled from the bottom at a fixed rate of 30 ml/min to give a uniform, fine N_2 bubble distribution. Increase in foam volume (capacity) was recorded after 30, 40 and 50 sec. Time (sec) required to drain 85 and 92% of the original volume was recorded (stability).

Foam capacity or overrun was defined as

$$\% \text{ Overrun} = 100 \, (V_f - V_0)/V_0 \qquad (3)$$

where V_f is volume of foam at any time and V_0 is the original liquid volume (2.0 ml).

Structural Properties. Surface activity, hydrophobicity before and after denaturation, and circular dichroism (CD) were measured.

Surface Activity. Surface adsorption was measured by the Whilhelmy plate method (35, 36) using a Cahn electrobalance. Surface pressure of protein solution (6 ml, 0.01%) in 20 mM citrate (pH 3-5), phosphate (pH 6-8) and borate (pH 9-10) buffers was measured by injecting 30 ul of 2% protein solution into the subphase while it was stirred continuously with a stirrer bar (10 x 1 mm) to prevent surface disruption. The kinetics were calculated (37-39). Surface pressure as a function of time was calculated (40). The area of interface (dA) cleared for penetration of a protein molecule was calculated from the initial slope of the plot of $\ln(d\pi/dt)$ vs. π as described by MacRitchie (41), Torenberg (38) and Waniska and Kinsella (40).

Hydrophobicity. Surface hydrophobicities of the protein (0.5%) in 25 mM phosphate buffer, pH 7.0, before and after heating ("exposed hydrophobicity") at 90°C for 10 min in 25 mM phosphate buffer containing 0.5% sodium dodecyl sulfate were measured fluorometrically using cis-parinaric acid (42) and 8-anilinonaphthalene sulfonate (43) as hydrophobic probes.

Circular Dichroism (CD). Structural changes of native and modified α-lactalbumin and β-lactoglobulin were determined as a function of pH, using a Jasco J-600 Spectropolarimeter (Japan Spectroscopic Co., CTD) at wavelengths of 190 to 350 nm. Measurement parameters were: band width of 1.0 nm, sensitivity of 20 mdeg/FS, time constant of 2.0 sec, step resolution of 0.2 nm and scan speed of 20 nm/min. The protein samples (0.01-0.07 nM) were prepared from dilution of stock protein (1 mg/ml) in a 10 mM citrate (pH 3-5), 10 mM phosphate (pH 6-8) or 10 mM borate buffer (pH 9-10). The diluted proteins were equilibrated in the buffer for 30 min at 23°C.

The CD properties of the protein solution were calculated based on molar ellipticity [θ]. The ellipticity in degrees (θ deg) of the protein solution is equal to:

$\theta = +H \times S$, where S = CD scale ($m°/cm^2 = x\ 10^{-3}$ deg/cm) and H = reading (cm).
Molecular ellipticity $[\theta]$ in deg cm^2 decimeter^{-1} of the protein solutions was calculated
according to Eq. 4.

$$[\theta] = (\theta \times 100)/Cl \tag{4}$$

where C is molar concentration (mol/l) and l is cell length (cm). $[\theta]$ can also be
calculated from

$$[\theta] = [\Psi] \times M/100 \tag{5}$$

where specific ellipticity ($[\Psi]$) is θ/Cl, with C, concentration, in g/mol and l, the cell
length, in decimeters. M is the average weight of amino acid residues (113 g/mol
residue).

Results and Discussion

α-Lactalbumin and Phosphorylated α-Lactalbumin.

The stability, functional
properties and structural properties of native α-lactalbumin (α-La) and phosphorylated
α-lactalbumin (α-La-phos) were compared under the same experimental conditions.
The α-La-phos contained 1.75% phosphorus (7.8 moles P/mol protein).

Solubility. Figure 1B shows that 0.1% α-La and α-La-phos were 92-96%
soluble above pH 4. However, the solubility of α-La-phos decreased markedly below
pH 4 compared to α-La, probably because of protonation of the phosphate groups.
Similar results below pH 4 were reported for phosphorylated casein (21, 22).

Emulsifying Activity and Stability. The emulsifying activity index
(E.A.I.) of α-La-phos was less than that of α-La below pH 2.5 (Fig. 1C), probably
because of its decreased solubility, and was better than α-La at pH 2.5-6 and above
pH 8. At pH 5, the pI of α-La, the E.A.I. of α-La-phos was 5 times better than that of
α-La, although both had the same solubility in this pH range (Fig. 1B). The emulsion
stability of α-La-phos was lower than that of α-La below pH 4, but much better than
α-La in the pH range 4 to 6 (Fig. 1D). These data indicate that α-La has poor
emulsifying activity and stability at the pI (5.1) of the protein, as expected from the
DLVO (Derjaguin, Landau, Verwey and Overbeek) theory. The marked improvement
for α-La-phos is expected since the pI is shifted to a lower pH (calculated pH of 3.6).
The emulsion properties of phosphorylated casein gave similar results as a function of
pH (21, 22) as reported here for α-La-phos.

Foaming Capacity and Stability. Except below pH 3.5, α-La had better
foaming capacity than did α–La-phos (Fig. 2A). Particularly at pH 6, the foaming
capacity of α-La is 3.6 times that of α-La-phos while at pH 3 to 5 and 7 to 10 the
foaming capacities are similar. It is known that foaming capacity of a protein is best
near the pI, especially at low salt concentrations. This expectation is confirmed by
these experiments. At pH ≤ 3.6, (calculated pI) the α-La-phos had better foaming
capacity than α-La, but not at pH ≥ 3.6. α-La had better foaming capacity near pH 5,
its pI. The foaming stability of α-La-phos was much better than that of α-La at pH 2-4
(Fig. 2B). At pH 5 and above, neither protein formed a stable foam. Similar
improvement of the functional properties of phosphorylated casein were reported (21,
22).

Structural Properties. Gel filtration chromatography of α-La on Superose
12 showed a single peak at 13.8 ml elution volume (data not shown), while the

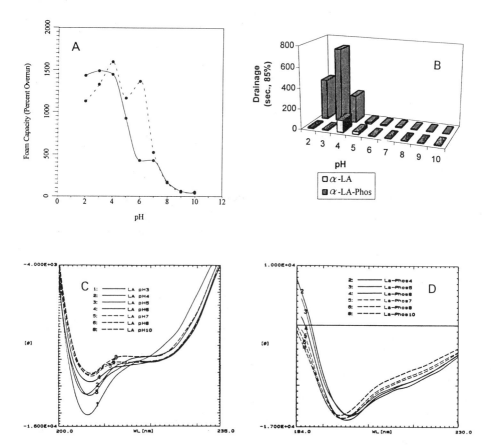

Figure 2.   Functional and structural properties of native (α-La) and
phosphorylated α-lactalbumin (α-La-phos). A. Foam capacity-pH profile of α-La
(--•--) and α-La-phos (—•—).   B. Foam stability-pH profile of α-La and α-La-
phos. C. CD profiles of α-La at pH 3-10. D. CD profiles of α-La-phos at pH 4-
10.

chromatogram of α-La-phos (Fig 1A) had a small shoulder (MW ~30,000) on the left side of the major peak. The major peak had a MW of ~15,000 compared to 14,200 for α-La while the shoulder had a MW of ~30,000 (apparently a dimer formed by cross linking of α-La by $POCl_3$). This point will be discussed later. But otherwise, the protein is intact (no fragmentation).

The CD patterns of the two proteins are compared in Figures 2C and 2D over the pH range of 4 to 10. The α-La-phos was too insoluble below pH 4 to do CD analysis. The CD patterns are different. This difference is better shown in Table I. There is less α-helix content and more β-sheet content at all pHs for α-La-phos than for α-La, indicating that there is a "loosing" of the α-helices of the secondary structure. However, this protein is far from being a totally denatured protein where all α-helices and β-sheets are destroyed.

The "expanded" nature of α-La-phos is also indicated by the use of the hydrophobic probes *cis*-parinaric acid (*cis*-PnA) and 8-anilinonaphthalene sulfonate (ANS) (Table II) and the average surface area cleared per protein molecule during the protein's adsorption and rearrangement (Table III). The surface hydrophobicity was higher for α-La-phos than for native α-La (Table II). In fact, the surface hydrophobicity of α-La-phos was higher than for denatured α–La by *cis*-PnA (201 vs. 162) but smaller by the ANS assay (63 vs. 83). The average surface area cleared per protein molecule during adsorption and rearrangement, and the number of amino acids at the interface, were appreciably higher (except at pH 5) for α-La-phos than for α-La (Table III). This correlates also with the effect of the two proteins in lowering the surface pressure (Table III).

β-Lactoglobulin and Phosphorylated β-Lactoglobulin. The solubility, functional properties and structural properties of β-lactoglobulin (β-Lg) and phosphorylated β-lactoglobulin (β-Lg-phos) were compared under the same experimental conditions. The β-Lg-phos had 2.66% phosphorus (15.7 moles P/mol protein).

Solubility. β-Lg and β-Lg-phos at 0.1% had 90 to 95% solubility from pH 4 to 10 (Fig. 3B). But the solubility of β-Lg-phos dropped drastically below pH 4, with essentially no solubility at pH 3. This is probably due to protonation of the phosphate groups, since β-Lg was very soluble in this pH range.

Emulsifying Activity and Stability. The E.A.I. for β-Lg-phos was lower than for β-Lg at pH 3 (because of poor solubility), but was better than β-Lg from pH 4 to 10 (Fig. 3C). The largest difference between β-Lg and β-Lg-phos was at pHs 6 and 10. Emulsion stability of β-Lg-phos was lower than that for β-Lg below pH 4 and above pH 7, but it was better (11 times) at pH 5-6 (near the pI of β-Lg) (Fig. 3D). This improvement is not due to improved solubilities of the two proteins (Fig. 3B), but rather a shift of the pI from about pH 5 for β-Lg to about pH 3.2 for β-Lg-phos. Overall, phosphorylation of β-Lg markedly improves its emulsifying properties.

Foaming Capacity and Stability. The foaming capacities of β-Lg-phos and β-Lg are about the same below pH 4, but above pH 4 the foaming capacity drops markedly for β-Lg-phos (Fig. 4A). The foam stability of β-Lg-phos is higher than β-Lg at pH 4 and below, but is less than that for β-Lg above pH 4 (Fig. 4B). Therefore, β-Lg-phos has better foaming properties than β-Lg at pH 2-4 but poorer foaming properties than β-Lg-phos at pH 5 to 10. Based on these results, we conclude that the additional negative charges of β-Lg-phos cause too many repulsive anionic forces at pH 5 and above. Only when the phosphate groups are protonated (pH 2-4) do the foaming properties of β-Lg-phos exceed those of β-Lg. The repulsive anionic forces could be masked by using higher salt concentrations.

Table I.   Secondary Structure Conformation of Native and Phosphorylated α-Lactalbumin by Circular Dichroism Measurements

	Secondary conformation				
pH	α-Helix (%)	β-Sheet (%)	β-Turn (%)	Random (%)	Unorder (%)
α-Lactalbumin					
3	33	17	5	45	50
4	30	20	5	45	50
5	28	20	5	45	50
6	28	16	10	48	58
7	27	16	10	48	58
8	27	21	7	45	52
9	26	23	8	43	51
10	27	23	8	42	50
Phosphorylated α-lactalbumin					
3	-	-	-	-	-
4	20	25	7	49	56
5	16	25	7	52	59
6	15	25	7	54	61
7	12	26	7	55	62
8	11	22	6	52	58
9	7	31	7	55	62
10	7	30	8	53	61

Table II.   Surface and Exposed Hydrophobicity of Proteins by the Fluorometric Probe Method (42, 43)

Protein	cis-PnA		ANS	
	Surface	Exposed	Surface	Exposed
α-La (native)	113	431	8	105
α-La-denatured	162	531	83	214
α-La-phos	201	591	63	117
β-Lg (native)	1627	848	54	371
β-Lg-phos	271	598	109	140
BSA-phos	61	-	-	-

Table III.   Effect of Native and Phosphorylated α-Lactalbumin on Surface
Properties at Air-Solution Interface

Treatment (pH)	Average area cleared per protein molecule (Å^2)	Apparent number of amino acid residues penetrating the surface	Surface pressure (mNm^{-1})
α-Lactalbumin			
3	112	7.5	30.6
4	122	8.1	31.6
5	170	11.3	30.8
6	108	7.2	27.0
7	88	6.0	24.2
8	99	6.6	22.0
9	149	9.9	16.3
10	168	11.2	15.0
α-Lactalbumin-phosphorylated			
3	256	17.0	23.6
4	207	13.8	22.8
5	147	9.8	20.0
6	190	12.7	15.2
7	264	17.6	12.6
8	343	22.9	11.4
9	352	23.5	10.4
10	278	18.5	10.3

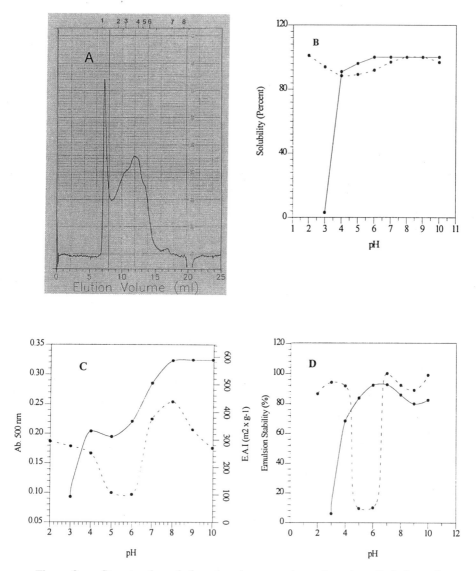

Figure 3.   Structural and functional properties of native (β-Lg) and phosphorylated β-lactoglobulin (β-Lg-phos). A. Superose 12 gel filtration chromatogram of β-Lg-phos. See Fig. 1A for standard MW markers. B. Solubility-pH profile of β-Lg (--•--) and β-Lg-phos (—•—). C. Emulsifying activity-pH profile of β-Lg (--•--) and β-Lg-phos (—•—). D. Emulsion stability-pH profile of β-Lg (--•--) and β-Lg-phos (—•—).

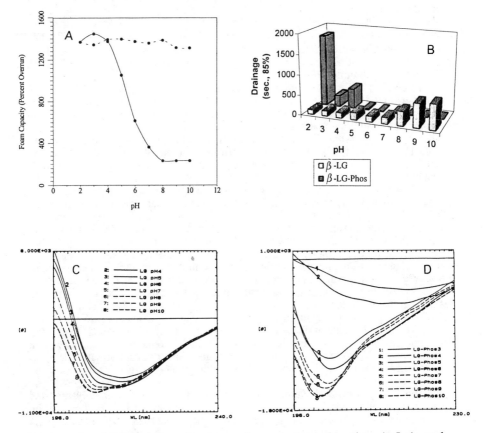

Figure 4.   Functional and structural properties of native (β-Lg) and phosphorylated β-lactoglobulin (β-Lg-phos). A. Foam capacity-pH profile of β-Lg (--•--) and β-Lg-phos (—•—). B. Foam stability-pH profile of β-Lg and β-Lg-phos. C. CD profiles of β-Lg from pH 4 to 10. D. CD profiles of β-Lg-phos from pH 3 to 10.

The solubility, emulsifying and foaming properties of phosphorylated casein as a function of pH were similar (21, 22) to those reported here for β-Lg-phos.

Structural Properties. There is evidence for considerable cross linking of β-Lg during phosphorylation (Fig. 3A). β-Lg eluted from a Superose 12 column in a sharp symmetrical peak at 13.0 ml using an FPLC system (data not shown). As shown in Figure 3A, the β-Lg-phos is a polydispersed product with a small amount of monomeric form (MW of 18,300), but the major portion of the protein was eluted in a molecular weight range between ∼200,000 (peak 1) and 40,000 (peak 2; perhaps dimeric β-Lg-phos). β-Lg-phos appears to be much more cross linked by POCl₃ than α-La, possibly because it may be in dimeric form under the conditions used for phosphorylation and because the extent of phosphorylation was higher (15.7 moles P/mol β-Lg vs. 7.8 moles P/mol α-La).

The CD profiles from 190 to 230 nm for β-Lg and β-Lg-phos are compared in Figures 4C and 4D from pH 4 to 10. Except for pH 2 and 3, the profiles for β-Lg-phos appear to show less randomness in the secondary structure than for β-Lg. But the calculated results show that the α-helix content is less for β-Lg-phos than for β-Lg at all pHs (3 to 10), while the β-sheet content is about the same, except at pH 9 and 10 (Table IV). There is also much less β-turn content in β-Lg-phos than in β-Lg. Therefore, β-Lg-phos appears to be a more "expanded" molecule than β-Lg by CD measurements.

The hydrophobic probe, *cis*-PnA, indicates that β-Lg-phos has much less surface hydrophobicity than β-Lg (271 vs. 1627) even when denatured by heat and SDS (848 vs. 598; Table II). This probably is a result of the intermolecular cross linking postulated above. The ANS probe shows that the surface hydrophobicity of β-Lg-phos is higher than β-Lg (109 vs. 54) while the heat and SDS denatured β-Lg has more surface hydrophobicity than β-Lg-phos. The *cis*-PnA and ANS results are different, since *cis*-PnA measures primarily hydrophobicity due to alkyl side chains, while ANS measures primarily hydrophobicity due to aromatic side chains (42, 43).

The average area cleared per protein molecule and the number of amino acid residues at the air-water surface is less for β-Lg-phos than for β-Lg (Table V). The effect of β-Lg-phos and β-Lg on surface pressure is about the same (Table V).

The overall results indicate that β-Lg-phos may be less "expanded" than β-Lg. This is probably the result of extensive cross linking due to the high phosphorylation of β-Lg-phos (15.7 moles P/mol protein). Perhaps a lower phosphorylation level as used for α-La-phos (7.8 moles P/mol protein) would give better functional properties.

Zein and Phosphorylated Zein. It is of interest to compare the newly reported results here on phosphorylation of α-La and β-Lg with those of Chobert *et al.* (23) on the phosphorylation of zein to improve its solubility properties. Zein is essentially insoluble in water but soluble in 60% aqueous acetone; phosphorylated zein (zein-phos) is more soluble in water, while α-La, α-La-phos, β-Lg and β-Lg-phos are readily soluble in water. Zein had very low nutritional quality while zein-phos(Trp, Lys) had half the nutritional value of casein, measured with *Tetrahymena thermophili*. The reader is referred to the published paper (23) for the experimental procedures. The solubility, foaming and emulsifying properties of zein, zein-phos, zein-phos(Lys), zein-phos(Trp) and zein-phos(Trp, Lys) were determined as described for α-La, α-La-phos, β-Lg and β-Lg-phos.

Solubility. The solubilities of zein, and zein-phos with 0.27, 2.60 and 9.5 moles phosphate/mol zein are shown in Figure 5A, as a function of pH (23). At pH 2.5, zein-phos(0.27) (with 0.27 mol phosphate/mol zein) was more water soluble than zein. Zein-phos(2.60) had about the same solubility as zein, while zein-phos (9.5) was less soluble than zein. But from pH 5 to 7.8, all zein-phos derivatives were

Table IV.   Secondary Structure Conformation of Native and Phosphorylated β-Lactoglobulin by Circular Dichroism Measurements

	Secondary conformation				
pH	α-Helix (%)	β-Sheet (%)	β-Turn (%)	Random (%)	Unorder (%)
β-Lactoglobulin					
3	20	45	15	20	35
4	20	45	15	20	35
5	20	45	15	20	35
6	15	47	13	25	38
7	15	45	10	30	40
8	15	42	10	33	43
9	10	42	8	40	48
10	10	40	5	46	51
Phosphorylated β-lactoglobulin					
3	15	45	10	30	40
4	15	45	9	31	41
5	5	44	5	46	51
6	5	44	2	49	51
7	5	40	0	55	55
8	5	38	0	57	57
9	4	34	0	62	62
10	3	34	0	63	63

Table V.   Effect of Native and Phosphorylated β-Lactalbumin on Surface Properties
            Air-Solution Interface

Treatment (pH)	Average area cleared per protein molecule ($\mathring{A}^2$)	Apparent number of amino acid residues penetrating the surface	Surface pressure (mNm^{-1})
β-Lactoglobulin			
3	-	-	18.6
4	430	26.7	23.1
5	318	21.2	24.0
6	332	22.1	20.0
7	260	17.3	19.5
8	181	12.1	18.8
9	339	22.6	14.2
10	181	12.1	16.0
β-Lactoglobulin-phosphorylated			
3	322	19.8	24.0
4	321	19.0	23.8
5	360	15.0	21.8
6	365	15.0	18.6
7	343	13.8	18.4
8	326	13.6	18.2
9	304	10.9	15.1
10	308	8.7	14.4

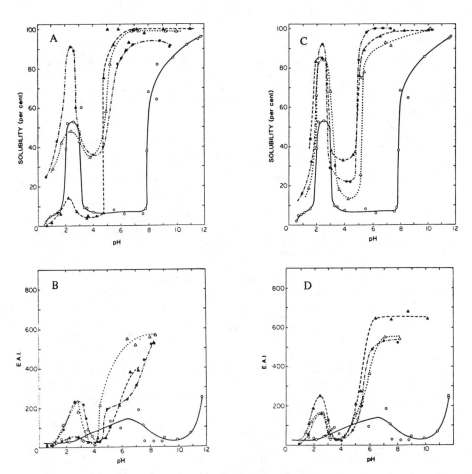

Figure 5.   Solubility-pH and emulsifying activities-pH profiles for zein and modified zeins at 0.1% protein and 23°C. A. Solubility-pH profiles of zein (o—o); zein-phos (0.27 mol P/mol zein) (•—•); zein-phos (2.6 mole P/mol zein) (Δ···Δ); zein-phos (9.5 mol P/mol zein (▲---▲). B. Emulsifying activity profiles for zein and zein-phos. Symbols used as in A. C. Solubility-pH profiles of zein (o—o), zein-phos(Trp) (1.07 mol P/mol zein and 0.61% Trp) (•—•); zein-phos(Lys) (1.20 mol P/mol zein and 0.46% Lys) (Δ···Δ); and zein-phos(Try, Lys) (0.98 mol P/mol zein, 1.05% Trp and 0.24% Lys) (▲---▲). D. Emulsifying-pH profiles for zein, zein-phos(Trp), zein-phos(Lys) and zein-phos(Trp, Lys). Symbols used and concentrations of phosphate, Trp and Lys as in C. Reproduced with permission from Ref. 23. Copyright 1987 American Oil Chemists Society.

very much more soluble than zein. Zein-phos(0.27) and zein-phos(2.60) were also much more soluble than zein at pH 3.5 to 5.

Emulsifying Activity. The emulsifying activities of the phosphorylated zeins were much better than zein at all pHs except near pH 4 (Fig. 5B). Zein-phos (0.27) had equal or better emulsifying activity than the more extensively phosphorylated zein (zein-phos(2.60) and zein-phos(9.5)).

Therefore, phosphorylation markedly improved the water solubility and emulsifying activity of zein. However, a low level of phosphorylation was better than more extensive modification.

Zein and Phosphorylated Zein with Covalently-bound Amino Acids.
Chobert *et al.* (23) reported that lysine and tryptophan are covalently bound to zein in the presence of POCl3, giving a phosphorylated, amino acid derivative of zein. The derivatives prepared were zein-phos, zein-phos(Trp), zein-phos(Lys) and zein-phos (Trp, Lys).

Solubility. The pH-solubility curves for zein, zein-phos(Trp), zein-phos (Lys) and zein-phos(Trp, Lys) are shown in Figure 5C. The degree of substitution is given in the figure legend. All the derivatives had better water solubility than zein; the differences are most apparent at pH 2.5 and above pH 4.5.

Emulsifying Activity. The emulsifying activity indices (E.A.I.) of zein, zein-phos(Trp), zein-phos(Lys) and zein-phos(Try, Lys) are shown in Figure 5D. Except at pH 3 to 5, the E.A.I. of the derivatives were much higher than those for zein and were similar to the E.A.I. for zein-phos(0.27) shown in Figure 5B. At pH 6 and above, the E.A.I. values were similar to those for bovine serum albumin, a highly functional protein.

Nutritional Quality of Zein and Modified Zein. Chobert *et al.* (23) determined the nutritional quality of zein and modified zeins, by measuring the growth rates of the protozoan, *T. thermophili*. *T. thermophili* have similar essential amino acid requirements as humans (19). The results are shown in Table VI. Casein was used as a standard protein in determining the relative rates of growth of the protozoa.

Zein had 4.5% the nutritional quality of casein. This was reduced to 1% by phosphorylation because of additional loss of Trp and Lys during modification. Zein-phos(Trp), at 0.61% Trp, had 22% the nutritional quality of casein, while zein-phos (Lys), at 0.46% Lys had 8.8% the nutritional quality of casein. Zein-phos(Trp, Lys), at 1.05% and 0.24% Trp and Lys, respectively, had 48.8% the nutritional quality of casein. A more highly phosphorylated zein, Mr1 (Table VI), with less Trp had only 24.4% the nutritional quality of casein.

Further research is needed to optimize the conditions for covalent incorporation of free amino acids into proteins by coupled phosphorylation/peptidization. We have not determined the mechanism of binding of the amino acids to proteins. It seems reasonable to suggest that POCl3 and amino acids initially form a mixed phosphate-carboxyanhydride, which then reacts with an ϵ-amino group of a protein (Pro-NH2) (44), as shown in Eq. 6.

$$\text{Pro-NH}_2$$
$$Cl_3PO + AA\text{-}COOH \rightarrow AA\text{-}COOPOCl_2 \rightarrow AA\text{-}CONH\text{-}Pro + Pro\text{-}NH\text{-}POCl_2 \qquad (6)$$

Pro-NH-POCl2 is hydrolyzed to Pro-NH-PO(OH)2. At the low pH of phosphorylation, the ammonium group of the amino acid would not be phosphorylated. It is known that both amino and imidazole nitrogens and hydroxyl

Table VI.   Zein and Modified Zeins and Their Effect on Growth Rate of *Tetrahymena thermophili*

Sample	Zein concn. (%)	Molar ratio[a] (POCl$_3$/zein)	P bound (mol P/mol) zein)	Amino acid bound[b] Trip	Lys	Relative growth[c]
Casein	-	-	-	-	-	100
Zein	-	-	0	0.16	0.09	4.5
A	4.75	100	2.60	0.09	0.00	1.1
D	1.20	100	0.27	0.09	0.00	1.1
6	5.00	400	9.5	-	-	-
T10	5.00	200	1.07	0.61(10)[d]	0.09	22.2
L40	5.00	200	1.20	0.09	0.46(40)[d]	8.8
M1	5.00	200	0.98	1.05(5)[d]	0.24(10)[d]	48.8
Mr1	5.00	400	12.3	0.63(5)[d]	0.22(10)[d]	24.4

[a]Molecular weight of zein, 38,000.
[b]Total, in percentage, including small amounts of the original zein.
[c]Growth relative to that on casein.
[d]Molar ratio of free amino acid to zein added to the reaction mixture.

SOURCE: Reprinted with permission from Ref. 23. Copyright 1987 American Oil Chemists Society.

groups of seryl, threonyl and tyrosyl residues of proteins react with $POCl_3$ and that the partitioning is influenced by reaction conditions and nature of the protein (10, 18, 21). It is also known that triethylamine and hexylamine facilitate phosphorylation of proteins (27). The P-Cl bonds are very easily hydrolyzed in water to form phosphorylated protein and HCl. Therefore, no toxic compounds are formed. Earlier in this paper we mentioned the inter- and intra-cross linking of α-La and β-Lg by $POCl_3$. It could occur by the same mechanism as shown in Eq. 6, replacing the amino acid with a second ∈-amino group of protein-bound lysine.

Recently, porcine insulin was phosphorylated with $POCl_3$ for use in treating diabetes mellitus (26). The pI of insulin-phos was < 4. It was used in pancreatectomized dogs to decrease blood glucose from 180 mg% (with regular insulin infusion) to 120-130 mg%. Chelators, e.g. EDTA, were used to increase the stability of phosphorylated insulin. If insulin-phos is approved eventually for use in humans, perhaps *in vitro* phosphorylated proteins will be accepted for use in foods some day.

Conclusions. This chapter, through new and published data, documents that α-La and β-Lg can be easily phosphorylated under mild conditions (0°C, neutral pH) with $POCl_3$ to give higher foaming and emulsifying properties than the original protein, especially near the pI. At a low level of phosphorylation, the solubility, foaming and emulsifying properties were improved also at pH 4 and below. However, extensive phosphorylation gave a less functional protein than the original protein at pH 4 and below. Phosphorylation increased the water solubility of zein, thereby improving its functional properties (23). Also, the nutritional quality of zein was greatly improved by covalent attachment of lysine and tryptophan to zein, catalyzed by $POCl_3$ (23).

Literature Cited

1.   Whitaker, J. R. In *Food Proteins: Improvement Through Chemical and Enzymatic Modification;* Feeney, R. E. and Whitaker, J. R., Eds.; Advances in Chemistry Series 160; American Chemical Society: Washington, D. C., 1977; pp. 95-155.

2.   Uy, R.; Wold, F. In *Chemical Deterioration of Proteins;* Whitaker, J. R. and Fujimaki, M., Eds.; ACS Symposium Series 123; American Chemical Society: Washington, D. C., 1980; pp. 49-62.

3.   Whitaker, J. R. In *Modification of Food Proteins: Food, Nutritional, and Pharmacological Aspects;* Feeney, R. E. and Whitaker, J. R., Eds.; Advances in Chemistry Series 198; American Chemical Society: Washington, D. C., 1982; pp. 57-87.

4.   Protein Phosphorylation, Chemical Abstracts, Vol. 122; American Chemical Society: Columbus, Ohio, 1995.

5.   Chung, S. L.; Ferrier, L. K. *J. Food Sci.* **1992**, *57*, 40-42.

6.   Kato, A. In *Interactions of Food Proteins;* Parris, N. and Barford, R., Eds.; ACS Symposium Series 454; American Chemical Society: Washington, D. C., 1991; pp. 15-24.

7.   Sen, L. C.; Gonzales-Flores, E.; Feeney, R. E.; Whitaker, J. R. *J. Agric. Food Chem.* **1977**, *25*, 632-638.

8.   Campbell, F.; Shih, F. F.; Marshall, W. E. *J. Agric. Food Chem.* **1992**, *40*, 403-406.

9.   Whitaker, J. R.; Feeney, R. E. In *Advances in Experimental Medicine and Biology, Vol. 86B, Protein Crosslinking B, Nutritional and Medical Consequences;* Freidman, M., Ed.; Plenum Publishing Corp.: New York, 1977; pp. 155-175.

10.   Matheis, G.; Whitaker, J. R. *Int. J. Biochem.* **1984**, *16*, 867-873.

11.   Heidelberger, M.; Davis, B.; Treffers, H. P. *J. Am. Chem. Soc.* **1941,** *63,* 498-503.
12.   Mayer, M.; Heidelberger, M. *J. Am. Chem. Soc.* **1946,** *68,* 18-25.
13.   Boursnell, J. C.; Dewey, H. M.; Wormall, A. *Biochem. J.* **1948,** *43,* 84-90.
14.   Ferrel, R. E.; Olcott, H. S.; Fraenkel-Conrat, H. *J. Am. Chem. Soc.* **1948,** *70,* 2101-2107.
15.   Salak, J.; Vodrazka, Z.; Cejka, J. *Collect. Czech. Chem. Commun.* **1965,** *30,* 1036-1040.
16.   *Modification Reactions;* Hirs, C.; Timasheff, S. N., Eds.; *Methods in Enzymology, Vol.* 25; Academic Press: Boca Raton, FL, 1972.
17.   Yoshikawa, M.; Sasaki, R.; Chiba, H. *Agric. Biol. Chem.* **1981,**45, 909-914.
18.   Woo, S. L.; Creamer, L. K.; Richardson, T. *J. Agric. Food Chem.* **1982,** *30,* 65-70.
19.   Matheis, G.; Penner, M.; Feeney, R. E.; Whitaker, J. R. *J. Agric. Food Chem.* **1983,** *31,* 379-387.
20.   Sung, H. Y.; Chen. H.-J.; Liu, T.-Y.; Su, J.-C. *J. Food Sci.* **1983,** *48,* 716-721.
21.   Woo, S. L.; Richardson, T. *J. Dairy Sci.* **1983,** *66,* 984-987.
22.   Matheis, G.; Whitaker, J. R. *J. Agric. Food Chem.* **1984,** *32,* 699-705.
23.   Chobert, J.-M.; Sitohy, M.; Whitaker, J. R. *J. Am. Oil Chem. Soc.* **1987,** *64,* 1704-1711.
24.   Chobert, J.-M.; Sitohy, M. Z.; Whitaker, J. R. *Sciences des Aliments.* **1989,** *9,* 749-761.
25.   Casella, M. L. A.; Whitaker, J. R. *J. Food Biochem.* **1990,** *14,* 453-475.
26.   Lougheed, W. D. PCT Int. Appl. WO 92 14,754, 1992.
27.   Sitohy, M.; Chobert, J.-M.; Haertle, T. *Milchwissenschaft.* **1994,** *49,* 610-615.
28.   Sitohy, M.; Chobert, J.-M.; Haertle, T. *J. Agric. Food Chem.* **1995,** *43,* 59-62.
29.   Davis, B. J. *Ann. N. Y. Acad. Sci.* **1964,** *121,* 404-427.
30.   Weber, K.; Osborn, M. In *The Proteins*; Neurath, H., Hill, R. L. and Boeder, C.-L., Eds.; The Proteins, Vol. I; Academic Press: Boca Raton, FL, 1975; pp. 179-223.
31.   Allen, R.J.L. *Biochem. J.* **1940,** *34,* 858-865.
32.   Lowry, O. H.; Rosebrough, N. J.; Farr, A. L.; Randall, R. J. *J. Biol. Chem.* **1951,** *193,* 265-275.
33.   Pearce, K. N.; Kinsella, J. E. *J. Agric. Food Chem.* **1978,** *26,* 716-723.
34.   Cameron, D. R.; Weber, M. E.; Idziak, E. S.; Neufeld, R. J.; Cooper, D. G. *J. Agric. Food Chem.* **1991,** *39,* 655-659.
35.   Graham, D. E.; Phillips, M. C. *J. Colloid Interface Sci.* **1979,** *70,* 403-414.
36.   Graham, D. E.; Phillips, M. C. *J. Colloid Interface Sci.* **1979,** *70,* 415-426.
37.   MacRitchie, F.; Alexander, A. E. *J. Colloid Sci.* **1963,** *18,* 453-457.
38.   Tornberg, E. *J. Colloid Interface Sci.* **1978,** *64,* 391-402.
39.   Tornberg, E. *J. Sci. Food Agric.* **1978,** *29,* 762-776.
40.   Waniska, R. D.; Kinsella, J. E. In *Food Proteins;* Kinsella, J. E. and Soucie, W. G., Eds.; American Oil Chemists Society: Champaign, Ill., 1989; pp. 100-131.
41.   MacRitchie, F. *Adv. Protein Chem.* **1978,** *32,* 283-326.
42.   Kato, A.; Nakai, S. *Biochim. Biophys. Acta .* **1980,** *624,* 13-20.
43.   Hayakawa, S.; Nakai, S. *J. Food Sci.* **1985,** *50,* 486-491.
44.   van Wazer, J. R. *Phosphorus and Its Compounds-Chemistry,* Vol. 1; Interscience: New York, NY, 1958.

Chapter 18

Improvement of Functional Properties of Food Proteins by Conjugation of Glucose-6-Phosphate

Takayoshi Aoki

Department of Biochemical Science and Technology, Faculty of Agriculture, Kagoshima University, Kagoshima 890, Japan

Phosphorylation of food proteins improves functional properties. However, chemical phosphorylation of food proteins is not easily accepted by consumers and enzymatic phosphorylation gives only low phosphate content to proteins. In this study, whey protein isolate (WPI), egg white protein, β-lactoglobulin, and ovalbumin were conjugated with glucose-6-phosphate (G6P) through the Maillard reaction in a controlled dry state. The phosphorus contents of protein-G6P conjugates, prepared by incubation at 50 °C and 65% relative humidity for 0.5-1 day were nearly 1%. The proteins became heat-stable and their emulsifying activity was increased by the conjugation with G6P. Protein-G6P conjugates solubilized calcium phosphate and inhibited the precipitation of calcium phosphate in the solution supersaturated with respect to calcium phosphate. Soluble calcium level in the small intestine and Ca absorption were higher in rats fed with WPI-G6P conjugates than in rats fed with soy protein.

Phosphorylation of food proteins improves functional properties such as water solubility, emulsifying activity, foaming properties, and gel-forming properties (1-2). It is well known that phosphoproteins interact with calcium and calcium phosphate. Casein solubilizes calcium phosphate. In milk, calcium and phosphate, present in excess of their solubilities, are not precipitated as a result of the interaction with casein through phosphate groups (3-5). Egg white riboflavin-binding protein (RfBP), which has 8 phosphate groups in the C-terminal region (6-7), also solubilizes calcium phosphate (8). Caseins have been shown to yield phosphopeptides upon digestion in the small intestine (9). These phosphopeptides also solubilize calcium phosphate and enhance calcium absorption in the intestinum ileum, increasing the calcium concentration by preventing the precipitation of calcium phosphate. Solubilization of calcium phosphate is needed for the enhancement of calcium absorption in the intestinum ileum. Accordingly, it is expected that food proteins with no phosphate groups can be altered to solubilize calcium phosphate by phosphorylation. However, excessive phosphorylation may not improve the

0097–6156/96/0650–0230$15.00/0

functional properties because egg yolk phosvitin, which is highly phosphorylated and contains about 10% phosphorus, does not enhance calcium absorption (*10*). Chemically phosphorylated proteins are not easily accepted as foods by consumers, although chemical phosphorylation with phosphorus oxychloride is considered to be a promising tool for changing the functional properties (*2*). Enzymatic phosphorylation is the most preferred method for food proteins with respect to food safety, but it gives only a low phosphate group content to proteins because protein kinase has a specificity for the amino acid sequences (*11-13*) and costs a great deal since it needs ATP.

Recently, some attempts have been made to improve the functional properties of enzyme and food proteins by the conjugation of glucose (*14*) or dextran (*15*) through the Maillard reaction. For example, Kato and Kobayashi (*15*) succeeded in making protein-dextran conjugates with excellent emulsifying properties. Glucose-6-phosphate (G6P) retains a reducing group and reacts with amino groups of proteins through the Maillard reaction (Figure 1). Thus, we can introduce phosphate groups to proteins by G6P modification. By the conjugation of proteins with G6P, the positive charge of amino groups is altered to the negative charge of the phosphate groups. In this study, using the Maillard reaction, we have conjugated food proteins such as egg white proteins and whey proteins with G6P to improve their functional properties.

MATERIALS AND METHODS

Ovalbumin (Ova) was a Grade VI from Sigma Ltd. (Tokyo, Japan). β–Lactoglobulin (β-Lg) was prepared from raw skim milk using the method of Armstrong et al. (*16*). Whey protein isolate (WPI) was obtained from Taiyo Kagaku Co., Ltd.. (Yokkaichi, Japan), and contained 89.7% protein, 0.5% fat, 1.3% ash, and 6.1% moisture. Egg white protein (EWP) was prepared as follows: egg white was separated from infertile eggs of White Leghorn hens was homogenized, acidified to pH 5.5, and then centrifuged at 1000g for 15 min. The supernatant obtained was dialyzed to remove the glucose and then lyophilized.

The proteins were dissolved in distilled water (0.5% w/v for the measurement of free amino group content and brown color, and 5% for sample preparation) with G6P-2Na-6H_2O (94% of the protein dry weight, equivalent to 50% glucose). The protein-G6P solutions were adjusted to pH 7.5 with 1 N NaOH and then lyophilized. The dried samples were incubated at 50 °C and 65% relative humidity using a saturated KI solution for 0-5 days. For the sample preparation, the samples were dissolved in 5% protein concentration, dialyzed, and lyophilized.

The phosphorus content of protein-G6P conjugates was measured by the method of Allen (*17*) after digestion in perchloric acid. Brown color was measured by the absorbance at 420 nm of the protein-G6P mixture (0.5% protein solution). Free amino groups were measured using the method of Böhlen et al. (*18*).

Gel filtration HPLC was carried out at room temperature (22-25 °C) with a Shimadzu LC-5A chromatograph equipped with a Shimadzu SPD-2A spectrophotometric detector, using a TSK-GEL G3000SW column. Elution buffers used were 0.1 M sodium phosphate buffer (pH 7.0) containing 0.3 M NaCl for the estimation of protein polymerization and simulated milk ultrafiltrate (SMUF), which was prepared using the method of Jenness and Koops (*19*), for the measurement of the cross-linking of protein by calcium phosphate.

The heat stability of protein samples was measured as follows: protein samples were dissolved at a protein concentration of 1 mg/mL with 50 mM Tris-HCl buffer (pH 7.0) and heated at 60-100 °C for 10 min. The aggregates were precipitated by centrifugation at 1500g for 10 min. The absorbance of the supernatant was measured at 280 nm to estimate the protein concentration of the solution.

The emulsifying property was measured by the method of Pearce and Kinsella (20).   Peanut oil (150 μL) was added to 100 μL of protein solution (1 mg of protein/mL), and each sample was well mixed and then emulsified by mixing with a homogenizer (Vortex Genie) at the highest revolution for 1 min.   Immediately after the mixture was emulsified, aliquots of the emulsion were diluted with 0.1% NaDodSO4, and the absorbance of the diluted emulsion was then determined in a 1-cm path-length cubette at a wavelength of 550 nm.

The test solutions for the solubilization of the calcium phosphate by protein-G6P conjugates were prepared using the procedure for artificial casein micelles (21).   To 2 ml of 4% protein solution, 200 μL of 0.2 M potassium citrate, 200 μL of 0.2 M CaCl2, and 240 μL of 0.2 M K2HPO4 were added, followed by the addition of 100 μL of 0.2 M CaCl2 and 50 μL of 0.2 M K2HPO4.   This addition of 100 μL of 0.2 M CaCl2 and 50 μL of 0.2 M K2HPO4 was repeated to give the prescribed concentrations of calcium and inorganic phosphate (Pi).   The addition intervals were 15 min, and all additions were accompanied by stirring at pH 6.7.   The volume was adjusted to 4 ml by measuring the weight of the solutions.   The prepared solutions were allowed to stand for 20 h at 25 °C and then centrifuged at 1000g for 15 min.   The precipitated calcium and Pi concentrations were estimated from their differences between the total and the supernatant fractions.   Calcium and Pi contents were measured in the 12% trichloroacetic acid filtrate by a Hitachi 208 atomic absorption spectrophotometer and the method of Allen (17).   Inhibitory effects of protein-G6P conjugates on the spontaneous precipitation of calcium phosphate were measured by the method of Hay et al. (22).

Male Wister rats, 27 days old, were introduced to examine the effect of calcium absorption.   The composition of the experimental diet was the same as that reported by Sato et al. (23).   The rats were trained to consume the diet within 1.5 h each day.   On the 9th day of meal feeding, the rats were divided into three groups: the whey protein, casein, and WPI-G6P protein diet groups.   The number of rats in each group was 6.   The rats were allowed to consume either the 20% soy protein diet or the 20% casein diet or the 20% WPI-G6P diet within 1.5 h.   The ileum was ligated at the point of 12 cm and 20 cm from the ileocecal junction, and 0.2 mL of ^{45}Ca was administered into the ligated loop.   After 1 h, the content of the ligated loop was washed out.   The soluble Ca in the ligated loop was estimated by determining the Ca concentration remaining in the supernatant after centrifugation at 7000g for 30 min.   The calcium absorption was estimated from the disappearance of the administered ^{45}Ca in the ligated ileum loop during 1 h.

RESULTS AND DISCUSSION

The Maillard Reaction of Proteins with G6P.   First the Maillard reaction of Ova with G6P was examined because Ova has been often used as a model protein of the Maillard reaction of proteins (15, 24, 25).   Figure 2 shows the changes in the relative value of the free amino group content of Ova incubated with G6P or glucose at 50 °C and 65% relative humidity.   The free amino group decreased quickly, and 50% of the free amino group was blocked with G6P or glucose after 1 day of incubation.   It was confirmed that there was no decrease in the free amino group of Ova incubated without sugars.   Ova has 20 lysine residues.   Therefore, the Ova-G6P conjugate incubated for 1 day was expected to already have more than 10 phosphate groups through lysine-ε-amino groups, assuming that no degradation of the reacted G6P molecules occurred.   In fact, the Ova-G6P conjugate incubated for 1 day contained 1.06% phosphorus.   This phosphorus content is a little higher than that of the whole bovine casein.   The electrophoretic mobility of Ova in native polyacrylamide gel electrophoresis increased with the increase in the incubation period, and the

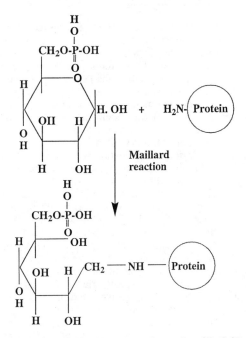

Figure 1. Maillard reaction of protein with G6P.

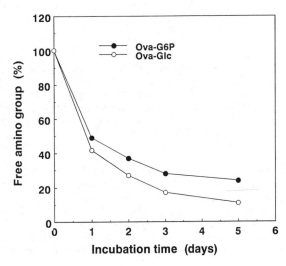

Figure 2. Decrease in the free amino group in Ova incubated with G6P or glucose (Glc).

increase in mobility was much larger in the Ova-G6P systems than in the Ova-glucose systems (26).

The brown color formation and the intermolecular covalent cross-linking are accompanied by the Maillard reaction of proteins with sugars. Figure 3 shows the brown color formation during the incubation of Ova with G6P or glucose. The brown color formation of the Ova-G6P systems was developed more strongly than that of the Ova-glucose system, especially at the later stage of incubation period. The absorbance at 420 nm of the Ova-G6P system was about 3 times that of the Ova-G6P system. The brown color development depended on the species of proteins. The absorbance at 420 nm of the β-Lg-G6P system incubated for 3 days was 2.2 times that of the Ova-G6P system. The brown color development of the WPI-G6P system was also faster than that of EWP-G6P system (27).

Protein polymerization during the incubation with G6P or glucose was examined by gel filtration HPLC. Figure 4 shows a typical elution profile for both Ova-G6P and Ova-glucose systems incubated for 3 days. Native Ova eluted at a retention time of about 38 min. Several additional peaks with shorter retention time appeared for samples incubated with sugars, and the number and height of the peaks with shorter retention time increased, especially in the Ova-G6P system. The peak height corresponding to the monomeric and polymeric forms of the Ova-G6P system was markedly lower and higher, respectively, than that of the Ova-glucose system. The peaks incubated by shaded and unshaded areas were regarded as the polymer and monomer fraction, respectively. Polymers formed in the Ova-G6P system incubated for 3 days were about 3 times that in the Ova-glucose system. The proportion of polymers formed in the β-Lg-G6P system incubated for 3 days was 56%, and this value was close to that of the Ova-G6P system incubated for 3 days.

Although the rate of the free amino group decrease in Ova-G6P system is slower than that of the glucose system (Figure 2), the browning reaction and protein polymerization at the later stage of reaction proceeded more rapidly in the Ova-G6P system than the Ova-glucose system, suggesting that the polar phosphate group of G6P accelerates the advanced step of the Maillard reaction. Therefore, the reaction time of proteins with G6P should be shortened as much as possible to avoid extra browning and cross-linking reactions.

Functional Properties of Protein-G6P Conjugates. In order to examine the heat stability of the Ova-G6P conjugate, 0.1% solutions of the Ova and Ova-G6P conjugate were heated at various temperatures (60-95 °C) for 10 min and then the soluble protein was determined. As shown in Figure 5, the solubility of Ova without treatment decreased as the heating temperature increased. There was little or no soluble Ova after heating at a temperature higher than 85 °C, whereas the Ova-G6P conjugate was soluble after heating at 95 °C for 10 min. Although β-Lg, EWP, and WPI were heat-unstable, their G6P conjugates were soluble after heating at 95 °C for 10 min.

The emulsifying activity of β-Lg phosphorylated with POCl3 was reported to become increased (28). The effect of G6P modification on emulsifying activity was studied by using Ovas incubated with G6P for various periods. As shown in Figure 6, the emulsifying activity of Ova-G6P conjugate increased as the incubation time increased and reached a maximum at 1 day. The activity of the Ova-G6P conjugate incubated for 1 day was about twice that of the untreated one. The emulsifying activity of the Ova-glucose conjugates was almost constant throughout the incubation time.

Amino groups are modified with G6P through the Maillard reaction. At neutral pH, for example pH 7.0, most of the ε-amino groups of lysine are positively charged because their pKa value is 10.0, and the net charge of one phosphoryl group gives a value of -1.8 when the values 1.5 for pK_1 and 6.4 for pK_2 are adopted (29).

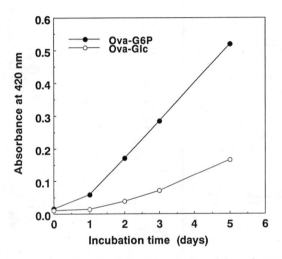

Figure 3. Brown color development of Ova-G6P and Ova-glucose systems.

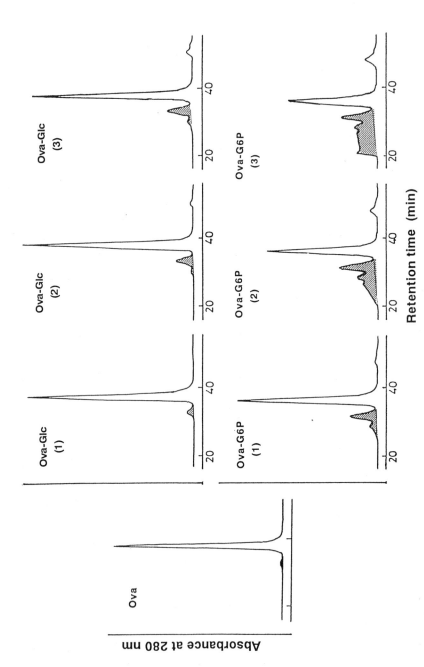

Figure 4. Elution profiles of Ovas incubated with G6P or Glc from a TSK-GEL G3000SW column. Typical profiles of samples incubated for 1, 2, and 3 days are shown.

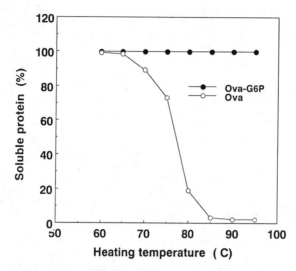

Figure 5. Heat stability of Ova and Ova-G6P conjugate.

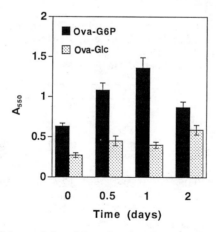

Figure 6. Emulsifying activity of Ovas incubated for 0-2 days with G6P or Glc.

Therefore, the increase in the number of net negative charges at pH 7 results in 2.8 by the modification of one ε-amino group with G6P.   Obviously this change of the net charge must make important   contributions to the improvement of the functional properties of proteins.

Interactions of Protein-G6P Conjugates with Calcium Phosphate.

Phosphoproteins interact with calcium and calcium phosphate and solubilize calcium phosphate.   Solubilization of calcium phosphate was examined using various protein-G6P conjugates incubated for 1 day.   In the absence of protein, more than two thirds of the calcium and half of the phosphate were precipitated at pH 6.7 in a solution containing 30 mM calcium, 22 mM Pi, and 10 mM citrate (Table I).   These salt concentrations are close to those of bovine milk.   In the presence of proteins without G6P, calcium and phosphate also precipitated, although slight solubilizing capabilities of these proteins were observed.   In the presence of 2% protein-G6P conjugates, no calcium phosphate precipitated at 30 mM calcium and 22 mM Pi, although a small amount of calcium phosphate precipitated at 40 mM calcium and 27 mM Pi.   In the presence of 2% RfBP, neither calcium nor phosphate precipitated even at 40 mM calcium and 27 mM Pi, although the phosphorus content of RfBP was lower than that of the protein-G6P conjugates.   The amounts of solubilized calcium and Pi were calculated using the WPI-G6P conjugate.   In the presence of a 2% WPI-G6P conjugate, 27.6 mM calcium and 17.5 mM Pi were estimated to be solubilized in the solution containing 40 mM calcium and 27 mM Pi.   Because the organic phosphate concentration was 6.9 mM in a 2% WPI-G6P conjugate, 4.0 mol calcium and 2.5 mol Pi were estimated to be solubilized per 1 mol organic phosphate if all the solubilized calcium and Pi are caused only by organic phosphates.   The calcium phosphate-solubilizing ability of the phosphate groups of the WPI-G6P conjugate was lower than that of RfBP and lower than that of the β-casein phosphopeptide reported by Naito and Suzuki (*30*).

Table I.   Phosphorus Content of Various Proteins, and Precipitated Ca and Pi in the Presence and Absence of Proteins
The test solution A contained 30 mM Ca, 22 mM Pi, and 10 mM citrate, and B contained 40 mM Ca, 27 mM Pi, and 10 mM citrate.

Protein sample	P content (%)	A		B	
		Precipitated		Precipitated	
		Ca (mM)	Pi (mM)	Ca (mM)	Pi (mM)
Ova	0.11	19.7	12.6	26.7	17.0
Ova-G6P	1.06	0	0	1.2	1.0
β-Lg	0.05	19.8	12.6	26.3	17.1
β-Lg-G6P	1.10	0	0	1.9	1.1
EWP	0.09	19.5	11.6	27.1	16.6
EWP-G6P	0.96	0	0	2.1	1.3
WPI	0.03	20.0	12.8	27.8	17.8
WPI-G6P	1.07	0	0	0.6	0.3
RfBP	0.82	0	0	0	0
In the absence of protein	–	20.9	13.6	28.2	17.8

As described before, phosphopeptides enhance calcium absorption, increasing the calcium concentration by preventing the precipitation of calcium phosphate.   We examined the inhibitory effect of β-Lg-G6P conjugate on the precipitation of calcium phosphate using the method of Hay et al. (*22*).   In the absence of protein, calcium phosphate precipitated spontaneously (Figure 7).   β-Lg had a slight inhibitory effect. When 0.025% β-Lg-G6P was present, almost no precipitate was formed after 24 h.

Calcium phosphate was solubilized in the presence of protein-G6P conjugates. However, 14.2 mM calcium and 9.2 mM Pi were sedimented at 100,000g in the solution containing 2% β-Lg-G6P conjugate, 30 mM calcium and 22 mM Pi, and 20.1% of the total protein was sedimented with calcium phosphate.   This indicates that the fine particles are composed of the complex of β-Lg-G6P conjugate and calcium phosphate.   In fact, particles of 30-150 nm diameter were observed by electron microscopy.   In the case of RfBP, almost no calcium phosphate sedimented at 100,000×g (data not shown).

We have already reported that cross-linking of RfBP by calcium phosphate (*8*). RfBP cross-linked by calcium phosphate formed at 12-20 mM calcium and 13-17 mM Pi and 10 mM citrate was dimer.   In order to examine the interactions of β-Lg-G6P conjugate with calcium phosphate in more detail, gel filtration HPLC of the β-Lg-G6P conjugate solution containing calcium, Pi and citrate was performed using simulated milk ultrafiltrate as the effluent.   As shown in Figure 8, the peak eluted at the void volume of the column appeared in the sample containing 30 mM calcium, 22 mM Pi, and 10 mM citrate.   This suggests that fine particles, which were observed by electron microscopy, appeared without formation of dimer.   The proposed scheme for the formation of the complex of β-Lg-G6P conjugate with calcium phosphate is shown in Figure 9.   At low calcium and Pi concentrations, even when calcium and Pi bind to β-Lg-G6P conjugate, the protein is not cross-linked.   When calcium and Pi concentrations become higher, relatively large particles of the β-Lg-G6P conjugate and calcium phosphate complex appear.   These particles disperse in the solution without precipitation.

Effects of Protein-G6P Conjugates on   Calcium Absorption.   The effect of the WPI-G6P conjugate on calcium absorption was examined using rats.   The proportions of soluble calcium in the ligated loop of the small intestine of the soy protein, casein, and WPI-G6P conjugate diet groups were 43.2 ± 8.7, 72.7 ± 8.0, and 77.7±9.7, respectively.   These values are means ± S.D. for 6 rats.   The soluble calcium in the small intestine was significantly higher in the WPI-G6P conjugate diet group than in the soy protein diet group.   This indicates that the WPI-G6P conjugate solubilizes calcium to prevent the precipitation of calcium phosphate in the small intestine.   The calcium absorption, which was estimated by the disappearance of [45]Ca in the ligated loop, of the soy protein, casein, and WPI-G6P conjugate diet groups was 50.7±17.6, 58.5±8.3, and 72.6±19.7, respectively.   The calcium absorption was also higher in the WPI-G6P conjugate diet group than in the soy protein diet group.   The effect of WPI without G6P on the calcium absorption was not examined in the present experiment.   However, Sato et al. (*23*) reported that soy protein did not solubilize calcium in the small intestine.   Accordingly, it is suggested that the protein-G6P conjugate has the ability to solubilize calcium phosphate and enhance calcium absorption in the small intestine.   Further studies are needed to evaluate the calcium absorption-stimulating function of protein-G6P conjugates.

CONCLUSION

Food proteins could be easily conjugated with G6P using the Maillard reaction. Functional properties such as heat stability and emulsifying properties of food proteins were markedly improved by the conjugation of G6P.   Protein-G6P

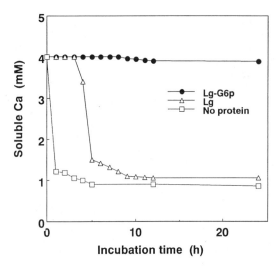

Figure 7. Inhibitory effects of β-Lg and β-Lg-G6P conjugate on precipitation of calcium phosphate from the solution supersaturated with respect to calcium phosphate. Protein concentration was 0.025%

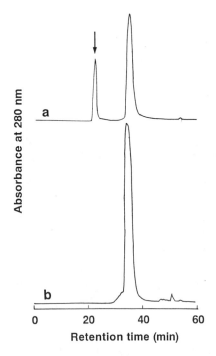

Figure 8. Elution profiles of β-Lg-G6P conjugate solution containing Ca, Pi, and citrate. a, β-Lg-G6P conjugate solution containing 30 mM calcium, 22 mM Pi, and 10 mM citrate; b, sample (a) was treated with EDTA.

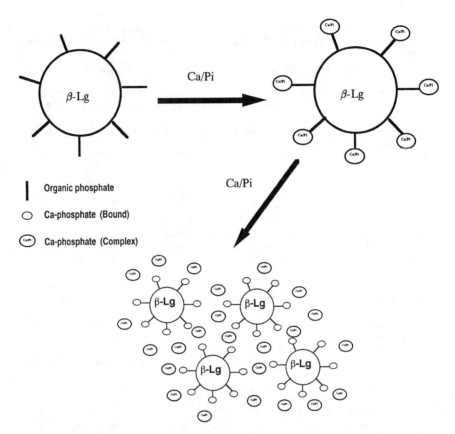

Figure 9. Scheme for formation of the complex of β-Lg-G6P conjugate with calcium phosphate.

conjugates have the ability to solubilize calcium phosphate and inhibit the spontaneous precipitation of calcium phosphate from the solution supersaturated with respect to calcium phosphate. When rats were fed with WPI-G6P conjugates, calcium was solubilized in the small intestine and calcium absorption in the intestinum ileum was enhanced.

The conjugation of food protein with G6P by the Maillard reaction improved not only functional properties such as emulsifying properties but also calcium absorption-stimulating property. The Maillard reaction naturally occurs in many food systems, therefore, it is superior to chemical phosphorylation in its simplicity and consumer acceptance. This method is considered to be useful for the phosphorylation of food proteins.

LITERATURE CITED

1. Matheis, G.; Whitaker, J.R. *J. Agric. Food Chem.* **1984**, *32*, 699-705.
2. Matheis, G. *Food Chem.*, **1991**, *39*, 13-26.
3. Schmidt, D.G. In *Developments in Dairy Chemistry-1 Proteins*; Fox, P.F., Ed.; Elsevier Applied Science Publishers Ltd: London, 1982, pp. 61-86.

4. Holt, C. In *Developments in Dairy Chemistry-3 Lactose and Minor Constituents*, Fox, P.F. Ed.; Elsevier Applied Science Publishers Ltd: London, 1985, pp. 143-181.
5. Aoki, T; Yamada, N.; Tomita, I.; Kako, Y.; Imamura, I. *Biochim. Biophys. Acta*, **1987**, *911*, 238-243.
6. Hamazume, Y.; Mega, T.; Ikenaka, T. *J. Biochem.*, **1984**, *95*, 1633-1644.
7. Fenselau, C.;Heller, D.N.; Miller, M.S.; White, H.B. *Anal. Biochem.*, **1985**, *150*, 309-314 .
8. Aoki, T.; Yamao, Y.; Yonemasu, E.; Kumasaki, Y.; Kako, K. *Arch. Biochem. Biophys.*, **1993**, *305*, 242-246.
9. Naito, H; Kawakami, A.; Imamura, T. *Agric. Biol. Chem.*, **1972**, *36*, 409-415.
10. Naito, H. In *Searches for the Diverse Role of Food Proteins.*; Taylor, T.G; Junkins, N.K. Ed.; Proceedings of the XIII International Congress of Nutrition, John Libbey, London, 1986, pp. 771-784.
11. Ross, L.F; Bhatnagar, D. *J. Agric. Food Chem.*, **1989**, *37*, 841-849.
12. Seguro, K.; Motoki, M. *Agric. Biol. Chem.*, **1990**, *54*, 1271, 1274.
13. Aluko, R.E.; Yada, R.Y. *Biosci. Biotech. Biochem.*, **1995**, *59*, 2207-2209.
14. Kato, Y; Matsuda, M: Nakamura, R. *Biosci. Biotech. Biochem.*, **1993**, *57*, 1-5.
15. Kato A.; Kobayashi, K. In *Microemulsions and Emulsions in Foods,* El-Nokaly M.; Cornell, D. Ed; American Chemical Society, Washington, D.C., 1991, pp. 213-229.
16. Armstrong, J.M.; McKenenzie, H.A.; Sawyer, W.H. *Biochim. Biophys. Acta*, **1967**, *147*, 60-72.
17. Allen, R.J.L. *Biochem. J.*, **1940**, *34, 858-868.
18. Böhlen, P.; Stein, S.; Dairmen, W. *Arch. Biochem. Biophys.*, **1973**, *155*, 213-220.
19. Jenness, R.; Koops, J. *Neth. Milk Dairy J.*, **1962**, *16*, 153-164.
20. Pearce, K.N.; Kinsella, J.E. *J. Agric. Food Chem.*, **1978**, *62*, 716-723.
21. Aoki, T. *J. Dairy Res.*, **1989**, *56*, 613-618.
22. Hay, D.I.; Moreno, E.C.; Schlesinger, D.H. *Inorg. Prespec. Biol. Med.*, **1979**, *2*, 271-185.
23. Sato, R.; Noguchi, T; Naito, H. *J. Nutr. Vitaminol.*, **1986**, *32*, 72-76.
24. Watanabe, K., Sato, Y.; Kato, Y. *J. Food Processing & Preservation.*, **1980**, *3*, 263-274.
25. Kato, Y.; Matsuda, T.; Kato, N.; Nakamura, R. *J. Agric. Food Chem.*, **1988**, *36*, 806-809.
26. Kato, Y.; Aoki, T.; Kato, N.; Nakamura, R.; Matsuda, T. *J. Agric. Food Chem.*, **1995**, *43*, 301-305.
27. Aoki, T.; Fukumoto, T.; Kimura, T.; Kato, Y.; Mastsuda, T. *Biosci. Biotech. Biochem.*, **1994**, 1727-1728.
28. Woo, S.L.; Richardson, T. *J. Dairy Sci.*, **1983**, *66*, 984-987.
29. Swaisgood, H.E. In *Developments in Dairy Chemistry-1 Proteins*; Fox, P.F., Ed.; Elsevier Applied Science Publishers Ltd: London, 1982, pp. 1-60.
30. Naito, H; Suzuki, H. *Agric. Biol. Chem.*, **1974**, *38*, 1543-1545.

Chapter 19

Novel Functional Properties of Glycosylated Lysozymes Constructed by Chemical and Genetic Modifications

Akio Kato[1], S. Nakamura[2], H. Takasaki[1], and S. Maki[1]

[1]Department of Biochemistry, Faculty of Agriculture, Yamaguchi University, Yamaguchi 753, Japan
[2]Department of Food Science, University of British Columbia, 6650 NW Marine Drive, Vancouver, British Columbia V6T 1W5, Canada

Maillard-type lysozyme-polysaccharide conjugates prepared in a controlled dry state revealed dramatic improvements of the functional properties of proteins, such as emulsifying properties and heat stability. In order to elucidate the molecular mechanism of the dramatic improvements of the functional properties of proteins by glycosylation, oligomannose- and polymannose-linked lysozymes were constructed by using genetic engineering. cDNA encoding hen egg white lysozyme was subjected to site-directed mutagenesis to have the signal sequence for asparagine-linked glycosylation (Asn-X-Thr/Ser) at the positions 19 and 49. The oligomannosyl and polymannosyl lysozymes were secreted in the yeast carrying mutant cDNAs. The polymannosyl lysozymes showed remarkable heat stability and excellent emulsifying properties. These functional properties of polymannosyl lysozymes were much higher than those of oligomannosyl lysozymes, suggesting the importance of the length of saccharide chains. The double polymannosyl lysozyme at both positions 19 and 49 showed better emulsifying properties than the single polymannosyl lysozymes.

As requirements for high quality proteins increase, many researchers have developed a lot of methods to improve the functional properties of proteins by using chemical (1-9) and enzymatic modifications (10-14) However, most of these methods are not used for food applications, because of the potential health hazard or the appearance of detrimental products. Therefore, the idea different from the conventional approaches is desirable for the improvement of functional properties of proteins in food systems.

Proteins have unique surface properties due to their large molecular size and their amphiphilic properties. However, the industrial applications of food proteins are limited, because proteins are generally unstable against heating, organic solvents and proteolytic attack,. Therefore, if proteins can be converted into stable forms, the applications of proteins will be greatly broadened to food processing. Glycosylation of proteins is expected to overcome their instability to heating and to further improve the functional properties. Thus, the glycosylation of proteins with monosaccharides or oligosaccharides were attempted to improve their functional properties (7-9).

0097–6156/96/0650–0243$15.00/0

However, the effects were not enough to improve for industrial use. Marshall (15) reported that soluble protein-dextran conjugation by coupling proteins to cyanogen bromide activated dextran was dramatically effective to heat stability of various enzymes. We also found that the soluble protein-dextran conjugates prepared by coupling with cyanogen bromide activated dextran showed excellent emulsifying properties in addition to heat stability (16). Based on these observation, we have developed the novel method to conjugate proteins with polysaccharides by spontaneous Maillard reaction occurred in the controlled dry-heating between the ε-amino groups in protein and the reducing-end carbonyl group in polysaccharide. Most notably, the resulting protein-polysaccharide conjugates showed the excellent emulsifying properties superior to commercial emulsifier (17-19). Among many chemical and enzymatic modifications of proteins to improve its functionality, this method could be one of the most promising approaches for food applications, because of the safety and the various advantages described below. In addition to the dramatic improvement of the emulsifying properties, this approach was efficient to the improvement of the solubility of insoluble gluten (20), the enhancement of antioxidant effect of ovalbumin (21), and the broadening of bactericidal effect of lysozyme (22,23). Therefore, this Maillard-type protein-polysaccharide conjugates can be used as proteinaceous food additives such as emulsifiers, antibacterial agents and antioxidants. However, the molecular mechanisms of the dramatic improvement of the functional properties are still unknown.

This chapter focuses on the properties of Maillard-type lysozyme-polysaccharide conjugates and describes the properties of glycosylated lysozymes constructed by genetic engineering in order to understand the molecular mechanism of remarkable improvements of the functional properties of proteins by the glycosylation.

Preparation of Maillard-type Protein-Polysaccharide Conjugates

Maillard-type protein-polysaccharide conjugates can be efficiently prepared during storage of the freeze-dried powders of protein-polysaccharide mixtures (molar ratio of 1: 5) at 60 °C for a given day under either 65 % or 79 % relative humidity in a desiccator containing saturated KI or KBr solution, respectively, in the bottom. The Maillard reaction between the ε-amino groups in protein and the reducing-end carbonyl group in polysaccharide is accelerated in the low water activity described above. The rate of reaction for the formation of the conjugates seems to depend on the conformation of proteins. The casein-polysaccharide conjugate is formed only within 1 day, while it takes long time (about 1-2 weeks) to form the lysozyme-polysaccharide conjugate. It seems likely that the rigid structure of protein may suppress and the unfolding structure may accelerate the formation of conjugates, because of the difference in the reactivity of the lysyl residues exposed outside between folded and unfolded proteins.

The binding mode was investigated using ovalbumin-dextran and lysozyme-dextran conjugates (18, 24). The SDS polyacrylamide gel electrophoretic patterns demonstrate a broad single band for protein and carbohydrate stains near the boundary between stacking and separating gels, indicating the formation of the conjugate between ovalbumin or lysozyme and dextran. The molecular weight and the decrease in free amino groups of protein-dextran conjugates suggest that about two moles of dextran are bound to ovalbumin or lysozyme, respectively. Since the binding ratio is expressed as average values, it is probable that proteins bound with one to three moles of dextran may also exist in the protein-dextran conjugates.The analysis of the low-angle laser light scattering combined with HPLC suggests that these protein-polysaccharide conjugates are ease to form oligomeric micelle structure in aqueous solution due to the amphiphilic property (25). The peptide analysis of the lysozyme-galactomannan conjugate showed that an active reducing-end group in polysaccharide was attached to ε-amino group in the lysine residues at position 1 and 97 in lysozyme (26). The limited number of bound polysaccharides may come from the steric hindrance of attached polysaccharide. This limitation is suitable for designing the functional properties of proteins, because the functions of proteins are deteriorated if most lysyl residues are masked by saccharides as observed in the conjugates of proteins with monosaccharides and oligosaccharides. This is the case for the rigid and folded proteins. In contrast, unfolded proteins attach polysaccharides more than folded ones. Casein bound about four polysaccharides per mole in the conjugates (19), because the lysyl residues of unfolded protein are exposed outside and easily reactive to the reducing-end carbonyl group in polysaccharide with a smaller steric hindrance than that of folded proteins.

Emulsifying Properties of Protein-Polysaccharide Conjugates

Both protein and polysaccharide have a role in the stabilization of oil in water emulsions. Proteins adsorb at the oil-water interface during emulsification to form a coherent viscoelastic layer. On the other hand, polysaccharides confer colloid stability through their thickening and gelation behavior in the aqueous phase. Therefore, the protein-polysaccharide conjugates are expected to exhibit the good emulsifying properties. As expected, the dramatic enhancements of emulsifying properties for lysozyme-polysaccharide, casein-polysaccharide and soy protein-polysaccharide conjugates were observed (Figure 1). The conjugates of proteins with polysaccharide revealed much better emulsifying activity and emulsion stability than the control mixtures of proteins with polysaccharides. The emulsifying property of the conjugate of lysozyme with galactomannan was the best of various proteins. The similar excellent emulsifying properties were obtained in the conjugates of lysozyme with dextran. The use of galactomannan is desirable for food ingredients, because it is not so expensive as dextran and already utilized as a thickener, binder and stabilizing agents in food. In order to evaluate the potential to industrial applications, the emulsifying properties of dried egg white (DEW)-polysaccharide conjugates were compared with commercial emulsifiers (27). The DEW-galactomannan conjugate was much better than those of commercial emulsifiers (sucrose-fatty acid ester and glycerin-fatty acid ester). In addition, the emulsifying properties of the conjugates were not affected in acidic conditions, in the presence of 0.2M NaCl and by heating of the conjugates. Since high salt conditions, acidic pH, and/or heating process are commonly encountered in industrial application, the DEW-galactomannan conjugate may be a suitable ingredient to be used in food processing. Since the commercial mannase hydrolysate (galactomannan) of guar gum is contaminated with considerable amounts of small molecular carbohydrates, thereby resulting in deterioration of emulsifying properties, the low-molecular weight of galactomannan should be removed prior to the preparation of the DEW-polysaccharide conjugate.

By screening various polysaccharides, galactomannan (Mw 15,000-20,000) obtained from mannase hydrolysate of guar gum was found to be suitable polysaccharides besides dextran. When glucose is attached to proteins in a similar manner, the function of proteins are unfavorably lowered and the detrimental effects (browning color development etc.) are observed.

In addition to the excellent emulsifying properties, the improvements of various functional properties of Maillard-type protein-polysaccharide conjugates were reported from a viewpoint of the development of new-type of food additives, medicines and cosmetics. The examples reported so far are shown in Table I.

Table 1. Improvement of Functional Properties of Protein-polysaccharide Conjugate Prepared by Maillard Reaction

Proteins	Polysaccharides	Functional properties	References
Ovalbumin	Dextran	Emulsifying properties	Agric.Biol.Chem., 54, 107-112 (1990)
Lysozyme	Dextran	Emulsifying properties Antimicrobial action	J. Agric. Food Chem., 39, 647-650 (1991)
Gluten	Dextran	Emulsifying properties Solubility	J. Agric. Food Chem., 39, 1053-1056 (1991)
Casein	Dextran Galactomannan	Emulsifying properties	Biosci. Biotech.Biochem. 56, 567-571 (1992)
Lysozyme	Galactomannan	Emulsifying properties Antimicrobial action	J. Agric. Food Chem., 40, 735-739 (1992)
Ovalbumin	Dextran Galactomannan	Emulsifying properties Antioxidant effect	J. Agric. Food Chem., 40, 2033-2037 (1992)
Egg-white	Galactomannan	Emulsifying properties	J. Agric. Food Chem., 41, 540-543 (1993)
Protamin	Galactomannan	Emulsifying properties Bactericidal action	J. Food Sci., 59, 428-431(1994)

Heat Stability of Maillard-type Lysozyme-Polysaccharide Conjugates

As shown in Figure 2, the heat stability of lysozyme was dramatically increased by the conjugation with polysaccharide. No coagulation was observed in the lysozyme-polysaccharide conjugates, while nonglycosylated lysozyme formed the insoluble aggregates during heating to 90°C with the transition temperature of 82.5 °C. The lytic activity of lysozyme in the conjugate was also recovered by cooling after heating to 90°C. These results suggest that the attachment of polysaccharide causes proteins to form a stable structure. Upon heating in aqueous solution, the protein molecule partially unfolds and results in the aggregates due to the heat-induced disruption of the delicate balance of various noncovalent interactions. This process may be reversible in the protein-polysaccharide conjugates, because of the inhibition of the unfolded protein-protein interaction due to the attached polysaccharide. This resistance to heating is favorable to food applications, because the heating is essential for pasteurization of food ingredients. Therefore, the approach of protein-polysaccharide conjugates can be useful for the developments of various functional food proteins

Antimicrobial Action of Lysozyme-Polysaccharide Conjugates

Many attempts have been made to develop food preservatives having superior antimicrobial effect without toxicities. For this purpose, hen egg white lysozyme may be one of the most promising ingredients. It is well known that lysozyme attacks only specific positions of glycosidic bonds between N-acetylhexosamines of the peptidoglycan layer in bacterial cell walls. However, since the cell envelope of these bacteria contains a significant amount of hydrophobic materials such as lipopolysaccharide (LPS) covered over the thin peptidoglycan layer, lysozyme fails to lyse Gram-negative bacteria when it is simply added to the cell suspension in the native form. As discussed in a previous paper (28), synergistic factors such as detergents and heat treatment destabilize and consequently solubilize the outer membranes which are

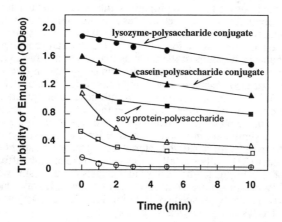

Figure 1. Comparison of Emulsifying Properties of Protein-Polysaccharide Conjugates.Solid symbols indicate Maillard-type protein-polysaccharide conjugates, and open symbols indicate protein-polysaccharide mixtures (without Maillard reaction). Galactomannan (Mw. 15-20 kDa) was used as a polysaccharide. Circle is for lysozyme-polysaccharide, triangle is for casein-polysaccharide, and square is for soy protein-polysaccharide.

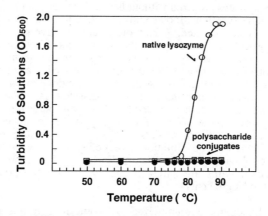

Figure 2.   Heat Stability of Lysozyme-Polysaccharide Conjugates. Open circle, native lysozyme; solid circle, lysozyme-dextran conjugate; open square, lysozyme-galactomannan conjugate. The turbidity of 0.1 % sample solutions in 1/15 M phosphate buffer (pH 7.4) was measured at the absorbance at 500 nm.

mainly composed of LPS. Therefore, the excellent surfactant activity of lysozyme-polysaccharide conjugate seems to cause to destroy the outer membrane of Gram-negative bacterial cells synergistically along with thermal stresses.

Because of the improved surface properties, the antimicrobial action of lysozyme-polysaccharide conjugates is expected. Despite the steric hindrance due to the attachment of dextran or galactomannan, the lytic activity of lysozyme was considerably preserved and the antimicrobial action of lysozyme-dextran conjugate for typical Gram-positive bacteria was almost the same as that of control lysozyme, as shown in Figure 3. Antimicrobial effect of lysozyme-dextran conjugate on five different Gram-negative bacteria was also observed. The living cells were dramatically decreased with heating time at 50°C in the presence of lysozyme-dextran conjugate and completely disappeared from the medium after 40 min. On the contrary, the bactericidal effects were not observed in the presence of native lysozyme as well as in the control medium (buffer alone). The similar antimicrobial effects of lysozyme-galactomannan conjugate were observed for five Gram-negative bacterial strains measured after heat-treatment at 50°C for 30 minutes (23). Although all strains tested were slightly affected by heating in the absence of the conjugate, the bactericidal effects on all strains were observed in the presence of lysozyme-galactomannan conjugate. Thus, it was concluded that the lethal effect was effectively induced by exposing the cells to lysozyme-galactomannan conjugate as well as lysozyme-dextran conjugate.

Lysozyme-Polysaccharide Conjugates as a Bifunctional Food Additive

Since lysozyme-polysaccharide conjugate is very stable to heat-treatment and capable of perturbing the outer membrane of Gram-negative bacteria, it is favorable to use for food applications requiring heat-treatments. In addition to the antimicrobial activity, lysozyme-galactomannan conjugate demonstrated better emulsifying properties than those of commercial emulsifiers under different conditions (in acidic pH or high salt solutions). The conjugates prepared without using chemicals can be applied for food formula as a safe multifunctional food additive. In addition, some therapeutic effect of galactomannan may be expected. Yamamoto et al.(29) have reported that the oral administration of galactomannan decreases the total content of lipids in the liver of rats. Since galactomannan is not so expensive as dextran, lysozyme-galactomannan conjugate can probably be used as a food preservative as well as emulsifier.

Animal dose test and bacterial mutagenesis test were done to confirm the safety of the protein-polysaccharide conjugates (21). It has been shown that these conjugates are nontoxic for oral administration to mice and that they are negative for the Ames test and *rec* assay. In addition, the effect of protein-polysaccharide conjugate on the proliferation of mammalian cell was also investigated to ensure the safety of the conjugates (27). There were no inhibition of the cell growth, suggesting no detrimental effects of the conjugates on the mammalian cell. Therefore, these protein-polysaccharide conjugates are safety and may be fruitful products as novel macromolecular food ingredients.

Polymannosyl Lysozyme Constructed by Genetic Engineering

In order to understand the molecular mechanism of the dramatic improvements of the functional properties of protein by the glycosylation, we attempted to construct the glycosylated lysozymes in yeast expression system using genetic engineering (30-32). In yeast cells, the proteins having Asn-X-thr/Ser sequence are N-glycosylated in the endoplasmic reticulum and the attached oligosaccharide chain can be elongated with further extension of a large polymannose chain in the Golgi apparatus. For this reason, we attempted to construct the yeast expression plasmid. For construction of the yeast expression vectors, the mutant cDNAs of hen egg white lysozyme having N-glycosylation signal sequence (Asn-X-Thr/Ser) at the molecular surface were inserted into *Sal* 1 site of pYG-100, as shown in Figure 4. The expression vectors were

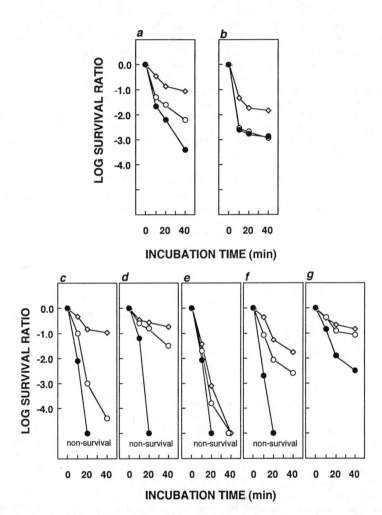

Figure 3. Antimicrobial activity of lysozyme-dextran conjugate for two Gram-positive (a,b) and five Gram-negative (c - g) bacteria. a, *B. cereus* IFO13690; b, *S. aureus* IFO 14462; c, *V. parahaemolyticus* IFO 13286; d, *E. coli* IFO 12713; e, *A. hydrophila* IFO 13286; f, *P. mirabilis* IFO 12668; g, *K. pneumoniiae* IFO 14438. (◇) Control (medium without the addition of lysozyme or conjugate); (○) native lysozyme; (●) lysozyme-dextran conjugate.

introduced into *S. cerevisiae* AH22. A large amount of polymannosyl lysozyme was predominantly expressed in the yeast carrying the lysozyme expression vectors, while a small amount of oligomannosyl lysozyme was also secreted in the yeast medium. Figure 5 shows the elution profiles of the mutant lysozyme G49N (glycine 49 was substituted with asparagine) on a CM Toyopearl (Panel A) , a Sephadex G-50 column (Panel B) and a concanavalin A-Sepharose column (Panel C). The glycosylated lysozymes were purified by these steps. The mutant lysozyme R21T (arginine 21 was substituted with threonine) was also constructed and expressed in a large amount of glycosylated lysozymes. In addition to these mutants having single glycosylation site, the mutant (R21T/G49N) having double glycosylation site was also constructed. The oligomannosyl and polymannosyl lysozymes were secreted in the yeast carrying cDNA of R21T, G49N and R21T/G49N mutants. The carbohydrate contents of these glycosylated lysozymes are shown in Table II. Fortunately the oligomannosyl and polymannosyl lysozymes were obtained enough to compare their functional properties.

Table II. Carbohydrate Composition of Glycosylated Lysozymes

Mutant lysozymes	Contents (moles/mole lysozyme)	
	N-acetylglucosamine	Mannose
G49N		
Polymannosyl	2	310
Oligomannosyl	2	18
R21T		
Polymannosyl	2	338
Oligomannosyl	2	14
R21T/G49N		
Polymannosyl	4	290

Heat Stability of Oligomannosyl and Polymannosyl Lysozymes

The enzymatic activity and the turbidity of glycosylated lysozymes during heating were measured as indications of the apparent heat stability (Figure 6). As expected, the thermal stability of polymannosyl lysozyme was much higher than that of oligomannosyl lysozyme. No coagulation was observed in the polymannosyl lysozyme on heating up to 95°C, whereas the wild-type lysozyme completely coagulated at more than 85°C. In addition, about 60 % of the residual enzymatic activities were retained in the polymannosyl lysozyme after heating up to 95°C, while about 15 % of residual enzymatic activities were retained in oligomannosyl lysozyme. This result suggests t hat the attachment of polysaccharide is much more effective for the improvements of thermal stability of proteins than that of oligosaccharide. It is probable that polysaccharide attached to lysozyme may confer to stabilize aqueous phase around protein molecule and to protect the intermolecular interaction with unfolded molecules, thereby causing the stabilization of protein structure against heating.

Emulsifying Property of Oligomannosyl and Polymannosyl Lysozymes

The emulsifying properties were greatly increased by the attachment of polymannose chains (Figure 7), as expected from the data of Maillard-type lysozyme-polysaccharide

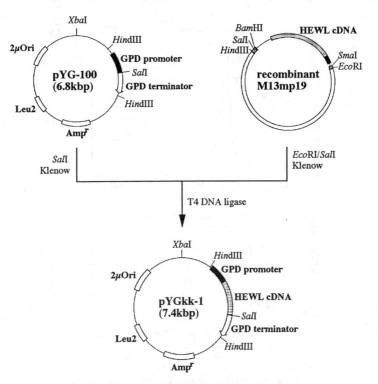

Figure 4   Flow sheet of construction of yeast expression plasmid, pYGkk-1. In the pYGkk-1: solid box, yeast glyceraldehyde 3-phosphate dehydrogenase (GPD) promoter ; shaded box, HEW prelysozyme cDNA; open arrow, GPD terminator. The plasmids are not drawn to the scale.

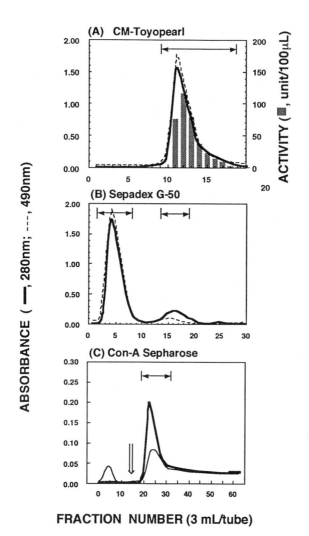

Figure 5.   Isolation of glycosylated lysozymes secreted from *S. cerevisiae* carrying G49N lysozyme cDNA. Panel A represents the cation-exchange chromatography of the yeast cultured supernatant using a CM-Toyopearl resin. The fractions indicated by a horizontal arrow were used for further experiments. The shadow bars indicate the lytic activity. Panel B represents the gel filtration pattern of the collected mutant lysozyme using a Sephadex G-50. Panel C represents the affinity chromatography of the separated mutant lysozymes using a concanavalin A-Sepharose. The thick and thin lines indicate the elution patterns of the large and small peaks in panel B, respectively. A vertical arrow marks the start of elution with 100mM α-methylmannoside. (Reproduced with permission from ref. 30. Copyright 1993 The American Society for Biochemistry and Molecular Biology)

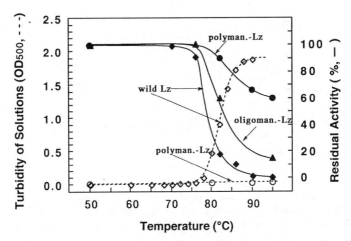

Figure 6. Thermal stability of polymannosyl and oligomannosyl lysozymes (G49N). The thermal stability was estimated by measuring the developed turbidity (dotted line) and the residual enzymatic activity (straight line) when 0.1% samples (for protein) were heated to 95°C at a rate of 1°C/min from 30°C in 1/15M sodium phosphate buffer, pH7.4. The turbidity was measured at the absorbance of 500nm (◇, wild-type lysozyme; ○, polymannosyl lysozyme).The residual enzymatic activity of the heated samples was measured by glycolysis of glycol chitin at 40°C (◆, wild-type lysozyme; ▲, oligomannosyl lysozyme; ●, polymannosyl lysozyme).

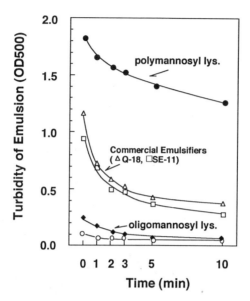

Figure 7. Comparison of emulsifying property of wild (○), oligo-mannosyl (◆), Polymannosyl Lysozymes (●) with commercial emulsifiers. (Reproduced with permission from ref. 31. Copyright 1993 Elsevier Science Publishers)

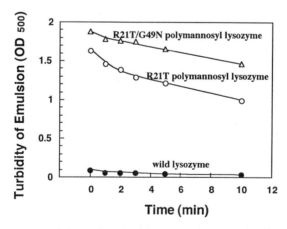

Figure 8. Comparison of emulsifying property of mutant lysozymes polymannosylated at single (R21T) and double (R21T/G49N) sites.

conjugate mentioned above. The polymannosyl lysozyme showed much higher emulsifying properties than the oligomannosyl lysozyme. This suggests that the length of polysaccharide is very important and critical for the emulsifying property of protein-polysaccharide conjugates. It is interesting that the polymannosyl lysozyme shows better emulsifying properties than commercial emulsifiers. It is probable that the role of polysaccharide in the stabilization of emulsion is considered as follows. The hydrophobic residues in protein molecule are anchored in the oil droplets during emulsion formation. The polysaccharide orients to the aqueous layer around the oil droplets after the emulsion preparation and accelerates the formation of thick steric stabilizing adsorbed layer around emulsion, thereby inhibiting the coalescence of oil droplets. Thus, it was evidenced on the molecular basis that the dramatic improvements of the thermal stability and emulsifying property of lysozyme are brought about by the attachment of polysaccharide but not oligosaccharide.

Effects of the Number of Glycosylation Sites on the Functional Properties of Polymannosyl Lysozyme

Another subject to be solved is the effect of the number of glycosylation site in proteins on the functional properties of polymannosyl lysozyme. This is possible to elucidate only by genetic protein glycosylation, because the chemical glycosylation of proteins including Maillard-type glycosylation can not control the number of glycosylation sites. In order to solve this problem, the single polymannosyl mutants (R21T, G49N) and double mutant (R21T/G49N) were successfully constructed by genetic engineering. As shown in Figure 8, the double polymannosyl lysozyme revealed much higher emulsifying properties than single polymannosyl lysozyme. This suggests that the formation of thick steric stabilizing adsorbed layer around emulsion is further enhanced and the coalescence of oil droplets was more effectively inhibited by increasing the number of glycosylation sites.

Conclusions

The novel and promising approaches were described here to improve the functional properties of hen egg white lysozyme. The conjugation of lysozyme with polysaccharides through Maillard reaction in a dry state was very effective to improve the thermal stability and emulsifying property of proteins. In addition, the lysozyme-polysaccharide conjugate showed the novel antimicrobial action against Gram-negative bacteria, and the ovalbumin-polysaccharide conjugate exhibited the enhanced antioxidant action. In order to understand the molecular mechanism of the dramatic improvement of functional properties of lysozyme, the oligomannosyl and polymannosyl lysozymes were constructed in yeast expression system using genetic engineering. The importance of the size of saccharide chain in the effective improvement of the functional properties was proved by the genetic modification of lysozyme. In addition, the double polymannosyl lysozyme showed much higher emulsifying properties than the single polymannosyl lysozyme.

Acknowledgments

This work was supported by Grant-in-Aid 02660143, 05660142, 07280219 for Scientific Research from the Minister of Education, Science, and Culture of Japan. I would like to thank Dr. Kobayashi for his kind cooperation to carry out this project.

Literature Cited

1. Haque, Z.; Kito, M. *J. Agric. Food Chem.* 1976, 24, 504-510.
2. Wu, C.H.; Nakai, S.; Powrie, W.P.*J. Agric. Food Chem.* 1983, 31, 1225-1230.
3. Matsudomi, N.; Sasaki, T.; Kato, A.; Kobayashi, K. *Agric. Biol. Chem.* 1985, 49, 1251-1256.
4. Kinsella, J. E. *CRC Crit. Rev. Food Sci. Nutr.* 1976, 7, 219-280.
5. Sen, L.C.; Lee, H.S.; Feeney, R. E.; Whitaker, J.R. *J. Agric. Food. Chem.* 1981, 29, 348-354.
6. Matheis, G.; Penner, M.H.; Feeney, R.E.; Whitaker, J. R. *J. Agric. Food. Chem.* 1983, 31, 379-387.
7. Marsh, J.W.; Denis,J; Wriston, Jr.J. C. *J. Biol. Chem.* 1977. 252, 7678-7684.
8. Krantz, M. J.; Holtzman, N. A.; Stowell, C. P.; Lee, Y.C. *Biochemistry* 1976, 15, 3963-3968.
9. Kitabatake, N.; Cuq, J. L.; Cheftel, J.C. *J. Agric. Food. Chem.*1985, 33, 125-130.
10. Watanabe, M. ; Shimada, A.; Yazawa, E.; Kato ,T.; Arai, S. *J. Food Sci.*1981 46, 1738-1740.
11. Watanabe, M.; Toyokawa, H.; Shimada, A.; Arai, S. *J. Food Sci.* 1981, 46, 1467-1469.
12. Kato, A.; Tanaka, A.; Lee, Y.; Matsudomi, N.; Kobayashi, K. *J. Agric. Food Chem.* 1987, 35, 285-288.
13. Nio, N.; Motoki , M.; Takinami, K. *Agric. Biol. Chem.* 1985 49, 2283-2286.
14. Motoki, M.; Seguro, K.; Nio, N.; Takinami, K. *Agric. Biol. Chem.* 1986, 50, 3025-3030
15. Marshall, J. J.; Rabinowitz, M. L. *J. Biol. Chem.* 1976, 251, 1081-1087.
16. Kato, A.; Murata, K.; Kobayashi, K. *J. Agric. Food. Chem.*1988, 36, 421-425.
17. Kato, A.; Sasaki, Y.; Furuta, R.; Kobayashi, K. *Agric. Biol. Chem.* 1990, 54, 107-112.
18. Kato, A.; Kobayashi, K. In *Microemulsions and Emulsions in Foods* ; El-Nokaly, M.; Cornell, D. Eds.; ACS Symposium Series 448, 1991, pp. 213-229.
19. Kato, A.; Mifuru,R.; Matsudomi, N.; Kobayashi, K. *Biosci. Biotech. Biochem.* 1992, 56, 567-571.
20. Kato, A.; Shimokawa, K.; Kobayashi, K. *J. Agric. Food Chem.* 1991, 39, 1053-1056.
21. Nakamura, S.; Kato, A.; Kobayashi, K. *J. Agric. Food Chem.* 1992, 40, 2033-2037.
22. Nakamura, S.; Kato, A.; Kobayashi, K. *J. Agric. Food Chem.* 1991, 39, 647-650
23. Nakamura, S.; Kato, A.; Kobayashi, K. *J. Agric. Food Chem.* 1992, 40, 735-739
24. Nakamura, S.; Kato, A.; Kobayashi, K. In *Food Proteins*; Schwenke, K.D.; Mothes, R. Eds.; VCH, Weinheim, 1993, pp. 29-39.
25. Kato, A.; Kameyama, K.; Takagi, T. *Biochim. Biophys. Acta* 1992, 1159, 22-28
26. Kato, A.; Sahara, S.; Shu, Y.W.; Nakamura, S. *J. Agric. Food Chem.* 1996 Submitted
27. Kato, A.; Minaki, K.; Kobayashi, K. *J. Agric. Food Chem.* 1993, 41, 540- 543.
28. Nakamura, K.;Mizushima, S. *Biochim. Biophys. Acta* 1975, 413, 371-393.
29. Yamamoto, T.; Yamamoto, S.; Miyahara, I.; Matsumura,Y.;Hirata, A.; Kim, M. *Denpun Kagaku* 1990, 37, 99-105.
30. Nakamura, S.; Takasaki, H.; Kobayashi, K.; Kato, A. *J. Biol. Chem.* 1993, 268, 12706-12712.
31. Nakamura, S.; Kobayashi, K.; Kato, A. *FEBS Letters* 1993, 328, 259-262.

Chapter 20

Crystallization and X-ray Analysis of Normal and Modified Recombinant Soybean Proglycinins

Three-Dimensional Structure of Normal Proglycinin at 6 Å Resolution

S. Utsumi, A. B. Gidamis[1], Y. Takenaka, N. Maruyama, M. Adachi, and B. Mikami

Research Institute for Food Science, Kyoto University, Uji, Kyoto 611, Japan

Glycinin, which is one of the dominant storage proteins of soybean seeds, is a suitable target to create an ideal food protein by protein engineering. To attain this, clarification of structure-function relationships of glycinin at the molecular level is essential. Therefore, we attempted to crystallize normal recombinant proglycinin and its modified versions having food functions different from those of the normal glycinin and proglycinin for X-ray analysis. We succeeded to crystallize all the samples, each under different suitable crystallization conditions. Among them, the crystals of the normal proglycinin and some of the modified proglycinins diffracted X-rays sufficiently for crystallographic analysis. The preliminary crystallographic data were obtained for these crystals. The three dimensional structure of the normal proglycinin was determined at 6 Å resolution.

Soybean proteins have a function to lower cholesterol level in human serum (1). Heat-induced gel forming and emulsifying abilities of soybean proteins are significant functional properties with respect to their utilization in food systems. However, the functional properties of soybean proteins are usually inferior to those of animal proteins. In addition, soybean proteins are deficient in the essential sulfur-containing amino acids. Therefore, improvement of functional properties and nutritional value of soybean proteins is one of major objectives in the food industry (2). Glycinin (11S globulin) is a suitable target for such improvement, since it is one of the dominant storage proteins in soybean seeds and superior to the other dominant protein, β-conglycinin (7S globulin), with regard to nutritional value as well as functional properties (3). In other words, we can create an ideal food protein from glycinin.

[1]Current address: Department of Food Science and Technology, Faculty of Agriculture, Sokoine University of Agriculture, P.O. Box 3006, Morogoro, Tanzania

In order to design modified glycinins with enhanced food functions, structural characteristics of glycinin and the relationships between the structure and the functional properties of glycinin should be considered. Clarification of these at molecular level is essential for definite design. For this, elucidation of higher-order structure of glycinin by X-ray crystallography is essential.

Expression of Soybean Proglycinin in *Escherichia coli*

Crystals of glycinin are required for X-ray crystallography. However, crystallization of glycinin has not yet been succeeded because of the presence of several molecular species having different subunit compositions (3). Recombinant proteins expressed from glycinin cDNA in microorganism are homogeneous. Therefore, we can expect to get easily crystals of the recombinant proteins (3).

Microorganisms such as *Escherichia coli* and *Saccharomyces cerevisiae* have no enzyme which is responsible for cleavage of proglycinin to a mature form (4,5). Therefore, the mature forms of glycinin having the structure of hexamers are not available from the *E.coli* (4) and yeast (5,6) expression systems, but proglycinin trimers are available. A trimer is smaller in size than a hexamer, indicating that elucidation of higher-structure of a proglycinin trimer is easier than that of a mature hexamer.

Five kinds of subunits were identified as a constituent of glycinin (3). These are classified into two groups: group I (A1aB1b, A1bB2, A2B1a) and group II (A3B4, A5A4B3) (3). The subunits belonging to group I have better nutritional value than those of group II. We constructed an *E. coli* expression system of A1aB1b cDNA using expression vector pKK233-2 and *E. coli* strain JM105 and achieved a high level expression, 20% of the total bacterial proteins, by devising culture conditions: cultivation at low speed at 37 °C was important (7). The expressed proteins were accumulated as soluble proteins without forming inclusion bodies. They were verified to assemble into trimers by a sucrose density gradient centrifugation of the cell extracts, to have the secondary structure similar to that of soybean glycinin by circular dichroism, and to exhibit calcium-induced precipitation, cryoprecipitation and functional properties (heat-induced gel-forming and emulsifying abilities) as the mature glycinin does (7). These facts indicate that X-ray crystallography of a proglycinin trimer is required as a step of that of the mature glycinin and is desired as a model of that of the mature glycinin.

Construction of Modified Soybean Proglycinins

In addition to normal proglycinin, we attempted to crystallize the modified proglycinins shown in Figure 1. Modifications introduced into these proglycinins were mainly based on the structural characteristics (conserved and variable regions) revealed from the comparison of the amino acid sequences of many glycinin-type proteins from legumes and nonlegumes (8). Open and shaded areas a conserved and a variable regions, respectively. Glycinin has five variable regions. Each variable region should be located on the surface of the protein since it has a strong hydrophilic nature, suggesting that such variable regions may tolerate modifications.

Deletion of a part of each variable region and insertion of an oligopeptide composed of four contiguous methionines into each variable region would result in a strengthening of the relative hydrophobisity and the destabilization of the proglycinin molecule (9). The substitution of a sulfhydryl residue involved in disulfide bond formation for another amino acid causes a change of the number and topology of free sulfhydryl residues and may result in destabilization of the molecule. As a result of these modifications, improvement of functional properties such as heat-induced gel-forming and emulsifying abilities would be expected based on the following

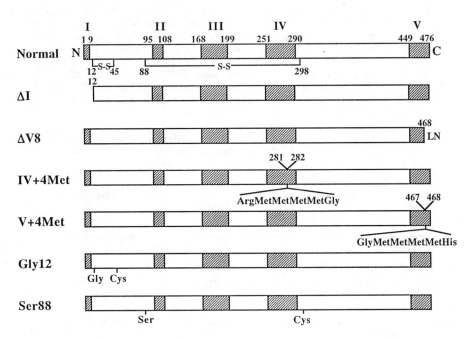

Figure 1. Schematic representation of the normal and modified proglycinin subunits. N and C represent NH2- and COOH-terminus, respectively. Open and hatched areas are conserved and variable regions, respectively. The numbers of residues from the NH2-terminus for the variable regions I—V are shown above the alignment. The positions of the disulfide bonds are indicated in the case of the normal proglycinin. A free cysteine residue resulted from the substitution of another cysteine residue with glycine or serine is indicated in Gly12 and Ser88. (Adapted from reference16)

relationships between the structure and the functional properties revealed at the subunit level:(i) heat-instability of the constituent subunits of glycinin is related to the heat-induced gel-forming ability (*10*); (ii) hydrophobicity is an important factor in the emulsifying properties (*11,12*); (iii) the surface properties of a protein depend on the conformational stability - the more unstable, the higher the emulsifying properties (*13*); (iv) the topology of free sulfhydryl residues is closely related to the heat-induced gel-forming ability (*3*).

We designed many modified proglycinins and evaluated their abilities to form a correct conformation similar to that of the normal proglycinins (*9,14*) Then, we observed that the modified proglycinin shown in Figure 1 can form a correct conformation. ΔI and ΔV8 lack the eleven residues on the NH$_2$-terminus and the eight residues on the COOH-terminus, respectively. ΔV8 has two extra amino acids derived from the universal terminator sequence, Leu-Asn, at its COOH-terminus. IV+4Met has Arg-Met-Met-Met-Met-Gly between Pro281 and Arg282. V+4Met has Glu-Met-Met-Met-Met-His between Pro467 and Gln468. Cys12 and Cys88 are substituted with Gly and Ser in Gly12 and Ser88, respectively. Emulsifying activities of the modified proglycinins are shown in Table I. All the modified proglycinins

Table I. Emulsifying Activities of the Soybean Glycinin and the Normal and Modified Proglycinins

Sample	Emulsifying activity [a] (%)	Sample	Emulsifying activity [a] (%)
Soybean glycinin	100	IV+4Met	127
Normal proglycinin	123	V+4Met	200
ΔI	125	Gly12	115
ΔV8	203	Ser88	133

[a] Emulsifying activity was expressed as relative value (%) compared with the soybean glycinin.
SOURCE: Adapted from references 9 and 14.

exhibited higher emulsifying activities than the native glycinin. Especially, ΔV8 and V+4Met exhibited twice the value of the native glycinin. All the modified proglycinins could form gels by boiling. The gel hardness are shown in Figure 2. ΔI, IV+4Met, V+4Met, and Ser88 formed gels having the hardness higher than that of the native glycinin, especially Ser88 could form hard gels at a low protein concentration where the native glycinin formed very soft gels (*9,14*). Gly12 exhibited different gel-forming profile from the others; Gly12 could form gel similarly to the normal proglycinin at higher protein concentration, but not at all at lower protein concentration (*14*). Elucidation of the higher structure of these modified proglycinins having the functional properties different from those of the native glycinin and the normal proglycinin is desired to clarify the relationships between the structure and the functional properties diversely.

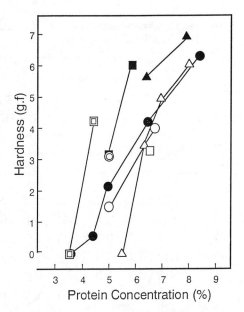

Figure 2. Hardness of gels from the soybean glycinin, the normal and modified proglycinins. Protein samples in 3.5 mM potassium phosphate buffer (pH 7.6) were boiled for 30 min. ●; soybean glycinin; ○, normal proglycinin; ■, ΔI; □, ΔV8; ▲, IV+4Met; ◉, V+4Met; △, Gly12; ▣, Ser88. (Adapted from references 9 and 14)

Crystallization of the Normal and Modified Proglycinins

Crystallization of the normal proglycinin was attempted under various conditions by the dialysis and the vapor diffusion methods. The biggest crystals were obtained by the dialysis against 0.1 M Tris-HCl buffer (pH 7.6) at 4 °C (Figure 3A) (15,16). Crystallization usually commenced within 48 hours and the crystals grew to more than 1.0 mm in length within one week. More than 70 % of the proglycinin in the protein solution formed crystals, supporting the suggestion that the proglycinin expressed in *E. coli* can form the correct conformation.

Crystallization of the modified proglycinins shown in Figure 1 was also attempted by the dialysis method (Table II) (16). All of the modified proglycinins

Table II. Crystallization Conditions and Maximum Crystal Length of the Modified Proglycinin

Proglycinins	Suitable temperature (°C)	Maximum crystal length (mm)							
		Concentration of Tris-HCl[a] (M)							
		0.05	0.1	0.125	0.135	0.15	0.16	0.175	0.2
ΔI	4	0.2	—[b]	0.2	1.5	0.2	ND	—	—
ΔV8	8	ND	0.1	0.1	ND	0.1	0.1	ND	ND
IV+4Met	8	ND	0.2	0.2	ND	1.0	ND	—	ND
V+4Met	8	ND	0.2	3.0	0.2	0.2	ND	ND	ND
Gly12	6	ND	0.5	2.0	ND	ND	ND	ND	ND
Ser88		ND	—	—	ND	—	ND	—	—

[a] pH 7.6.   [b] No crystals.   ND, not done.
SOURCE: Reprinted with permission from reference 16.

except for Ser88 formed crystals during dialysis, although the concentration of Tris and the temperature most suitable for the crystallization were different among the modified proglycinins. Crystallization usually started within 48 h. Crystals of IV+4Met, V+4Met, and Gly12 grew to longer than 1.0 mm within one week and those of ΔI grew to this long in more than four weeks (Figure 3B,D,F,G). ΔV8 did not form large crystals under the conditions used here. Ser88 had a tendency to form a precipitate during dialysis under the conditions used here. Crystallization of ΔV8 and Ser88 by the hanging drop vapor diffusion method with polyethylene glycol 6000 (PEG) as the precipitant was also attempted (16). Crystals of Ser88 appeared in the protein-PEG solutions containing 8.2 to 9.6 % PEG within two to three days. The largest crystals having the length of 2.5 mm were obtained in 9.0 % PEG in one week (Figure 3E). ΔV8 formed only small crystals in the protein-PEG solutions containing 3.4 to 6.0 % PEG within two to three days, and did not grow to a size suitable for X-ray diffraction (Figure 3C). The differences in the suitable conditions for

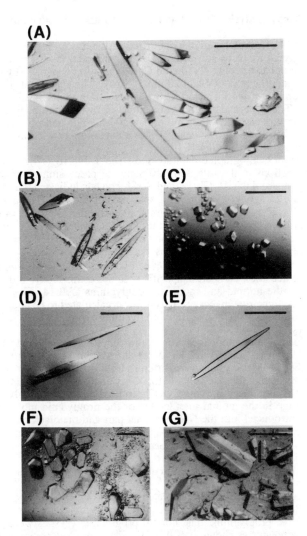

Figure 3.   Photomicrographs of crystals of the normal and modified proglycinins.   (A) normal proglycinin; (B) ΔI; (C) ΔV8; (D) Gly12; (E) Ser88; (F) IV+4Met; (G) V+4Met. The bar in each panel represents 1 mm. (Reproduced with permission from reference 16. Copyright 1994 Japan Society for Bioscience, Biotechnology, and Agrochemistry)

crystallization of the normal and modified proglycinins presumably reflects the assumption that the normal and modified proglycinins have different structural characteristics each other that may affect its functional properties.

Preliminary X-ray analysis of the Normal and Modified Proglycinins

The crystals of the normal and the modified (ΔI, Gly12, Ser88, IV+4Met, and V+4Met) proglycinins were subjected to X-ray diffraction experiments, using a Nonius precession camera with Ni-filtered CuKα radiation generated by a Rigaku X-ray generator operating at 35 kV and 20 mA (*15,16*). The crystals of the normal, ΔI, Gly12, Ser88, and V+4Met diffracted X-rays to resolution limits of 2.9 - 4.1 Å on still photographs. These crystals were resistant to X-ray radiation damage and the diffraction did not decay even after more than 100 h. The exception was IV+4Met crystals, which diffracted X-rays weakly up to a maximum resolution limit of 4.8 Å, and the diffraction decayed within 10 h.

Preliminary crystallographic data are shown in Table III (*16*). The systematic absence of reflections in the $h0l$ and $hk0$ zones of precession photographs of the normal proglycinin and the ΔI, Gly12, and Ser88 crystals indicated that these crystals belonged to the tetragonal system, with space group $P4_1$ or $P4_3$. The unit cell dimensions of these crystals were very similar to each other; thus a=b=around 115 Å and c=around 146 Å. On the other hand, the V+4Met crystals were of the monoclinic space group $P2$ with unit cell dimensions a=118.7 Å, b=78.1 Å, c=109.9 Å, and β=119°. The densities of all the crystals, as measured in a ficoll density gradient, were found to be 1.16 g/cm^3 at 25 °C. With unit cell volumes calculated from the unit cell dimensions, the densities, and partial specific volumes estimated from the amino acid compositions of the normal and modified proglycinins (*7,9,14,17*), the number of protomers per asymmetric unit for each crystal was calculated to be about three. This indicates that these normal and modified proglycinins expressed in *E. coli* are surely forming trimers.

Of the normal and modified proglycinin crystals which diffracted X-rays sufficiently (ΔI, Gly12, Ser88, and V+4Met), the system (Table III) and the shape (Figure 3) of the V+4Met crystals were different from those of the others. ΔV8 could not form crystals having the size suitable for X-ray analysis. These facts suggest that the modifications introduced into the COOH-terminal part of the proglycinin cause a substantial change in the mutual interactions of the proglycinin molecules, but the modifications introduced into the NH$_2$-terminal part and the disulfide bonds did not affect such interactions substantially. In other words, the COOH-terminal part plays an important role for protein-protein interaction.

Three-Dimensional Structure of the Normal Proglycinin at 6 Å Resolution

X-ray crystallography of proglycinin-related proteins has not yet been carried out. Therefore, we have to employ heavy atom isomorphous replacement method to do X-ray analysis of the proglycinin. The normal proglycinin crystals were treated with heavy atom compounds as shown in Table IV. The X-ray diffraction data collected on a Siemens multiwire area detector for both native and heavy atom derivative crystals were evaluated using the program XENGENS (*18*). Most of the data processing was performed using PHASIT from the PHASES program package developed by Furey (*19*). By using the PHASIT program, the diffraction data between the native and the derivative crystals were scaled to select the derivatives that gave significant intensity changes for the determination of heavy atom positions: *p*-chloromercuribenzene sulfonic acid (PCMBS), methyl mercury hydroxide (CH$_3$HgOH) and potassium gold chloride (KAuCl$_4$) were usable.

Table III. Preliminary X-ray Crystallographic Data for the Normal and Modified Proglycinins

Proglycinins	Maximum resolution (Å)	Crystal system	Space group	Unit cell dimensions (Å)			Density (gcm^{-3})	Protomers/ asymmetric unit	V_m[a] (Å^{3}Da^{-1})
				a	b	c			
Normal	2.9	Tetragonal	$P4_1/P4_3$	115.2	115.2	147.1	1.16	3.17	3.05
ΔI	3.5	Tetragonal	$P4_1/P4_3$	115.9	115.9	145.1	1.16	3.25	3.12
Gly12	3.4	Tetragonal	$P4_1/P4_3$	114.9	114.9	146.1	1.16	3.13	3.02
Ser88	3.0	Tetragonal	$P4_1/P4_3$	114.3	114.3	145.7	1.16	3.09	2.98
V+4Met	4.1	Monoclinic	$P2$	118.7	78.1	109.9	1.16	2.87	2.76

a Ratio of unit cell volume to unit protein mass.
SOURCE: Reprinted with permission from ref. 16. Copyright 1994.

**Table IV. Summary of Conditions for Crystal Treatment with
Heavy Atom Derivatives**

Heavy atom derivatives	Heavy atom concentration[a] (mM)	Soaking time at 20 °C (hours)	Heavy atom derivatives	Heavy atom concentration (mM)	Soaking time at 20 °C (hours)
CH_3HgOH	1.0	15	PCMBS	1.0	18
$Pr(NO_3)_3$	0.8	24	K_2PtCl_4	5.0	18
$UO_2(CH_3COO)_2$	0.5	20	$KAuCl_4$	2.0	15

[a] Heavy atom derivatives prepared in 0.1 M Tris-HCl buffer (pH 7.6) at 20 °C.

The data obtained from PCMBS (Hg1), CH3HgOH (Hg2), and KAuCl4 (Au1) derivative crystals were used to solve the protein structure to 6 Å resolution as described (20). The heavy metal derivative sites were deduced from the isomorphous difference-Patterson maps and were further confirmed with difference-Fourier synthesis by inspection of the relevant Harker sections. Three binding sites for individual heavy atom derivatives were identified. The phase ambiguity was resolved using two Hg derivatives and one Au derivative. From the space group of the crystals, each Patterson-map section contains four asymmetric units. There are three heavy atom sites in each section corresponding to the number of protomers per asymmetric unit of the crystals. Positions of these heavy atom sites were determined from these Patterson maps and used for further phase calculations. Table V summarizes the final refinement statistics of the heavy atom parameters at 6 Å resolution as refined by using PHASIT program. The overall mean figure of merit was 0.607 for 4348 independent reflections. The solvent flattening technique and non-crystallographic averaging (21) were applied on the isomorphous replacement phases to improve the overall figure of merit to 0.85 for calculating the electron density map.

An electron density map at 6 Å resolution shown in Figure 4 was calculated from the data based on the best phases obtained from PCMBS, CH3HgOH, and KAuCl4 derivatives. This electron density map clearly distinguishes the protein region from that of solvent. The protein region indicates the boundaries of the proglycinin molecule composed of three protomers related by a three-fold axis symmetry, which coincides with the axis relating the heavy atom binding sites. The dimensions are 93 x 93 x 36 Å. The dimensions of the proglycinin trimer are quite similar to those of phaseolin, 90 x 90 x 35 Å, and canavalin, 80 x 80 x 40 Å, which are 7S globulins of kidney bean (22) and jack bean (23), respectively. The back born of phaseolin can be superimposed on the electron density map of the proglycinin trimer (Figure 5). However, we can not solve the structure of the proglycinin trimer by molecular replacement method using the structural data of phaseolin and canavalin. Therefore, it is suggested that the structure of the proglycinin trimer is fairly similar to those of phaseolin and canavalin but not very similar. Our data support the suggestion by Lawrence et al. (24) that the 7S globulins and the 11S globulins are derived from the common ancestral gene and share a common structure.

Although crystallization of 11S globulins and their proforms of legume seeds has not yet been achieved, hemp edestin, brazil nuts excelsin, and cucurbitin have been

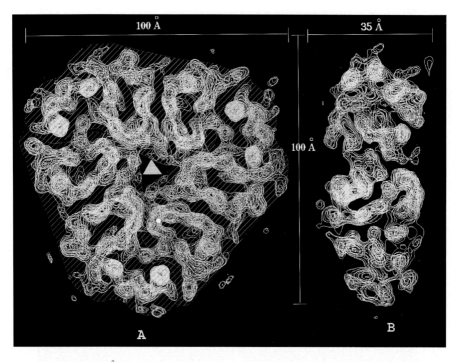

Figure 4. 10 Å-Thick slices of electron density around soybean proglycinin trimer. The map is calculated after solvent flattening and non-crystallographic symmetry averaging. A. View from a three-fold axis (Δ). Green, red and blue hatched areas correspond to a proglycinin protomer. B. Side view of A.

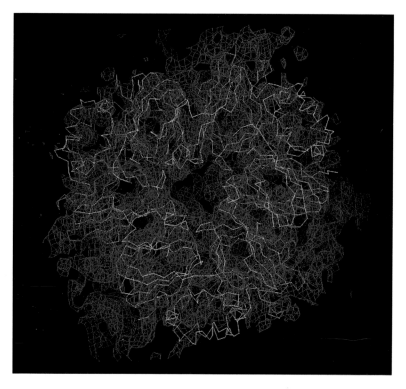

Figure 5. Best fitting of phaseolin trimer to the electron density of proglycinin trimer. Phaseolin and proglycinin trimer are drawn by blue and red color, respectively.

Table V. Final Results of the Refinement Statistics of the Heavy Atom Parameters at 6 Å Resolution

Heavy atom derivatives	Site	Fractional Coordinates			B^a ($Å^2$)	Occ.[b]	RC[c] (%)	RK[d] (%)	Pp[e]
		X	Y	Z					
PCMBS	Hg1-1	0.347	0.867	0.152	10.25	0.32			
	Hg1-2	0.988	0.782	0.100	8.27	0.34	58.6	5.8	1.80
	Hg1-3	0.130	0.537	0.000	46.05	0.32			
CH_3HgOH	Hg2-1	0.344	0.863	0.161	10.44	0.31			
	Hg2-2	0.991	0.779	0.110	7.94	0.30	58.6	5.5	1.68
	Hg2-3	0.126	0.533	0.003	8.18	0.26			
$KAuCl_4$	Au1-1	0.350	0.869	0.164	31.19	0.32			
	Au1-2	0.986	0.780	0.110	15.62	0.29	63.9	5.6	1.55
	Au1-3	0.128	0.537	0.011	19.23	0.31			

[a] Isotropic temperature factor.   [b] Occupancy.   [c] R Cullis.
[d] R Kraut.   [e] Phasing power.

crystallized (*25,26*). Comprehensive X-ray analyses of these crystals had not yet been undertaken since because of the unavailableness of their primary structures and the low-qualities of these crystals. Recently, Patel et al. succeeded to get crystals of edestin having reasonable stability against X-ray radiation and resolution limit of 3.5 Å (*27*). They reported that edestin has an open ring structure with the diameter of 145 Å and the thickness of 90 Å. The thickness of edestin molecule is close to twice of that of the proglycinin determined by us, but the diameter is much bigger than that of the proglycinin. We do not know whether the difference in the diameter is derived from the processing of the proform to the mature form.

Literature Cited

1. Kito, M.; Moriyama, T.; Kimura, Y.; Kambara, H. *Biosci. Biotech. Biochem.* **1993**, *57*, 354-355.
2. Kinsella, J. E. *J. Am. Oil Chem. Soc.* **1979**, *56*, 242-258.
3. Utsumi, S. In *Advances in Food and Nutrition Research;* Kinsella, J. E., Ed.; Academic Press: SanDiego, California, 1992, Vol 36; pp. 89-208.
4. Utsumi, S.; Kim, C. -S.; Sato, T.; Kito, M. *Gene* **1988**, *71*, 349-358.
5. Utsumi, S.; Sato, T.; Kim, C. -S.; Kito, M. *FEBS Lett.* **1988**, *233*, 273-276.
6. Utsumi, S.; Kanamori, J.; Kim, C. -S.; Sato, T.; Kito, M. *J. Agric. Food Chem.* **1991**, *39*, 1179-1186.
7. Kim, C. -S.; Kamiya, S.; Kanamori, J.; Utsumi, S.; Kito, M. *Agric. Biol. Chem.* **1990**, *54*, 1543-1550.

8. Wright, D. J. In *Developments in Food Proteins;* Hudson, B. J. F., Ed.; Elsevier: London, 1988, Vol. 6, pp. 119-178.
9. Kim, C. -S.; Kamiya, S.; Sato, T.; Utsumi, S.; Kito, M. *Protein Eng.* **1990,** *3,* 725-731.
10. Nakamura, T.; Utsumi, S.; Kitamura, K.; Harada, K.; Mori, T. *J. Agric. Food Chem.* **1984,** *32,* 647-651.
11. Nishimura, T.; Utsumi, S.; Kito, M. *J. Agric. Food Chem.* **1989,** *37,* 1266-1270.
12. Utsumi, S.; Kito, M. *Comments Agric. Food Chem.* **1991,** *2,* 261-278.
13. Kato, A.; Yutani, K. *Protein Eng.* **1988,** *2,* 153-156.
14. Utsumi, S.; Gidamis, A. B.; Kanamori, J.; Kang, I. J.; Kito, M. *J. Agric. Food Chem.* **1993,** *41,* 687-691.
15. Utsumi, S.; Gidamis, A. B.; Mikami, B.; Kito, M. *J. Mol . Biol.* **1993,** *233,* 177-178.
16. Gidamis, A. B.; Mikami, B.; Katsube, T.; Utsumi, S.; Kito, M. *Biosci. Biotech. Biochem.* **1994,** *58,* 703-706.
17. Utsumi, S.; Kohno, M.; Mori, T.; Kito, M. *J . Agric. Food Chem.* **1987,** *35,* 210-214.
18. Howard, A. J.; Gilliland, G. L.; Finzel, B. C.; Poulos, T. L.; Ohlendorf, D. H.; Salemme, F. R. *J. Appl. Chrys.* **1987,** *20,* 383-387.
19. Furey, W. Biocrystallographic Lab., VA Medical Center, Univ. Pittsburg, Pa 15240, U.S.A.
20. Mikami, B.; Sato, M.; Shibata, T.; Hirose, M.; Aibara, S.; Katsube, Y.; Morita, Y. *J. Biochem.* **1992,** *112,* 541-546.
21. Wang, B. C. *Methods Enzymol.* **1985,** *115,* 90-112.
22. Lawrence, M. C.; Suzuki, E.; Varghese, J. N.; Davis, P. C.; Van Donkelaar, A.; Tulloch, P. A.; Colman, P. M. *EMBO J.* **1990,** *9,* 9-15.
23. Ko, T. -P.; Ng, J. D.; McPherson, A. *Plant Physiol.* **1993,** *101,* 729-744.
24. Lawrence, M. C.; Izard, T.; Beuchat, M.; Blagrove, R. J.; Colman, P. M. *J. Mol. Biol.* **1994,** *238,* 748-776.
25. Schepman, A. M. H.; Wichertjes, T.; Van Bruggen, E. F. J. *Biochim. Biophys. Acta* **1972,** *271,* 279-285.
26. Colman, P. M.; Suzuki, E.; Van Donkelaar, A. *Eur. J. Biochem.* **1980,** *103,* 585-588.
27. Patel, S.; Cudney, R.; McPherson, A. *J. Mol. Biol.* **1994,** *235,* 361-363.

Chapter 21

Some Characteristics of a Microbial Protein Cross-Linking Enzyme: Transglutaminase

Katsuya Seguro, Noriki Nio, and Masao Motoki

Food Research and Development Laboratories, Ajinomoto Company, Inc.,
1–1 Suzuki-cho, Kawasaki-shi, Kanagawa 210, Japan

Some characteristics of a transglutaminase derived from a variant of *Streptoverticillium mobaraense* (MTGase) were investigated. MTGase catalyzes the crosslinking of most food proteins, such as caseins, soybean globulins, gluten, actin, myosins, and egg proteins, through the formation of ε-(γ-glutamyl)lysine bond, as well as the well-known guinea pig liver enzyme. Molecular weight as determined by SDS-PAGE and mass spectrometry was 38,000 and 37,824, respectively. This indicates MTGase is a simple, monomeric enzyme. Inhibitory effects by thiol-modifying agents as well as copper, lead, and zinc ions on MTGase indicate that MTGase is a thiol-enzyme. For digestion of the MTGase-catalyzed crosslinked proteins, γ-glutamyltransferase was found to cleave the ε-(γ-glutamyl)lysine bond into lysine and glutamate. Lysine in the ε-(γ-glutamyl)lysine bond was, actually, utilized as an essential amino acids in rats.

In tertiary and quaternary structures of protein molecules, bonds and interactions such as hydrogen bond, disulfide bond, electrostatic, and van der Waal's hydrophobic interactions, greatly contribute to their formation and maintenance. Especially in food, not only these, but also other bonds, particularly those found in connective tissues, are very important in the expression of textures or how food feels in the mouth. Among such bonds and interactions, ε-(γ-glutamyl)lysine crosslinking is considered noteworthy. To illustrate, transglutaminase (protein-glutamine γ-glutamyltransferase, EC 2.3.2.13) primarily catalyzes an acyl transfer reaction between the γ-carboxyamide group of peptide-bound glutaminyl residue and a variety of primary amines (1-6). If no amines are present in the reaction system, transglutaminase catalyzes the hydrolysis of the γ-carboxyamide group of the glutaminyl residue, resulting in deamidation (1-7). When the ε-amino group of peptide-bound lysyl residue acts as a substrate, peptide chains are covalently connected through ε-(γ-glutamyl)lysine bonds (Figure 1). A typical example of such transglutaminase-catalyzed protein crosslinking reaction is the termination of bleeding in wound healing (blood coagulation) by Factor XIIIa, an activated form of plasma transglutaminase (8-9). Another example of such a reaction is the setting phenomenon or "suwari" of salted, ground fish protein pastes in Japanese kamaboko

0097–6156/96/0650–0271$15.00/0

manufacturing. Contribution of fish intrinsic transglutaminase in suwari has been well evidenced by many researchers (10-14). Transglutaminase has been believed to modify functional properties of protein substrates (15), as demonstrated in the above cases. In an attempt to develop new foods with unique protein utilization and processing methodology, we have been interested in the modification of the functional properties of food proteins through formation of ε-(γ-glutamyl)lysine bonds by transglutaminase.

Feasibility study on transglutaminase modification.

In the early 1980s, parallel to our study, the possibility of modification of functional properties in milk caseins and soybean globulins was demonstrated using transglutaminase derived from guinea pig liver (16-20) or bovine plasma (21). In these studies, crosslinking proteins of different origins as well as incorporation of amino acids or peptides to improve nutritive deficiency were shown. Concurrently, we were investigating the feasibility of food protein modification for industrial utilization using the guinea pig liver enzyme (22-30), and whey proteins, beef, pork, chicken and fish actomyosins as substrates, which could be gelled, for the transglutaminase reaction. Subsequently, improvement in solubility, water-holding capacity and thermal stability was demonstrated. Proteins in an oil-in-water type emulsion could also serve as the substrate, were gelled by transglutaminase. Moreover, when protein solutions were cast on a flat surface, a transparent, water-resistant, and slowly-digestible protein film could be prepared. Based on these results, transglutaminase was considered potentially useful in creating proteins with new, unique functional properties. However, its limited supply and the unfamiliarity of guinea pig liver as food-use hindered its commercialization. In all cases, mass production of transglutaminase had been eagerly desired.

Microbial transglutaminase derived from a variant of *Streptoverticillium mobaraense*

Discovery of microbial transglutaminase. In order to achieve commercialization of transglutaminase, its constant supply or mass production was a desperate, absolute prerequisite. Therefore, screening for transglutaminase in about 5000 microorganisms was then carried out in collaboration with Amano Pharmaceutical Co.. As a result, microorganisms that produced transglutaminase-like enzymes were screened using the hydroxamate assay (31). These microorganisms excreted the enzyme into the cultural broth, and the capability of one enzyme, excreting the highest activity in the broth, to form ε-(γ-glutamyl)lysine bond in proteins, a critically essential property for transglutaminase, was investigated. The result demonstrated that this enzyme actually was transglutaminase (32). This enzyme was thus first named "bacterial transglutaminase (BTGase)", however, finding that it was taxomically classified as a variant of *Streptoverticillium mobaraense* (33), the name, "microbial" transglutaminase (MTGase)", was accordingly adopted.

Characteristics of MTGase. Since MTGase is secreted into the cultural broth and cell disruption is unnecessary, its purification proved rather easy and subsequently, its commercialization has been much accelerated. Physicochemical properties, such as molecular weight and secondary structure, and enzymatic properties of MTGase had already been reported (31, 34), however, most of these had been determined in the presence of reducing agents in order to compare to those of the guinea pig liver enzyme. However, reducing agents would noticeably change such properties as thermal stability and sensitivity to metals. Therefore, these properties have been newly determined in the absence of the reducing agents.

Physicochemical Properties. Molecular weight (MW) and isoelectric point (pI) of MTGase were previously reported to be approximately 40,000 on both SDS-polyacrylamide electrophoresis and gel-permeation chromatography (31) and 8.9, respectively. More precise analysis, performed recently, has confirmed that MTGase has a MW of 38,000 on SDS-PAGE. Protein sequencing by automated Edman method and mass spectrometry revealed the primary structure of MTGase, demonstrating that MTGase is comprised of 331 amino acid residues (34). cDNA sequencing of the gene taken from the producing-microorganism coincided well and further revealed that MTGase would have a signal peptide (18 amino acid residues) in its amino terminal (33). Data obtained in both protein and cDNA sequencing analyses indicate that MTGase has a single cysteine residue. Based on its 331 amino acid residue, the calculated MW is 37,842, which concides well with experimentally obtained MW 38,000 on both SDS-PAGE and gel-permeation chromatography. Therefore, MTGase is a monomeric, simple protein (not a glycoprotein or lipoprotein, etc.), although there were two potential glycosylation sites (-Thr-Xxx-Asn-) in the primary structure.

Enzymatic Properties. The optimum pH of MTGase activity was around 5 to 8. Even at pH 4 or 9, MTGase still showed some activity (31). MTGase is thus considered stable at wide pH ranges (Figures 2a & 2c). The thermal properties have been measured in the absence of the reducing agent (Figures 2b & 2d). The optimum temperature for enzymatic activity was 60°C, and MTGase fully sustained its activity even at 50°C for 10 min. On the other hand, it lost activity within a few minutes on heating to 70°C. MTGase expressed its activity at 10°C, and still retained some activity at near freezing-point temperature. Concerning substrate specificity, to date most food proteins, such as legume globulins, wheat glutens, egg yolk and albumin proteins, actins, myosins, fibrins, caseins, α–lactalbumin, and β–lactoglobulin, as well as many other albumins, could all be crosslinked by MTGase (35-42).

Ordinarily, transglutaminases, including the well-characterized guinea pig liver enzyme, absolutely require Ca^{2+} for expression of enzymatic activity (1-7, 43). However, MTGase is totally independent of Ca^{2+}, and in this aspect, MTGase is quite unique from other mammalian enzymes. Such a property is very useful in the modification of functional properties of food proteins, since many food proteins, caseins, soybean globulins and myosins, are susceptible to Ca^{2+}, and are easily precipitated in the presence of Ca^{2+}. Sensitivity toward other cations has also been investigated (Figure 3). Cu^{2+}, Zn^{2+}, Pb^{2+}, and Li^+ notably inhibited MTGase. Since MTGase was inhibited by thiol-modifying agents (31), it is reasonable that such heavy metals as Cu^{2+}, Zn^{2+}, Pb^{2+}, Fe^{2+} bind to and block the thiol group of the single cysteine residue. This evidence strongly supports the idea that the cysteine residue could be an active center.

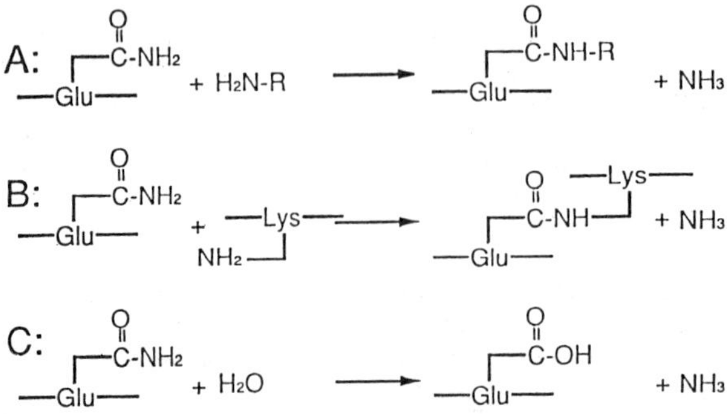

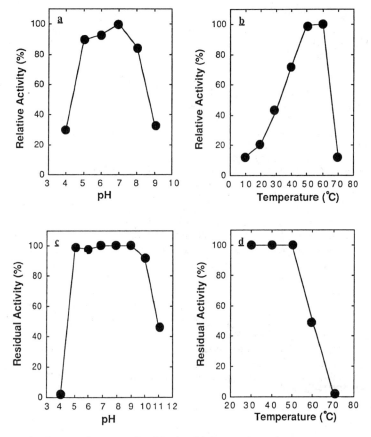

Figure 1. Catalytic action of transglutaminase.
A:, incorporation of a primary amine into γ–carboxyamide group of peptide-
bound glutaminyl residue; B:, crosslinking through formation of ε–(γ–
glutamyl)lysine bond between the γ–carboxyamide group and ε–amino group of
peptide-bound lysyl residue; C:, deamidation of the γ–carboxyamide group.

Figure 2. Enzymatic properties of microbial transglutaminase.
a, pH-activity profile; b, temperature-activity profile; c, pH stability profile, d,
thermal stability profile.

Crosslinking of protein substrates and incorporation of amino acids/peptides. MTGase is also capable of concentrating solutions of such proteins as soybean protein, milk proteins, gelatin, beef, pork, chicken and fish myosins, into a gels (35-42), as does guinea pig liver transglutaminase. The gelled soybean globulin further hardened by heating at 100°C, for 15 min, results in a protein with new gel properties. Caseins, non-heat setting proteins, were also gelled by MTGase without heating, and gelatin, a cold-setting protein, was also gelled. However, in this case, the gelled gelatin no longer melted on heating at 100°C. More than two different proteins can be covalently conjugated by MTGase to produce new proteins with novel functionalities, as with the guinea pig transglutaminase. For instance, conjugation of caseins or soybean globulins with ovomucin, one of the egg white glycoproteins, improved emulsifying activity (44). Also casein/gelatin conjugation yielded novel proteins highly soluble in acidic pH regions (45). MTGase is also able to incorporate amino acids or peptides covalently into substrate proteins. This reaction could improve nutritive values of caseins or soybean proteins, in which methionine and lysine would be limiting amino acids.. For practical application, all amino acids, except lysine, must have their α–carboxyl group amidated, esterified or decarboxylated. Lysine, whose ε-amino group is a primary amine, is a good substrate of MTGase. In the case of peptides, lysylpeptides should be used. For instance, lysylmethionine or methionyllysine could be incorporated in to caseins to improve methionine deficiency. Likewise, lysylarginine or arginyllysine can be incorporated into caseins to improve arginine deficiency in infant prawns. Recently, practical application of transglutaminases, including our MTGase, on food processing have been intensively reviewed (46, 47). It is recommended to consult those specific paper.

Bioavailability of crosslinked proteins.

Distribution of naturally-occurring ε-(γ–glutamyl)lysine. Since application of MTGase-catalyzed modification on food proteins is vast, intense attention must be focused on nutritional efficiency of such crosslinked proteins. It is obvious that between MTGase-modified and native proteins, only the quantity of the ε-(γ–glutamyl)lysine bond differs, all else being totally the same. The digestibility and bioavailability of the ε-(γ–glutamyl)lysine moiety in the proteins have been investigated. First, distribution of naturally-occurring ε-(γ–glutamyl)lysine bond was measured among raw materials and processed foods (47, 48). Such raw food materials as meats, tongues and fish and shellfish, and most processed foods, kamaboko, ham, stewed beef, fried chicken, grilled pork, and hamburger were found to contain more or less certain amounts of ε-(γ–glutamyl)lysine bond., with the exception of dairy products (Figure 4). The ε-(γ–glutamyl)lysine level was rather higher in processed, especially cooked, foods than in raw materials. ε-(γ–Glutamyl)lysine is also found naturally occurring in eggs of various fish, such as red salmon, lumpfish, herring, sardine, and Alaska pollack (Kumazawa, Y. *Fisheries Sci.*, in press). Among these are caviar, salted fish eggs, popularly consumed, in Western countries.

Mechanism of ε-(γ–glutamyl)lysine formation. It is highly likely that intrinsic transglutaminases form ε-(γ–glutamyl)lysine bonds in protien during the cooking process, since a variety of living organisms contain transglutaminase in their tissues and organs (1-6, 49-56). Temperature elevation in food materials during cooking is

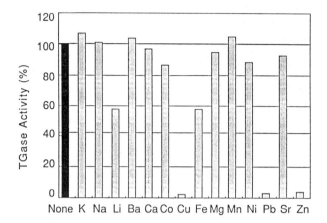

Figure 3. Effect of various cations on microbial transglutaminase activity. Concentrations of each cation were 0.5 mM in the final assaying system. Counter-ions for each cation were chloride ions.

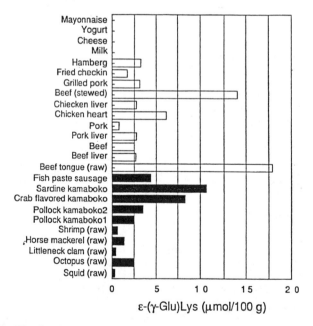

Figure 4. Distribution of naturally-occurring ε–(γ–glutamyl)lysine. Solid bars, marine materials, (raw or cooked); open bars, livestock materials, (raw or cooked).

actually very slow, such that intrinsic transglutaminases are able to remain active for some time. It is thus quite reasonable to conclude that cooked or processed foods contain greater amounts of the ε-(γ-glutamyl)lysine moiety than raw materials due to the catalytic activity of intrinsic transglutaminases. That the dairy products contained no ε-(γ-glutamyl)lysine bond could support such an idea, since milk itself does not contain any intrinsic transglutaminases. Besides the catalytic action of the intrinsic transglutaminases, heating was reported to induce formation of ε-(γ-glutamyl)lysine bonds due to chemical dehydration between the γ-carboxyl group of glutamate and ε-amino group of lysine (57-59). Thus in the presence of transglutaminase, cooking itself results in ε-(γ-glutamyl)lysine formation in proteins. In this aspect, mankind has been ingesting the ε-(γ-glutamyl)lysine moiety from ancient eras, since the discovery of the fire and cooking. Consequently, the safety of ε-(γ-glutamyl)lysine moiety has been unintentionally, and automatically demonstrated by long-lasting consumption of the ε-(γ-glutamyl)lysine moiety in cooked foods.

Degradation *in vitro* of ε-(γ-glutamyl)lysine. A kidney enzyme, γ-glutamylamine cyclotransferase, was reported to cleave an ε-(γ-glutamyl)lysine dipeptide, yielding free lysine and 5-oxo-proline (synonym; pyroglutamate) (60). Ordinary mammalian gastrointestinal digestive enzymes cleave proteins into amino acids but leave the ε-(γ-glutamyl)lysine dipeptide uncleaved after ingestion of crosslinked proteins. The uncleavable ε-(γ-glutamyl)lysine dipeptide may be absorbed through intestinal brush-border membranes to the kidney. The ε-(γ-glutamyl)lysine dipeptide could then be cleaved into lysine and 5-oxo-proline by γ-glutamylamine cyclotransferase. Since lysine is an essential amino acid for humans and other mammalians, the presence of such a pathway would satisfy nutritional requirements in the body.

5-Oxo-proline, however, should be converted to glutamate by an ATP-dependent enzyme, 5-oxo-prolinase. If a large amount of the crosslinked proteins were ingested, the amounts of ε-(γ-glutamyl)lysine dipeptide and subsequently, 5-oxo-proline, would increase and thus a large amount of ATP would be consumed in the kidney. It is feared that these would burden kidney function. Recently, γ-glutamyltransferase (EC 2.3.2.2), present mainly in intestinal brush-border membranes, kidney, and blood (61), has been found to cleave the ε-(γ-glutamyl)lysine dipeptide into lysine and glutamate (62), and in this reaction, no ATP is consumed to generate glutamate. The ε-(γ-glutamyl)lysine moiety, therefore, could be effectively metabolized into lysine by the two enzymes. Hopefully, the lysine generated would be nutritionally beneficial in the body.

Degradation *in vivo* of ε-(γ-glutamyl)lysine. Many researchers have demonstrated that the ε-(γ-glutamyl)lysine dipeptide could be metabolized in rats, and that lysine was integrated in rat tissues (63-66). On the other hand, the bioavailability of the ε-(γ-glutamyl)lysine moiety in the crosslinked proteins has not yet been demonstrated. This may be because it would be impractical to use guinea pig liver transglutaminase for large-scale preparation of crosslinked proteins for feeding animals. Thus with abundantly available microbial transglutaminase, crosslinked caseins were prepared in kilogram-scale volumes, and were fed to rats to evaluate the nutritive value of lysine in the ε-(γ-glutamyl)lysine moiety. In the results, rats fed the crosslinked caseins grew normally, when compared with rats fed the native caseins (67). It is thus suggested that the ε-(γ-glutamyl)lysine moiety in the crosslinked caseins is cleaved and the lysine in the moiety metabolically utilized in the body.

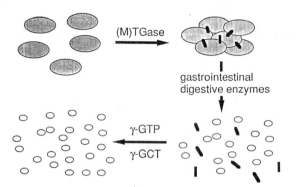

Figure 5. Fate of transglutaminase-catalyzed crosslinked proteins.
The large circles represent a protein molecule, small circles a single amino acid, solid line ε–(γ–glutamyl)lysine. TGase, transglutaminase; γ–GTP, γ–glutamyltransferase; γ–GCT, γ–glutamylamine cyclotransferase.

Concluding Remarks. MTGase can catalytically form the ε–(γ–glutamyl)lysine bond in many food proteins, and such a crosslink drastically alters protein functionalities. Thus, its vast application using ε–(γ–glutamyl)lysine crosslinks can bring forth such promising results as novel foods and processing methodologies. The ε–(γ–glutamyl)lysine moiety in the crosslinked proteins was demonstrated to be cleaved by both γ–glutamylamine cyclotransferase and γ–glutamyltransferase. The fate of the crosslinked proteins and lysine in the ε–(γ–glutamyl)lysine moiety is of much concern and both were found metabolizable in the body (Figure 5). Firstly, the safety of the ε–(γ–glutamyl)lysine moiety has been spontaneously proved by long-established, habitual intakes of cooked foods by mankind.

References

1. Folk, J. E., Chung, S. I., *Adv. Enzymol.*, **1973**, *38*, 109-191.
2. Folk, J. E., Finlayson, J. S., *Adv. Protein Chem.*, **1977**, *31*, 1-133.
3. Folk, J. E., Adv. *Enzymol. Relat. Areas Mol. Biol.*, **1983**, *54*, 1-56.
4. Lorand, L., Conrad, S. M., *Mol. Cell Biochem.*, **1984**, *58*, 9-35.
5. Greenberg, C. S., Birckbichler, P. J., Rice, R. H., *FASEB J.* **1991**, *5*, 3071-3077.
6. Aeschlimann, D., Paulsson, M., *Thromb. Haemost.*, **1994**, *71*,   402-415.
7. Mycek, M. J., Waelsch, H., *J. Bio. Chem.*, **1960**, *235*, 3513-3517.
8. Chung, S. I., Lewis, M. S., Folk, J. E., *J. Biol. Chem.*, **1974**, *249*, 940-950.
9. Hornyak, T. J., Bishop, P. D., Shafer, J. A., *Biochemistry*, **1989**, *28*, 7326-7332.
10. Kimura, I., Sugimoto, M., Toyoda, K., Seki, N., Arai, K., Fujita, T., *Nippon Suisan Gakkaishi*, **1991**, *57*, 1389-1396.
11. Tsukamasa, Y., Shimizu, Y., *Nippon Suisan Gakkaishi*, **1991**, *57*, 535-540.
12. Kamath, G. G., Lanier, T. C., Foegeding, E. A., Hamann, D. D., *J. Food Biochem.*, **1992**, *16*, 151-172.
13. Sato, K., Tsukamasa, Y., Imai, C., Ohtsuki, K., Shimizu, Y., Kawabata, M., *J. Agric. Food Chem.*, **1992**, *40*, 806-810.
14. Wang, J., Kimura, I., Satake, M., Seki, N., *Fisheries Sci.*, **1994**, *60*, 107-113.
15. Whitaker, J. R., *Food Proteins-Improvement through Chemical and Enzymatic Modification;* Feeney, R. E., Whitaker, J. R. Ed.; American Chemical Society; Washington, DC, 1977, pp. 95-105.

16. Ikura, K., Kometani, T., Yoshikawa, M., Sasaki, R., Chiba, H., *Agric. Biol. Chem*, **1980**, *44*, 1567-1573.
17. Ikura, K., Kometani, T., Sasaki, R., Chiba, H., *Agric. Biol. Chem*, **1980**, *44*, 2979-2984.
18. Ikura, K., Yoshikawa, M., Sasaki, R., Chiba, H., *Agric. Biol. Chem*, **1981**, *45*, 2587-2592.
19. Ikura, K, Goto, M., Yoshikawa, M., Sasaki, R., Chiba, H., *Agric. Biol. Chem*, **1984**, *48*, 2347-2354
20. Ikura, K., Okumura, K., Yoshikawa, M., Sasaki, R., Chiba, H., *Agric. Biol. Chem*, **1985**, *49*, 1877-1878.
21. Kurth, L, Rogers, P. J., *J. Food Sci.*, **1984**, *49*, 573-576, 589.
22. Nio, N., Motoki, M., *J. Food Sci.*, **1983**, *48*, 561-566.
23. Nio, N., Motoki, M., Takinami, K., *Agric. Biol. Chem*, **1984**, *48*, 1257-1261.
24. Nio, N., Motoki, M., Takinami, K, *Agric. Biol. Chem*, **1985**, *49*, 2483-2286.
25. Nio, N., Motoki, M., Takinami, K., *Agric. Biol. Chem*, **1986**, *50*, 851-855.
26. Nio, N., Motoki, M., Takinami, K., *Agric. Biol. Chem*, **1986**, *50*, 1409-1412.
27. Motoki, M., Seguro, K., Nio, N., Takinami, K., *Agric. Biol. Chem*, **1986**, *50* , 3025-3030.
28. Motoki, M., Nio, N., Takinami, K., *Agric. Biol. Chem*, **1987**, *51*, 237-239.
29. Motoki, M., Aso, H., Seguro, K, Nio, N., *Agric. Biol. Chem*, **1987**, *51*, 993-996.
30. Motoki, M., Aso, H., Seguro, K, Nio, N., *Agric. Biol. Chem*, **1987**, *51*, 997-1002.
31. Ando, H., Adachi, M., Umeda, K., Matsuura, A., Nonaka, M., Uchio, R, Tanaka, H., Motoki, M., *Agric. Biol. Chem*, **1989**, *53*, 2613-2617.
32. Nonaka, M., Tanaka, H., Okiyama, A., Motoki, M., Ando, H., Umeda, K., Matsuura, A., *Agric. Biol. Chem*, **1989**, *53*, 2619-2623.
33. Washizu, K., Ando, K., Koikeda, S., Hirose, S., Matsuura, A., Takagi, H., Motoki, M., Takeuchi, K., *Biosci. Biotech. Biochem.*, **1994**, *58*, 82-87.
34. Kanaji, T., Ozaki, H., Takao, T., Kawajiri, H., Ide, H., Motoki, M., Shimonishi, Y., *J. Biol. Chem.*, **1993**, *268*, 11565-11572.
35. Sakamoto, H., Kumazawa, Y., Motoki, M., *J. Food Sci.*, **1994**, *59*, 866-871.
36. Nonaka, M., Sakamoto, H., Toiguchi, S., Kawajiri, H., Soeda, T., Motoki, M., *J. Food Sci.*, **1992**, *57*, 1214-1218.
37. Huang, Y.-P., Seguro, K., Motoki, M., Tawada, K., *J. Biochem.*, **1992**, *112*, 229-234.
38. Seguro, K., Kumazawa, Y., Ohtsuka, T., Toiguchi, S., Motoki, M., *J. Food Sci.*, **1995**, *60*, 305-311.
39. Kang, I. J., Matsumura, Y., Ikura, K., Motoki, M., Sakamoto, H., Mori, T., *J. Agric. Food Chem.*, **1994**, *42*, 159-165.
40. Sakamoto, H., Kumazawa, Y., Toiguchi, S., Seguro, K., Soeda, T., Motoki, M., *J. Food Sci.*, **1995**, *60*, 300-304.
41. Kim, E., Motoki, M., Seguro, K., Muhlrad, A., Reisler, E., *Biophyis. J.*, **1995**, *69*, 2024-2032.
42. Nonaka, M., Matsuura, Y., Motoki, M., *Biosci. Biotech. Biochem.*, **1996**, *60*, 131-133.
43. *Enzyme Nomenclature;* Nomenclature Committee of the International Union of Biochemistry and Molecular Biology; Academic Press, Inc., San Diego, California, 1992, p. 201.
44. Kato, A., Wada, T., Kobayashi, K., Seguro, K., Motoki, M., *Agric. Biol. Chem*, **1991**, *55*, 1027-1031.
45. Neilsen, P. M., *Food Biotechol.*, **1995**, *9*, 119-156.
46. Zhu, Y., Rinzema, A., Tramper, J., Bol, J., *App. Microbiol. Biothechnol.*, **1995**, *44*, 277-282.
47. Sakamoto, H., Kumazawa, Y., Kawajiri, H., Motoki, M., *J. Food Sci.*, **1995**, *60*, 416-419.

48. Seguro, K., Kumazawa, Y., Kawajiri, H., Sakamoto, H., Motoki, M., *Institute of Food Technologists Annual Meeting Technical Program: Book of Abstacts*, 1994, Abstract 11-9, page 23.
49. Icekson, I., Apelbaum, A., *Plant Physiol.*, **1987**, *84*, 972-974.
50. Margosiak, S. A., Alavarets, M. E., Louie, D., Kuhen, G. D., *Plant Physiol.*, **1990**, *92*, 88-96.
51. Ramanujam, M. V., Hageman, J. H., *FASEB J.*, **1990**, *168*, A3630.
52. Seki, N., Uno, H., Lee, N.-H., Kimura, I., Toyoda, K, Fujita, T., Arai, K.,*Nippon Suisan Gakkaishi*, **1990**, *56*, 125-132.
53. Signorini, M., Beninati, S., Bergamini, C. M., *J. Plant Physiol.* **1991**, *137,* 547-552.
54. Klein, J. D., Guzman, E., Kuehen, G. D., *J. Biocteriol.*, **1992**, *174*, 2599-2605.
55. Kumazawa, Y., Numazawa, T., Seguro, K., Motoki, M., *J. Food Sci.*, **1995**, *60*, 715-726.
56. Araki, H., Seki, N., *Nippon Suisan Gakkaishi*, **1993**, *59*, 711-719.
57. Asquith, R. S., Otterburn, M. S. Sinclair, W. J., *Angew. Chem. internat. Edit.* **1974**, *13*, 514-520.
58. Hurrell, R. F., Carpenter, K. J., Sinclair, W. J., Otterburn, M. S., Asquith, R. S., *Br. J. Nutr.* **1976**, *35*, 383-395.
59. Otterburn, M., Healy, M., Sinclair, *W., Adv. Exp. Med. Biol.*, **1977**, *86B*, 239-262.
60. Fink, M. L, Chung, S. I., Folk, J. E., *Proc. Natl. Acad. Sci. U.S.A.*, **1980**, *77*, 4564-4568.
61. Meister, A., Tate, S. S., Griffith, O. W., *Method Enzymol.*, **1981**, *77*, 237-253.
62. Seguro, K., Kumazawa, Y., Ohtsuka, T., Ide, H., Nio, N., Motoki, M., Kubota, K., *J. Agric. Food Chem.*, **1995**, **43**, 1977-1981.
63. Raczynski, G., Snochowski, M., Buraczewski, S., *Br. J. Nutr.*, **1975**, *34*, 291-296.
64. Finot, P.-A., Mottu, F., Bujard, E., Mauron, J., *Nutritional Imprevement of Food and Foods Proteins;* Friedman, M. Ed,; Plenum, London, 1978, pp. 549-570.
65. Iwami, K., Yasumoto, K., *J. Sci. Food Agri.,* **1986**, *37*, 495-503.
66. Friedman, M., Finot, P.-A., *J. Agric. Food Chem.*, **1990**, *38*, 2011-2020.
67. Seguro, K., Kumazawa, Y., Kuraishi, C., Sakamoto, H., Motoki, M., J. Nutr. accepted for publication.

Chapter 22

Effect of the Bovine β-Lactoglobulin Phenotype on the Properties of β-Lactoglobulin, Milk Composition, and Dairy Products

Jeremy P. Hill, Mike J. Boland, Lawrence K. Creamer, Skelte G. Anema, Don E. Otter, Geoff R. Paterson, Ruth Lowe, Rose L. Motion, and Wayne C. Thresher

New Zealand Dairy Research Institute, Private Bag 11029, Palmerston North, New Zealand

The milk protein bovine β-lactoglobulin exists in several genetic variants, the structure of each a consequence of one or more point mutations that alter the amino acid sequence. The A variant of β-lactoglobulin differs from the B variant (Gly^{64}Asp and Ala^{118}Val) and is associated with a major increase in β-lactoglobulin concentration and a number of other compositional differences in the milk. Such changes are less marked, or unreported, for the other known variants. The denaturation temperatures of the A, B, and C (Gln^{59}His) variant proteins at neutral pH follow the order: C > A > B. However, the effects of heat on β-lactoglobulin-containing materials, such as whey protein concentrate, whey protein isolate, and concentrated milk, are influenced by the different ratios of the various proteins present as well as the intrinsic properties of the β-lactoglobulins. Although the differences in the effects of the variants measured in the laboratory appear to be relatively slight, the technical and economic impacts can be much greater.

It has been known for some time that the milk from some cows behaved differently from the milk of other cows during processing. After the effects of composition (e.g. fat and protein contents) had been taken into account, some differences remained. The discovery of so-called genetic variation within the milk proteins in the 1950s and 1960s (1-8) suggested that some cow-to-cow or breed-to-breed differences in processing and product characteristics could be related to these slight variations in protein structure (9-14). Not surprisingly, the genetic differences in κ-casein have been found to affect renneting characteristics (15) and cheesemaking characteristics (12-14), whereas differences in β-lactoglobulin affect denaturation-related characteristics such as the heat stability of milks (9,12-14) and the gelation of β-lactoglobulin solutions (16,17).

Aschaffenburg and Drewry (2) also noted that the A variant of β-lactoglobulin was associated with a higher yield of the protein in the milk. It is now clear that κ-casein also affects the quantity of protein expressed, and there is some evidence to suggest that this phenomenon is not limited to β-lactoglobulin and κ-casein alone. It

0097–6156/96/0650–0281$15.00/0

has also been found that there are small but consistent differences in the yields of non-protein milk components that are correlated with β-lactoglobulin variant type (12-14,18). Because many of the properties of milk protein mixtures are dependent on the proportions of the protein components as well as the properties of each protein in the mixture, it is difficult to segregate the "variant" effects on milk and product properties into those that derive from the altered protein structure and those that derive from the altered composition of the milk or product.

In an attempt to clarify the situation in New Zealand, where the national herd is about 18% Jersey, 57% Holstein/Friesian, and 16% Jersey-Friesian cross, we have explored the relationship between β-lactoglobulin variant type (A, B, or C) and some milk and protein characteristics. The C variant was unreported in New Zealand cattle until techniques for easy detection were developed (19-20) for routine use.

Methods of Detection of β-Lactoglobulin Variants

Early studies used electrophoresis of milk or whey, from individual cows, under near-native conditions using paper (1), agar gels (21), or starch gels (3) as the supporting matrix. Polyacrylamide gels are now used almost exclusively for both electrophoresis (PAGE) and isoelectric focusing (IEF) (15). Reverse phase column chromatography (22), capillary electrophoresis (19) and IEF (23) have all been used to determine β-lactoglobulin genetic variants. The latter technique was recently used in the identification of a new β-lactoglobulin variant in the rare Hungarian Grey cattle breed (24).

Recently electrospray ionization mass spectrometry has been used in the identification of a new variant form of β-casein, that contained an uncharged amino acid substitution (25); this is an example of one of the so-called silent variants of Ng-Kwai-Hang and Grosclaude (26). This technique should be especially useful because substitutions involving mass differences would be expected to occur more often than those involving charge differences, and we are currently using it in combination with amino acid sequencing to search for new variant forms of β-lactoglobulin.

Effect of Genetic Variant Type on Milk Composition

As the various phenotypes of β-lactoglobulin are associated with differences in milk composition (13,14,18), it is important that the frequency of such phenotypes is known.

The frequencies of the two major variants of β-lactoglobulin for Friesian cattle in New Zealand are similar to those found elsewhere (Table I). A more recent detailed investigation into the β-lactoglobulin frequencies in 2752 Jersey and Jersey-cross cattle (Hill, J. P.; Lowe, R.; Paterson, G. R.; Winkleman, A.; unpublished results; 1995) showed frequencies of 0.30 for A, 0.62 for B and 0.082 for C.

Table I. β-Lactoglobulin Variant and Phenotype Frequencies in Friesian Cattle

Country	No. of cows	AA	AB	BB	Reference
Canada	8,000	0.11	0.49	0.35	(27)
Italy	2,000	0.16	0.48	0.36	(28)
New Zealand	3,500	0.19	0.51	0.30	(29)
Australia	600	0.12	0.44	0.34	(18)[a]

[a] Included some Jersey cattle.

Table II. Relative Differences in the Compositions of the Milks Produced from β-Lactoglobulin AA and BB Phenotype Cows in New Zealand over Two Seasons

Component	Average Relative Differences[a,b] (%)	
	1992/93 $n^c = 18$	1993/94 $n^c = 21$
Total Protein	1.61[*]	1.21[**]
Casein	7.97[***]	5.03[***]
Whey Protein	-18.07[***]	-14.52[***]
Fat	8.90[***]	2.94[**]
Total Solids	3.18[***]	2.28[**]
β-Lactoglobulin	-29.56[***]	-26.26[***]
Sodium	-7.77[***]	-4.64[***]

[a] Relative difference between the concentration of a component X in a bulk milk from β-lactoglobulin phenotype AA and BB milks $\{([X_{BB}] - [X_{AA}])/[X_{AA}]\}$ and expressed as a percentage.

[b] [*] $p < 0.1$, [**] $p < 0.05$, [***] $p < 0.01$.

[c] n = number of bulk samples analyzed in each season, where each bulk milk sample was collected from approximately 200 β-lactoglobulin A or 200 β-lactoglobulin B phenotype cows.

In one large-scale investigation, Hill et al. (*30*) established a herd of 200 β-lactoglobulin A and 200 β-lactoglobulin B Friesians, selected from over 3,000 cows, and were able to segregate the milks at each milking so that bulk milks of each variant type were available. The compositions of the milks were determined and the results are reported in Table II. The A variant of β-lactoglobulin was associated with a 25-30% increase in β-lactoglobulin concentration in herd milks (Table II), corresponding changes in the whey protein fraction, a smaller inverse change in the casein fraction and a barely significant change in total protein content. Surprisingly, the AA variant was associated with a statistically significant decrease in both fat and total solids contents.

Our findings (*29-31*) that the milk produced by β-lactoglobulin BB phenotype cows contained more fat (Table II) is supported by a number of studies (*18,28,32-37*). However, there are also many studies that did not find such a relationship (*33,38-45*). No doubt, the relatively small effect requires a high level of analytical skill and a large sample population to obtain definitive results so that, as further trials are completed, a clearer picture will emerge. It is possible that the gene for β-lactoglobulin A promotes β-lactoglobulin synthesis at the expense of the synthesis of other milk proteins including β-lactoglobulin B or C; the caseins, and the enzymes involved with fat or protein synthesis, as the amino acid supply is limiting.

β-Lactoglobulin Structure and Properties

Primary Structure. Until about 1975, β-lactoglobulin was one of the most thoroughly investigated proteins (*46,47*) and was considered to be typical of most globular proteins. More recently, β-lactoglobulin has again attracted considerable interest, partly because of its well-known ability to bind hydrophobic molecules. The amino acid sequence and the disulfide cross-links of β-lactoglobulin have been known for some time, but tertiary structural information has become available only recently.

A	Leu	Leu	**Gln**	Lys	Trp	Gln	Asn	**Asp**	Glu	Cys	....Ser	Leu	**Val**	Cys	Gln
B	Leu	Leu	**Gln**	Lys	Trp	Gln	Asn	**Gly**	Glu	Cys	....Ser	Leu	**Ala**	Cys	Gln
C	Leu	Leu	**His**	Lys	Trp	Gln	Asn	**Gly**	Glu	Cys	....Ser	Leu	**Ala**	Cys	Gln
	57	58	59	60	61	62	63	64	65	66	116	117	118	119	120

Figure 1. Selected amino acid sequences of β-lactoglobulin A, B, and C (derived from Ref. 47).

The A and C variants of β-lactoglobulin differ from the B variant by two and one amino acid substitution respectively (Figure 1) with $Gly^{64}Asp$ and $Ala^{118}Val$ for B→A and $Gln^{59}His$ for B→C. The polypeptide from Leu^{58} to Glu^{65} is relatively unstructured and is likely to be exposed to solvent (48) suggesting that the substitutions at positions 59 and 64 would have less significance in terms of the internal structure and more in terms of the protein surface and intermolecular reactions, e.g. dimer formation. Conversely, the substitution of Val by the smaller Ala at position 118, on the inner surface of the β-barrel (49), may affect β-barrel structure and lipid binding but not protein reactivity.

Secondary Structure. Secondary structure estimations by infrared, optical rotatory dispersion, and circular dichroism spectroscopy together indicate that β-lactoglobulin contains about 10% α-helix and 45% β-sheet structures (50). These data were used together with sequence-based prediction algorithms to predict the likely positions of these secondary structures in the sequence (51). Examination of the predictions for all three variants showed no significant differences among the A, B, and C variant proteins. Dong et al. (52) used deuterium exchange experiments to measure the accessibility of the exchangeable protons on the A and B variant proteins and found that the A variant appeared to have greater structural mobility at pH 7.4 and at pH 7.0 after heating at 70°C. Nevertheless, the circular dichroism spectra of the A and B variant proteins did not differ significantly (52), supporting earlier studies (53,54) on the A, B, and C variants.

Tertiary Structure. Although the two preliminary high resolution structures (49,55) differ in detail, both show that the protein has an eight-stranded β-barrel and a three-turn helix that lies parallel to three of the β-strands. The reactive thiol, which is very important for catalyzing the disulfide interchange reactions central to the heat-induced reactions involving β-lactoglobulin, is situated within a group of hydrophobic residues between this helix and the β-strands. This structure is consistent with many of the known properties of β-lactoglobulin, such as the accessibility of the thiol under denaturing but not native conditions. This structure has been confirmed more recently by a thorough X-ray crystallographic study (Sawyer, L.; personal communication;1995) and a partial nuclear magnetic resonance structure of β-lactoglobulin at pH 2 has shown that the amino acids in the β-barrel regions have slowly exchanging amide protons (48).
One of the X-ray crystallographic studies (55) reported that the A and B variant proteins gave electron density maps that were almost indistinguishable.

Protein Charge. The isoelectric points for the A, B, and C variants in 0.1 M KCl are 5.26, 5.34, and 5.33 respectively (46), reflecting the extra carboxyl group present in the A variant (Figure 1). Titration studies have been assessed by McKenzie (46) and the charge versus pH curves are indistinguishable between pH 2 and pH 4.5. At higher pH, the curves diverge and at pH 7.0 and above the A and C variant proteins carry one more and one less negative charge than the B variant protein respectively, reflecting the different amino acid compositions of the variants (Figure 1). All the variant proteins undergo the so-called Tanford (*56*) transition and contain an "anomalous" carboxyl group that has an apparent pKa of 7.3 (instead of 4.7, Ref. 46).

Self-association. Self-association reactions differ, with the A variant forming both dimers and octomers under appropriate conditions whereas the B and C variants do not form octomers (*57*) and form dimers more readily than the A variant (*58*).

The self-association of β-lactoglobulin has been studied by several groups using ultracentrifugation (*46,57-59*) and light scattering (*50,57,60*) techniques; in no case could the monomer-dimer equilibrium constants be determined unambiguously. Difficulties arose from: (a) the slow formation of polymers during equilibrium sedimentation studies; (b) the unrecognized effects of the pressure dependence of dissociation; (c) the pH-dependent conformation changes that occurred. Nevertheless, the overall picture is that the A, B, and C variants all form dimers in the pH range from 3.5 to 6.5. Within this range (pH ~ 3.7 to 5.1), the A variant dimer tetramerizes, particularly at low temperatures. The relative ease of dimer dissociation has been estimated for the A, B, and C variants. At pH 7.5, 20.5°C, 0.1 ionic strength, K_d was < 5 μM for the C variant, whereas K_d was 8 μM for the B variant and 60 μM for the A variant. At pH 2.7, K_d was 130 μM for the A variant and 50 μM for the B variant (*61*). Recently Thresher et al. (*62*) have covalently bonded the A and B variants of β-lactoglobulin to an inert matrix and then measured the association strength between variant β-lactoglobulins passed down the column and the bonded proteins (Table III). From these results, it can be concluded that β-lactoglobulin B dimer dissociates less than the A variant protein and that the homodimers are more stable than the heterodimers. A similar technique has been used to separate β-lactoglobulin from milk and whey (*63,64*).

Ligand binding. Small organic molecules such as pentane and palmitic acid bind to β-lactoglobulin (*47*). Retinol, which is physiologically important, also binds to β-lactoglobulin which has raised an interest in whether β-lactoglobulin has a lipid- or retinol-binding function in milk (*47*). An early report (*65*) indicated that β-lactoglobulin B binds butane and pentane more strongly than β-lactoglobulin A, and more recently it has been reported (*66*) that retinoic acid binds to β-lactoglobulin B ($K_d = 1.75 \times 10^{-7}$ M) a little more strongly than to β-lactoglobulin A ($K_d = 1.96 \times 10^{-7}$ M).

Hydrolysis. In an early report (*67*), it was noted that the A variant protein was hydrolyzed by trypsin 3.5 times faster than the B variant protein at pH 7.0. This difference was confirmed using trypsin and chymotrypsin (*68*) and immobilized

Table III. The Association of Soluble β-Lactoglobulin A and B Variants with Immobilized β-Lactoglobulin A and B Variants in SMUF at 20°C

Immobilized β-Lactoglobulin	Soluble β-Lactoglobulin	K_d (μM)
A	A	1.5
A	B	2.1
B	A	1.8
B	B	0.72
A[a]	A	104

[a] Ref. 64.

trypsin (69-71). Kalan et al. (72) found that the release of C-terminal amino acids by carboxypeptidase was slower from β-lactoglobulin C than from β-lactoglobulin A or B. Schimdt and van Markwijk (73) suggested that the differences in the rates of hydrolysis by papain and trypsin of the A and B variant proteins were a consequence of the two variant proteins having different conformations; in addition, it was possible that the Gly^{64}Asp substitution affected a nearby cleavage site. The concept of different substrate conformations was supported by Huang et al. (70) who also reported (71) that the products of β-lactoglobulin A hydrolysis by immobilized trypsin were different from those of β-lactoglobulin B hydrolysis. At high (100 or 300 MPa) pressures (68), the rates of hydrolysis of the A and B variant proteins by either trypsin or chymotrypsin increased slightly and became more similar, indicating that the conformations of the two variant proteins had become comparable.

The hydrolysis by trypsin and chymotrypsin of the C variant of β-lactoglobulin has been compared with that of the A and B variants using size exclusion (74) and reverse phase (Motion, R. L.; Hill, J. P.; unpublished results; 1994) HPLC analysis. The rate of hydrolysis of the β-lactoglobulin variants, followed the order A > B > C. In the case of the β-lactoglobulin A and B variants the intermediates and products of the hydrolysis appeared to be identical (Figure 2). From these results we concluded that the orders of bond cleavage were the same in β-lactoglobulin A and B (cf. Ref. 71), but that these bonds were more rapidly hydrolyzed in β-lactoglobulin A. In contrast, β-lactoglobulin C gave different peptide patterns, indicating that the cleavage rates of the individual peptide bonds were different from those for β-lactoglobulin A or B.

Hydrolysis Inhibition. In a study on plasmin activity in milk, Schaar (75) reported that the levels of proteose peptones were significantly lower in the milks produced from β-lactoglobulin AA phenotype cows than in those produced from β-lactoglobulin BB phenotype cows. β-Lactoglobulin is known to inhibit the hydrolysis of casein by plasmin and it is probable that the higher concentrations of β-lactoglobulin found in β-lactoglobulin AA phenotype milks are more inhibitory than the lower β-lactoglobulin concentrations found in β-lactoglobulin BB phenotype milks (see Table II). Interestingly, Farrell and Thompson (76) found that the hydrolysis of p-nitrophenyl phosphate by phosphoprotein phosphatase was inhibited by 40, 22, and 12% by β-lactoglobulin A, B, and C respectively.

Effects of Heat on β-Lactoglobulin

Protein Denaturation. Most proteins tend to adopt a conformation that leads to a minimal energy state for the system. Under conditions close to physiological, the stable protein structure is the so-called native structure. In environments that differ from physiological (e.g. high or very low salt concentrations, chaotropic solutes such as urea, higher or lower temperature, higher or lower pH, high pressure, etc.), the protein usually adopts a different conformation. Under some conditions, irreversible (or very slowly reversing) reactions take place so that reversing the environmental changes does not necessarily reverse the conformational shifts. For example, when β-lactoglobulin solution is heated, the protein changes conformation, the thiol group becomes available for reaction and the protein can become involved in a series of complex inter- and intraprotein thiol-disulfide interchange reactions that are not readily reversed. If other proteins are present, they may be incorporated into complex entities with β-lactoglobulin (e.g. Ref. 77). Many such reactions involving other whey proteins or κ-casein have been invoked to explain some effects of dairy processing and the functional behavior of dairy products.

Purified Proteins in Buffer. Two different approaches can be taken to study the effect of heat on proteins. The protein can be heated in solution, the solution cooled and the consequences of the heat treatment determined. Alternatively the protein solution can be examined at an elevated temperature to determine changes in the state of the protein during reaction. Using the first approach gives results such as those shown in Figure 3 (*78*) in which the C variant protein has reacted less than the A variant which has reacted less than the B variant after heating at temperatures between about 70 and 80°C for 12.5 min. An example of the second approach is shown in Figure 4 in which the cysteine of β-lactoglobulin A reacts faster at 60°C than that of β-lactoglobulin B which reacts faster than that of β-lactoglobulin C at 60°C. This order for the β-lactoglobulin reactivities is different from that shown for the irreversible unfolding at about 72°C (Figure 3) and may be related to the changes in the reactivities at the different temperatures. It has been found (*79*) that at pH 7.5, presumably at about 20°C, the availability of the thiol is low and the order of reactivity is A ~ B >> C and when SDS is bound to the proteins, the thiol reactivity is further decreased and the order is B > A > C. Thus the C variant may be consistently less reactive. Gough and Jenness (*80*) showed that the B variant had apparently unfolded more than the A variant as measured by optical rotation, thiol availability and solubility at pH 5.0. Sawyer (*81*) used a turbidimetric method to show that, at 6 mg/mL in dilute Tris-citrate buffer at pH 7.5, the C variant solution was most turbid and the A variant solution was least turbid after heating at 97.5°C for 10 min. Sawyer (*81*) also noted that heating the A variant in 0.1 M phosphate buffers at pH 7.0 gave higher molecular weight products at 75 than at 97.5°C. The loss of β-lactoglobulin structure at low pH (~ 2.0) with heat or chaotropic solutes (compounds that disrupt water structure) is reversible and the B variant is less stable than the A variant (*82*).

Purified Proteins in Permeate. We (Hill, J. P.; Lowe, R.; unpublished results; 1995) have examined the effect of heat at 80°C on purified β-lactoglobulin variants dissolved in ultrafiltrate at 10 mg/mL using native-PAGE and SDS-PAGE on reduced and non-reduced samples. Under these conditions, the loss of native β-lactoglobulin followed the order B > C ~ A whereas the formation of disulfide-linked aggregates of β-lactoglobulin followed the order B > C > A. At 100 mg/mL protein concentration, the loss of native β-lactoglobulin was much faster than at 10 mg/mL and the protein formed SDS-dispersible aggregates that did not involve interprotein disulfide bonds. The loss followed the order A > B > C, which suggests that the reaction is concentration dependent as well as environment and variant dependent.

Whey. Parris et al. (*83*) observed some interesting differences in the heat-induced aggregation behavior of whey proteins in sweet whey samples prepared from bulk whole milk or milk containing β-lactoglobulin A or B variants. Two types of aggregate were formed upon heating, soluble aggregates and insoluble aggregates, with an increase in the relative amount of the latter type of aggregate being observed with increasing concentrations of calcium. Sweet whey prepared from milk containing β-lactoglobulin B had a tendency to form a greater proportion of soluble than insoluble aggregates upon heating, whereas sweet whey prepared from milk containing β-lactoglobulin A had a tendency to form a greater proportion of insoluble than soluble aggregates upon heating. Parris et al. (*83*) suggested that one reason for the difference in the aggregation properties of the β-lactoglobulin A and B wheys could be a consequence of the Asp[64]Gly substitution (Figure 1), the lower charge on β-lactoglobulin B leading to lower calcium binding and therefore the formation of a lower proportion of insoluble aggregates. Alternatively, as Parris et al. (*83*) also suggested, the difference in the proportion of soluble and insoluble aggregates that form in the β-lactoglobulin A and B type wheys could be due to differences in the reactivity of thiol groups within the variant proteins.

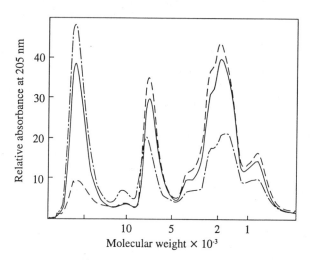

Figure 2. Size exclusion chromatography profiles of β-lactoglobulin A, B, and C following hydrolysis for 60 min by trypsin at pH 8.0 in phosphate buffer at 25°C. Symbols: (— —), β-lactoglobulin A; (———), β-lactoglobulin B; (— - —), β-lactoglobulin C.

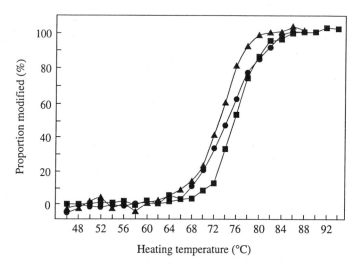

Figure 3. Denaturation of β lactoglobulin A, B and C in phosphate buffer at pH 6.7. Samples (3 mg/ml) were heated rapidly, held for 12.5 min, and cooled rapidly. The circular dichroism signal at 293 nm was determined and the results were normalized. Symbols: ●, β-lactoglobulin A; ▲, β-lactoglobulin B; ■, β-lactoglobulin C.

Heat-induced Gelation of β-Lactoglobulin. The heat-induced gelation characteristics of β-lactoglobulin A and B have been found to be different. Huang et al. *(17)* examined the gelation of the purified β-lactoglobulin A and B variants as 7% solutions in TES buffer at pH 7.0. They found that although both variants formed viscoelastic gels, the gelation point was lower and the initial gelation rate was higher for β-lactoglobulin A than for β-lactoglobulin B. Huang et al. *(17)* concluded that, because the β-lactoglobulin A and B gels exhibited different rheological properties in stress relaxation experiments, the gel matrix structures formed from β-lactoglobulin A and B in the heat-induced gels must involve different molecular interactions between the partially unfolded chains of the protein. In a similar study, McSwiney et al. *(16)* found that the gel stiffness of β-lactoglobulin A type gels was greater than that of β-lactoglobulin B type gels in imidazole-NaCl buffer, particularly at concentrations above 5%. This study also found that, whereas the gel stiffness of β-lactoglobulin B type gels was markedly affected by pH, being lower at lower pH values, the gel strength of β-lactoglobulin A type gels appeared to be independent of pH.

It was also noted *(16)* that prior to gelation some of the β-lactoglobulin formed high molecular weight aggregates that could be dispersed by SDS but did not involve disulfide bond formation. This is similar to effects noted above for heating without gel formation (Hill, J. P.; Lowe, R.; unpublished results; 1995)

Heat Coagulation of Milk. The effect of heat on milk is a very important area to understand and control. Almost all milk undergoes some form of heat treatment at some stage during its processing. One of the areas of intense interest is the heat stability of milk, i.e. the absence of coagulation, gelation, or sedimentation as a consequence of heating. The literature in this area is extensive and includes many review articles *(84-86)*. The heat stability of milk can be measured in a number of ways *(86)*, but can be defined as the length of time that elapses between placing a container of milk in an oil bath at a definite temperature and the onset of coagulation as indicated by flocculation, gelation or changes in protein sedimentability *(84)*.

Anema and McKenna *(87)* examined the effect of heat on milk reconstituted from whole milk powder and found that the reaction kinetic parameters for the depletion of β-lactoglobulin A and B (Figure 5) and α-lactalbumin *(87)* from the milk were very similar to those obtained previously for fresh milk *(88)*, with β-lactoglobulin B being depleted more rapidly than β-lactoglobulin A in the temperature range 70-90°C. This result indicates that processing does not affect the subsequent behavior of the heat-labile proteins towards heat treatment. However, at higher temperatures, β-lactoglobulin A was depleted faster than β-lactoglobulin B (Figure 5). This reversal at about 100°C supports the results of Lyster *(89)* but not those of Dannenberg and Kessler *(88)*.

Because the heat stability of normal-concentration milk at 140°C is dependent upon a number of different factors, the effect of β-lactoglobulin variants on this property has not always shown a clear effect. However, Rose *(90)*, Feagan et al. *(91)* and McLean et al. *(9)* all found the maximum HCT (heat coagulation time) in the HCT-pH curve to be affected by β-lactoglobulin phenotype and followed the general trend AA > AB > BB. Recently Robitaille *(92)* found that the effect of β-lactoglobulin phenotype on HCT-pH$_{max}$ was significant (AA > BB) only in milks that contained the AA phenotype of κ-casein.

The shape of the HCT-pH curve can also vary with β-lactoglobulin variant phenotype *(90,91)*. Feagan et al. *(91)* found that β-lactoglobulin AA, AB, AC, and CC phenotype milks always gave HCT-pH profiles that had a marked maximum near pH 6.7-6.8 and a minimum close to pH 6.8-7.0 (Type A curves), whereas β-lactoglobulin BB and BC phenotype milks generally gave profiles in which the HCT increased steadily over the range pH 6.4-7.2 (Type B curves). Rampilli et al. *(93)* found that Type B HCT-pH curves were produced upon heating samples of β-lactoglobulin BB phenotype milks, but only in combination with the κ-casein AB phenotype. In combination with the

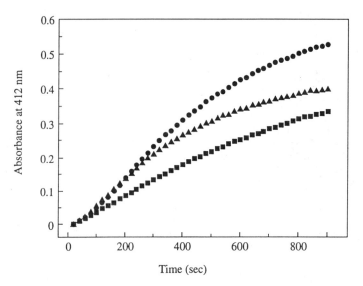

Figure 4. Relative rates of reaction of β-lactoglobulin A, B, and C with DTNB (5,5'dithiobis-(2-nitrobenzoic acid): Ellman's reagent) at 60.9°C in SMUF (simulated milk ultrafiltrate) at pH 6.6. The DTNB and β-lactoglobulin concentrations were both 46 mM. Symbols: ●, β-lactoglobulin A; ▲, β-lactoglobulin B; ■, β-lactoglobulin C.

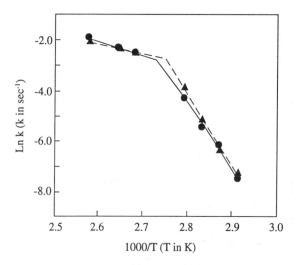

Figure 5. Plots of the rates of denaturation of β-lactoglobulin A and B in heated reconstituted whole milk versus inverse temperature. For experimental details, see Ref. 87. Symbols: ●, β-lactoglobulin A; ▲, β-lactoglobulin B.

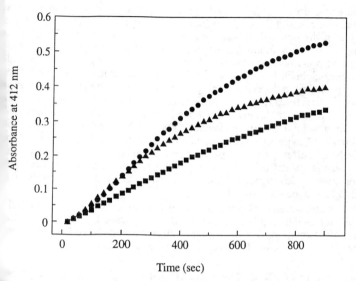

4. Relative rates of reaction of β-lactoglobulin A, B, and C with DTNB hiobis-(2-nitrobenzoic acid): Ellman's reagent) at 60.9°C in SMUF ted milk ultrafiltrate) at pH 6.6. The DTNB and β-lactoglobulin ations were both 46 mM. Symbols: ●, β-lactoglobulin A; ▲, β-lactoglob- ■, β-lactoglobulin C.

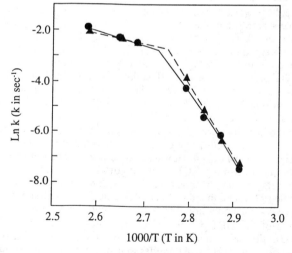

Plots of the rates of denaturation of β-lactoglobulin A and B in heated ed whole milk versus inverse temperature. For experimental details, see ymbols: ●, β-lactoglobulin A; ▲, β-lactoglobulin B.

Purified Proteins in Buffer. Two different approaches can be taken to study the effect of heat on proteins. The protein can be heated in solution, the solution cooled and the consequences of the heat treatment determined. Alternatively the protein solution can be examined at an elevated temperature to determine changes in the state of the protein during reaction. Using the first approach gives results such as those shown in Figure 3 (78) in which the C variant protein has reacted less than the A variant which has reacted less than the B variant after heating at temperatures between about 70 and 80°C for 12.5 min. An example of the second approach is shown in Figure 4 in which the cysteine of β-lactoglobulin A reacts faster at 60°C than that of β-lactoglobulin B which reacts faster than that of β-lactoglobulin C at 60°C. This order for the β-lactoglobulin reactivities is different from that shown for the irreversible unfolding at about 72°C (Figure 3) and may be related to the changes in the reactivities at the different temperatures. It has been found (79) that at pH 7.5, presumably at about 20°C, the availability of the thiol is low and the order of reactivity is A ~ B >> C and when SDS is bound to the proteins, the thiol reactivity is further decreased and the order is B > A > C. Thus the C variant may be consistently less reactive. Gough and Jenness (80) showed that the B variant had apparently unfolded more than the A variant as measured by optical rotation, thiol availability and solubility at pH 5.0. Sawyer (81) used a turbidimetric method to show that, at 6 mg/mL in dilute Tris-citrate buffer at pH 7.5, the C variant solution was most turbid and the A variant solution was least turbid after heating at 97.5°C for 10 min. Sawyer (81) also noted that heating the A variant in 0.1 M phosphate buffers at pH 7.0 gave higher molecular weight products at 75 than at 97.5°C. The loss of β-lactoglobulin structure at low pH (~ 2.0) with heat or chaotropic solutes (compounds that disrupt water structure) is reversible and the B variant is less stable than the A variant (82).

Purified Proteins in Permeate. We (Hill, J. P.; Lowe, R.; unpublished results; 1995) have examined the effect of heat at 80°C on purified β-lactoglobulin variants dissolved in ultrafiltrate at 10 mg/mL using native-PAGE and SDS-PAGE on reduced and non-reduced samples. Under these conditions, the loss of native β-lactoglobulin followed the order B > C ~ A whereas the formation of disulfide-linked aggregates of β-lactoglobulin followed the order B > C > A. At 100 mg/mL protein concentration, the loss of native β-lactoglobulin was much faster than at 10 mg/mL and the protein formed SDS-dispersible aggregates that did not involve interprotein disulfide bonds. The loss followed the order A > B > C, which suggests that the reaction is concentration dependent as well as environment and variant dependent.

Whey. Parris et al. (83) observed some interesting differences in the heat-induced aggregation behavior of whey proteins in sweet whey samples prepared from bulk whole milk or milk containing β-lactoglobulin A or B variants. Two types of aggregate were formed upon heating, soluble aggregates and insoluble aggregates, with an increase in the relative amount of the latter type of aggregate being observed with increasing concentrations of calcium. Sweet whey prepared from milk containing β-lactoglobulin B had a tendency to form a greater proportion of soluble than insoluble aggregates upon heating, whereas sweet whey prepared from milk containing β-lactoglobulin A had a tendency to form a greater proportion of insoluble than soluble aggregates upon heating. Parris et al. (83) suggested that one reason for the difference in the aggregation properties of the β-lactoglobulin A and B wheys could be a consequence of the Asp^{64}Gly substitution (Figure 1), the lower charge on β-lactoglobulin B leading to lower calcium binding and therefore the formation of a lower proportion of insoluble aggregates. Alternatively, as Parris et al. (83) also suggested, the difference in the proportion of soluble and insoluble aggregates that form in the β-lactoglobulin A and B type wheys could be due to differences in the reactivity of thiol groups within the variant proteins.

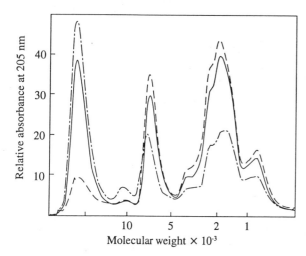

Figure 2. Size exclusion chromatography profiles of β-lactoglobulin A, B, and C following hydrolysis for 60 min by trypsin at pH 8.0 in phosphate buffer at 25°C. Symbols: (— —), β-lactoglobulin A; (———), β-lactoglobulin B; (— – —), β-lactoglobulin C.

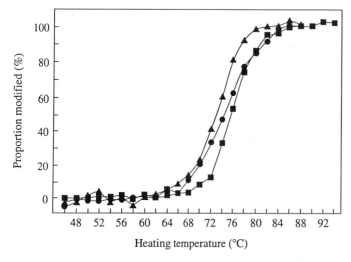

Figure 3. Denaturation of β-lactoglobulin A, B and C in phosphate buffer at pH 6.7. Samples (3 mg/ml) were heated rapidly, held for 12.5 min, and cooled rapidly. The circular dichroism signal at 293 nm was determined and the results were normalized. Symbols: ●, β-lactoglobulin A; ▲, β-lactoglobulin B; ■, β-lactoglobulin C.

Heat-induced Gelation of β-Lactoglobulin. The heat-induced of β-lactoglobulin A and B have been found to be different. Hu the gelation of the purified β-lactoglobulin A and B variants buffer at pH 7.0. They found that although both variants form gelation point was lower and the initial gelation rate was high than for β-lactoglobulin B. Huang et al. (*17*) concluded that, be A and B gels exhibited different rheological properties in stres the gel matrix structures formed from β-lactoglobulin A and F must involve different molecular interactions between the partia protein. In a similar study, McSwiney et al. (*16*) found β-lactoglobulin A type gels was greater than that of β-lact imidazole-NaCl buffer, particularly at concentrations above ! that, whereas the gel stiffness of β-lactoglobulin B type gels pH, being lower at lower pH values, the gel strength of β- appeared to be independent of pH.

It was also noted (*16*) that prior to gelation some of t high molecular weight aggregates that could be dispersed b disulfide bond formation. This is similar to effects noted ab formation (Hill, J. P.; Lowe, R.; unpublished results; 1995)

Heat Coagulation of Milk. The effect of heat on milk understand and control. Almost all milk undergoes some fo stage during its processing. One of the areas of intense in milk, i.e. the absence of coagulation, gelation, or sedime heating. The literature in this area is extensive and includes The heat stability of milk can be measured in a number of ￢ as the length of time that elapses between placing a contai definite temperature and the onset of coagulation as indic or changes in protein sedimentability (*84*).

Anema and McKenna (*87*) examined the effect of h whole milk powder and found that the reaction kinetic p β-lactoglobulin A and B (Figure 5) and α-lactalbumin similar to those obtained previously for fresh milk (*88*), depleted more rapidly than β-lactoglobulin A in the tem result indicates that processing does not affect the subseq proteins towards heat treatment. However, at higher temp depleted faster than β-lactoglobulin B (Figure 5). This ￢ the results of Lyster (*89*) but not those of Dannenberg a

Because the heat stability of normal-concentrat upon a number of different factors, the effect of β-lactog has not always shown a clear effect. However, Rose (*90* et al. (*9*) all found the maximum HCT (heat coagulatio be affected by β-lactoglobulin phenotype and followed t￢ Recently Robitaille (*92*) found that the effect of β-lactog was significant (AA > BB) only in milks that containec

The shape of the HCT-pH curve can also v: phenotype (*90,91*). Feagan et al. (*91*) found that β-lac phenotype milks always gave HCT-pH profiles that had 6.8 and a minimum close to pH 6.8-7.0 (Type A cur and BC phenotype milks generally gave profiles in ￢ over the range pH 6.4-7.2 (Type B curves). Ramp HCT-pH curves were produced upon heating sample: milks, but only in combination with the κ-casein AB ￢

Figure
(5,5'di
(simul
concen
ulin B;

Figure 5.
reconstitu
Ref. 87.

κ-casein AA phenotype, Type A HCT-pH curves were produced. Other β-lactoglobulin/κ-casein phenotype combinations gave Type A curves, but combinations involving the κ-casein BB phenotype were not investigated by Rampilli et al. (*93*). In a similar study by Robitaille (*92*) Type A curves were produced by all the β-lactoglobulin/κ-casein combinations investigated (β-lactoglobulin AA, AB, BB with κ-casein AA, AB).

We have recently examined the heat stability of pooled β-lactoglobulin AA and BB phenotype milks throughout a New Zealand dairy season (Paterson, G. R.; Anema, S. G.; Hill, J. P.; unpublished results; 1995). Very few of our milk samples gave the typical Type B heat stability pattern described by Rose (*90*). However, the Type A patterns were dependent on the β-lactoglobulin phenotype, with AA milk generally giving a higher maximum heat stability peak followed (at higher pH) by a relatively broad minimum followed by an increase in heat stability. The BB milks gave a lower heat stability maximum followed by a narrow pH minimum. The AB milks were intermediate in behavior. Because the quantity of β-lactoglobulin is greater for AA milk, it is not clear whether this effect was dependent on the proportion of β-lactoglobulin in the milk protein or on the thermal reactivity of the different variant proteins.

Milks are commonly heat treated to make them stable towards sterilizing heat treatments (~ 125°C) after concentration. McLean et al. (*9*) found that concentrated milks from cows with the β-lactoglobulin BB phenotype withstood high temperatures longer, which was the opposite trend from that for unconcentrated milks that had not been given a prior heat treatment. However van den Berg et al. (*94*), using a higher preheat treatment temperature and bulk milks, found that concentrates from β-lactoglobulin AA phenotype milks withstood sterilizing temperatures for longer.

Conclusions

There is considerable scope to obtain milk with appropriate properties through the selection and management of the natural genetic variation in dairy cattle. The effects of the A and B variants of β-lactoglobulin have been studied in detail, but work on the less common β-lactoglobulin variants has yet to be done. The development of modern separation and analysis techniques will surely lead to the discovery of new variants of β-lactoglobulin, with the potential for beneficial properties.

Acknowledgements

We would like to thank the staff of the Massey University Farms, staff at Tui Co-operative Dairy Company, staff at the Livestock Improvement Corporation, and Errol Conaghan and other members of the Analytical Chemistry Section (New Zealand Dairy Research Institute). We are grateful for the financial support of the New Zealand Dairy Board and the Foundation for Research, Science and Technology, contract numbers DRI 401, 402, and 403.

Literature Cited

1. Aschaffenburg, R.; Drewry, J. *Nature* **1955**, *176*, 218-219.
2. Aschaffenburg, R.; Drewry, J. *Nature* **1957**, *180*, 376-378.
3. Bell, K. *Nature* **1962**, *195*, 705-706.
4. Aschaffenburg, R. *Nature* **1961**, *192*, 431-432.
5. Thompson, M. P.; Kiddy, C. A.; Pepper, L.; Zittle, C. A. *Nature* **1962**, *195*, 1001-1002.
6. Woychik, J. H. *Biochem. Biophys. Res. Commun.* **1964**, *16*, 267-271.

7. Neelin, J. M. *J. Dairy Sci.* **1964**, *47*, 506-509.
8. Woychik, J. H. *J. Dairy Sci.* **1965**, *48*, 496-497.
9. McLean, D. M.; Graham, E. R. B.; Ponzoni, R. W.; McKenzie, H. A. *J. Dairy Res.* **1987**, *54*, 219-235.
10. Sherbon, J. W.; Ledford, R. A.; Regenstein, J.; Thompson, M. P. *J. Dairy Sci.* **1967**, *50*, 951.
11. Southward, C. R.; Dolby, R. M. *J. Dairy Res.* **1968**, *35*, 25-30.
12. Jakob, E.; Puhan, Z. *Int. Dairy J.* **1992**, *2*, 157-178.
13. Jakob, E.; Puhan, Z. *Bull. Int. Dairy Fed.* **1995**, *304*, 2-25.
14. Jakob, E. *Bull. Int. Dairy Fed.* **1994**, *298*, 17-27.
15. Jakob, E. (1993). *Beziehungen zwischen dem genetischen Polymorphismus der Milchproteine und der Labfähigkeit von Milch.* Dissertation ETH No. 10224, Swiss Federal Institute of Technology: Zurich, **1993**.
16. McSwiney, M.; Singh, H.; Campanella, O.; Creamer, L. K. *J. Dairy Res.* **1994**, *61*, 221-232.
17. Huang, X. L.; Catignani, G. L.; Foegeding, E. A.; Swaisgood, H. E. *J. Agric. Food Chem.* **1994**, *42*, 1064-1067.
18. McLean, D. M.; Graham, E. R. B.; Ponzoni, R. W.; McKenzie, H. A. *J. Dairy Res.* **1984**, *51*, 531-546.
19. Paterson, G. R.; Otter, D. E.; Hill, J. P. *J. Dairy Sci.* **1995**, *78*, 2637-2644.
20. Lowe, R.; Anema, S. G.; Paterson, G. R.; Hill, J. P. *Milchwissenschaft* **1995**, *50*, 663-666.
21. Aschaffenburg, R. *J. Dairy Sci.* **1965**, *48*, 128-132.
22. Visser, S.; Slangen, C. J.; Rollema, H. S. *J. Chromatogr.* **1991**, *548*, 361-370.
23. Krause, I.; Buchberger, J.; Weiss, G.; Pfluger, M.; Klostermeyer, H. *Electrophoresis* **1988**, *9*, 609-615.
24. Baranyi, M.; Bosze, Z.; Buchberger, J.; Krause, I. *J. Dairy Sci.* **1993**, *76*, 630-636.
25. Visser, S.; Slangen, C. J.; Lagerwerf, F. M.; van Dongen, W. D.; Haverkamp, J. *J. Chromatogr.* **1995**, *711*, 141-150.
26. Ng-Kwai-Hang, K. F.; Grosclaude, F. In *Advanced Dairy Chemistry - Volume 1*; Fox, P. F., Ed.; Elsevier Applied Science: London, **1992**; pp 405-455.
27. Ng-Kwai-Hang, K. F.; Monardes, H. G.; Hayes, J. F. *J. Dairy Sci.* **1990**, *73*, 3414-3420.
28. Aleandri, R.; Buttazzoni, L. G.; Schneider, J. C.; Caroli, A.; Davoli, R. *J. Dairy Sci.* **1990**, *73*, 241-245.
29. Hill, J. P. *J. Dairy Sci.* **1993**, *76*, 281-286.
30. Hill, J. P.; Paterson, G. R.; Lowe, R.; Wakelin, M. *Proc. N.Z. Soc. Anim. Prod.* **1995**, *55*, 94-96.
31. Hill, J. P.; Paterson, G. R. *Proc. N.Z. Soc. Anim. Prod.* **1994**, *54*, 293-295.
32. Comberg, G.; Meyer, H.; Groning, M. *Zuchtungskunde* **1964**, *36*, 248-255.
33. Graml, R.; Buchberger, J.; Klostermeyer, H.; Pirchner, F. *Z. Tierz. Zuechtungsbiol.* **1985**, *102*, 355-370.
34. Gelderman, H.; Pieper, U.; Roth, B. *Theor. Appl. Genet.* **1985**, *70*, 138-146.
35. Ng-Kwai-Hang, K. F.; Hayes, J. F.; Moxley, J. E.; Monardes, H. G. *J. Dairy Sci.* **1986**, *69*, 22-26.
36. Rahali, V.; Menard, J. L. *Lait* **1991**, *71*, 275-297.
37. Bovenhuis, H.; van Arendonk, J. A. M.; Korver, S. *J. Dairy Sci.* **1992**, *75*, 2549-2559.

38. Brum, E. W.; Rausch, W. H.; Hines, H. C.; Ludwick, T. M. *J. Dairy Sci.* **1968**, *51*, 1031-1038.

39. Hoogendoorn, M. P.; Moxley, J. E.; Hawes, R. O.; MacRae, H. F. *Can. J. Anim. Sci.* **1969**, *49*, 331-341.

40. Cerbulis, J.; Farrell, H. M., Jr. *J. Dairy Sci.* **1975**, *58*, 817-827.

41. Lin, C. Y.; McAllister, A. J.; Ng-Kwai-Hang, K. F.; Hayes, J. F. *J. Dairy Sci.* **1986**, *69*, 704-712.

42. Haenlein, G. F. W.; Gonyon, D. S.; Mather, R. E.; Hines, H. C. *J. Dairy Sci.* **1987**, *70*, 2599-2609.

43. Aaltonen, M.-L.; Antila, V. *Milchwissenschaft* **1987**, *42*, 490-492.

44. Gonyon, D. S.; Mather, R. E.; Hines, H. C.; Haenlein, G. F. W.; Arave, C. W.; Gaunt, S. N. *J. Dairy Sci.* **1987**, *70*, 2585-2598.

45. van Eenennaam, A.; Medrano, J. F. *J. Dairy Sci.* **1991**, *74*, 1730-1742.

46. McKenzie, H. A. In *Milk Proteins. Chemistry and Molecular Biology*; McKenzie, H. A., Ed.; Academic Press: New York, NY, **1971**, Vol. II; pp 257-330.

47. Hambling, S. G.; McAlpine, A. S.; Sawyer L. In *Advanced Dairy Chemistry - Volume 1*; Fox, P. F., Ed.; Elsevier Applied Science: London, **1992**; pp 141-190.

48. Molinari, H.; Ragona, L.; Varani, L.; Musco, G.; Consonni, R.; Zetta, L.; Monaco, H. *FEBS Letts* **1996**, *381*, 237-243.

49. Papiz, M. Z.; Sawyer, L.; Eliopoulos, E. E.; North, A. C. T.; Findlay, J. B. C.; Sivaprasadarao, R.; Jones, T. A.; Newcomer, M. E.; Kraulis, P. J. *Nature* **1986**, *324*, 383-385.

50. Timasheff, S. N.; Susi, H.; Townend, R.; Stevens, L.; Gorbunoff, M. J.; Kumosinski, T. F. In *Conformation of Biopolymers*; Ramachrandran, G. N., Ed.; Academic Press: New York, NY, **1967**, Vol. 1; pp 173-196.

51. Creamer, L. K.; Parry, D. A. D.; Malcolm, G. N. *Arch. Biochem. Biophys.*, **1983**, *227*, 98-105.

52. Dong, A.; Matsuura, J.; Allison, S. D.; Chrisman, E.; Manning, M. C.; Carpenter, J. F. *Biochemistry* **1996**, *35*, 1450-1457.

53. Timasheff, S. N.; Mescanti, L.; Basch, J. J.; Townend, R. *J. Biol. Chem.* **1966**, *241*, 2496-2501.

54. Townend, R.; Kumosinski, T. F.; Timasheff, S. N. *J. Biol. Chem.* **1967**, *242*, 4538-4545.

55. Monaco, H. L.; Zanotti, G.; Spadon, P.; Bolognesi, M.; Sawyer, L.; Eliopoulos, E. E. *J. Mol. Biol.* **1987**, *197*, 695-706.

56. Tanford, C.; Bunville, L. G.; Nozaki, Y. *J. Amer. Chem. Soc.* **1959**, *81*, 4032-4036.

57. Timasheff, S. N. In *Symposium on Foods: Proteins and Their Reactions*; Schultz, H. W.; Anglemier, A. F., Eds; AVI: Westport, CT, **1964**; pp 179-208.

58. McKenzie, H. A.; Sawyer, W. H. *Aust. J. Biol. Sci.* **1972**, *25*, 949-961.

59. Georges, C.; Guinand, S.; Tonnelat, J. *Biochim. Biophys Acta* **1962**, *59*, 737-739.

60. Georges, C.; Guinand, S. *J. Chim. Phys.* **1960**, *57*, 606-614.

61. Timasheff, S. N.; Townend, R. *J. Amer. Chem. Soc.* **1961**, *83*, 470-473.

62. Thresher, W. C.; Knighton, D. R.; Otter, D. E.; Hill, J. P.; *Brief Communications: 24th International Dairy Congress, Melbourne, 1994;* International Dairy Federation: Brussels, **1994**; p 101.

63. Chiancone, E.; Gattoni, M. *J. Chromatogr.* **1991**, *539*, 455-463 .

64. MacLeod, A.; Ozimek, L. *Milchwissenschaft* **1995**, *50*, 303-307 .

65. Wishnia, A; Pinder, T. W., Jr. *Biochemistry* **1966**, *5*, 1534-1542.

66. MacLeod, A.; Fedio, W. M.; Chu, L.; Ozimek, L. *Milchwissenschaft* **1996**, *51*, 3-7.

67. Monnot, M. *Biochim. Biophys. Acta* **1964**, *93*, 31-39.
68. van Willige, R. W. G.; FitzGerald, R. J. *Milchwissenschaft* **1995**, *50*, 183-186.
69. Chen, S. X.; Hardin, C. C.; Swaisgood, H. E. *J. Prot. Chem.* **1993**, *12*, 613-625.
70. Huang, X. L.; Catignani, G. L.; Swaisgood, H. E. *J. Agric. Food Chem.* **1994**, *42*, 1276-1280.
71. Huang, X. L.; Catignani, G. L.; Swaisgood, H. E. *J. Agric. Food Chem.* **1994**, *42*, 1281-1284.
72. Kalan, E. B.; Greenberg, R.; Walter, M. *Biochemistry* **1965**, *4*, 991-997.
73. Schmidt, D. G.; van Markwijk, B. W. *Neth. Milk Dairy J.* **1993**, *47*, 15-22.
74. Motion, R. L.; Hill, J. P. *Brief Communications: 24th International Dairy Congress, Melbourne, 1994.* International Dairy Federation: Brussels, **1994**; p 122.
75. Schaar, J. *J. Dairy Res.* **1985**, *52*, 369-378.
76. Farrell, H. M., Jr.; Thompson, M. P. *Protoplasma* **1990**, *159*, 157-167.
77. Gezimati, J.; Singh, H.; Creamer, L. K. *This volume.* **1996**.
78. Manderson, C. G.; Hardman, M. J.; Creamer, L. K. *J. Dairy Sci. Suppl. 1* **1995**, *78*, 132.
79. Phillips, N. I.; Jenness, R.; Kalan, E. B. *Arch. Biochem. Biophys.* **1967**, *120*, 192-197.
80. Gough, P.; Jenness, R. *J. Dairy Sci.* **1962**, *45*, 1033-1039.
81. Sawyer, W. H. *J. Dairy Sci.* **1968**, *51*, 323-329.
82. Alexander, S. S.; Pace, C. N. *Biochemistry* **1971**, *10*, 2738-2743.
83. Parris, N.; Anema, S. G.; Singh, H.; Creamer, L. K. *J. Agric. Food Chem.* **1993**, *41*, 460-464.
84. Fox, P. F.; Morrissey, P. A. *J. Dairy Res.* **1977**, *44*, 627-646.
85. Fox, P. F. In *Advanced Dairy Chemistry - Volume 1*; Fox, P. F., Ed.; Elsevier Applied Science: London, **1992**; pp 189-228.
86. Singh, H.; Creamer, L. K. In *Advanced Dairy Chemistry - Volume 1*; Fox, P. F., Ed.; Elsevier Applied Science: London, **1992**; pp 621-656.
87. Anema, S. G.; McKenna, A. B. *J. Agric. Food Chem.* **1996**, *44*, 422-428.
88. Dannenberg, F.; Kessler, H. G. *J. Food Sci.* **1988**, *53*, 258-263.
89. Lyster, R. L. J. *J. Dairy Res.* **1970**, *37*, 233-243.
90. Rose, D. *J. Dairy Sci.* **1962**, *45*, 1305-1311.
91. Feagan, J. T.; Bailey, L. F.; Hehir, A. F.; McLean, D. M.; Ellis, N. J. S. *Aust. J. Dairy Technol.* **1972**, *27*, 129-134.
92. Robitaille, G. *J. Dairy Res.* **1995**, *62*, 593-600.
93. Rampilli, M.; Cattaneo, T.; Caroli, A.; Bolla, P.; San Martino, A.; Sciocco Saita, A. In *Milk Proteins: Nutritional, Clinical, Functional and Technological Aspects*; Barth, C. A.; Schlimme, E., Eds.; Springer-Verlag: New York, **1988**; pp 251-255.
94. van den Berg, G.; Escher, J. T. M.; de Koning, P. J.; Bovenhuis, H. *Neth. Milk Dairy J.* **1992**, *46*, 145-168.

Author Index

Affiliation Index

Subject Index